Reaction Kinetics

Reaction Kinetics

Michael J. Pilling and
Paul W. Seakins

School of Chemistry, University of Leeds

OXFORD   NEW YORK   TOKYO
OXFORD UNIVERSITY PRESS

Oxford University Press, Great Clarendon Street, Oxford OX2 6DP

Oxford   New York

Athens   Auckland   Bangkok   Bogota   Bombay   Buenos Aires
Calcutta   Cape Town   Dar es Salaam   Delhi   Florence   Hong Kong   Istanbul
Karachi   Kuala Lumpur   Madras   Madrid   Melbourne   Mexico City
Nairobi   Paris   Singapore   Taipei   Tokyo   Toronto   Warsaw

and associated companies in
Berlin   Ibadan

Oxford is a trade mark of Oxford University Press

Published in the United States
by Oxford University Press Inc., New York

First published 1995
Reprinted 1996 (with corrections), 1997 (with corrections)

A catalogue record for this book is available from the British Library

Library of Congress Cataloging in Publication Data
Pilling, Michael J.
Reaction kinetics / Michael J. Pilling and Paul W. Seakins.
1. Chemical kinetics.   I. Seakins, Paul W.   II. Title.
QD502.P54 1995   541.3'94—dc20   94-46221
ISBN 0 19 855527 X (Pbk)

Printed in Great Britain by Biddles Ltd,
Guildford & King's Lynn

Copyright and figure acknowledgements

Chapter 2 Fig. 2.11 reproduced with permission from Bohland, F., Temps, F., and Wagner, H. Gg., (1986) *Berichte Bunsengesellschaft für Physikalische Chemie*, **90**, 468. Fig. 2.14 reproduced from Fendorf *et al.* (1993) *Journal of American Soil Science*, **57**, 57–62. Fig. 2.23 reproduced with permission from Markwalder, B., Guzel, P., and van den Bergh, H. (1993) *Journal of Physical Chemistry*, **97**, 5260. Copyright American Chemical Society.

Chapter 3 Fig. 3.2 reproduced with permission from Atkins, P. W. (1990), *Physical Chemistry* (4th edn), Oxford University Press.

Chapter 4 Fig. 4.5 reproduced with permission from Sinha, A., Haizo, M. C., and Crim, F. F., (1991) *Journal of Chemical Physics*, **94**, 4928. Fig. 4.7 reproduced with permission from Mayotte, D. H., Polanyi, J., and Woodall, K. B. (1972) *Journal of Chemical Physics*, **57**, 1547. Figs 4.15, 4.16, and 4.17 adapted from Porter and Karplus (1964) *Journal of Chemical Physics*, **40**, 1105. Fig. 4.20 from R. D. Levine and R. B. Bernstein, *(Molecular Reaction Dynamics and Chemical Reactivity)*, Oxford University Press, adapted from Neumark, D. M., Wodtke, A. M., Robinson, G. N., Hayden, C. C., Shobatake, K., Sparks, R. K., Schafer, T. P., and Lee, Y. T. (1985) *Journal of Chemical Physics*, **82**, 3067–77.

Chapter 5 Fig. 5.6 reproduced with permission from Seakins *et al.*, (1993) *The Journal of Physical Chemistry*, **97**, 4457. Copyright American Chemical Society. Figs 5.8 and 5.9 reproduced with permission from Lovejoy, E. R., Kim, S. K., and Moore, C. B., (1992) *Science*, **246**, 1541–2.

Chapter 6 Figs 6.4, 6.6, and 6.9 adapted with permission from Rice, S. A. (1984) *Diffusion limited reactions*, in *Comprehensive Chemical Kinetics*, vol. 25, Elsevier, Amsterdam. Fig. 6.8 reproduced with permission from Chuang, T. J., Hoffman, G. W., and Eisenthal, K. B. (1974) *Chemical Physics Letters*, **25**, 201–5. Fig. 6.11 from Papanikolas, J. M., Gord, J. R., Levinger, N. E., Ray, D., Vorsa, V., and Lineberger, W. C. (1991). *The Journal of Physical Chemistry*, **95**, 8028. Copyright American Chemical Society.

Chapter 7 Figs 7.6, 7.9, and 7.10 adapted with permission from Ertl, G. (1982) *Berichte Busengesellschaft für Physikalische Chemie*, **86**, 425. Fig. 7.11 reproduced with permission from Roth *et al.* (1992) *Journal of Chemical Physics*, **97**, 6880–9.

Chapter 8 Figs 8.7 and 8.8 reproduced with permission from Wayne, R. P. (1992) *Chemistry of Atmospheres* (2nd edn), Oxford University Press. Fig. 8.12 adapted from Blow *et al.* (1967) *Nature*, **214**, 652.

Chapter 9 Fig. 9.5 reproduced from Lightfoot, P. D. (1986) D.Phil Thesis, Oxford.

Chapter 10 Fig. 10.4 reproduced with permission from Dougherty, E. and Rabitz, H. (1980). *Journal of Chemical Physics*, **72**, 6571.

Chapter 11 Figs 11.2, 11.8, and 11.10 reproduced with permission from Scott, S. K. (1993). *Chemical Chaos*, pp. 13, 14, 15, 40, 286. Oxford University Press. Fig. 11.4 reproduced with permission from Proudler *et al.* (1991), *Philosophical Transactions of the Royal Society*, (London) A, **337**, 211.

Chapter 12 Figs 12.11 and 12.17 reproduced with permission from Fletcher, T. R., Woodbridge, E. L., and Leone, S. R. (1988) *The Journal of Physical Chemistry*, **92**, 5387. Copyright American Chemical Society. Fig. 12.15 adapted from Zewail, A. H. (1988) *Science*, **242**, 1645. Fig. 12.16 adapted from G. R. Fleming, (1986). *Chemical applications of ultrafast spectroscopy*, Oxford University Press. Fig. 12.9 reproduced with permission from Mourdaunt, D. H. *et al.* (1993) *Journal of Chemical Physics*, **98**, 2054. Fig. 12.20 reproduced with permission from Ball, S. M., Hancock, G., Murphy, I. J., and Rayner, S. P. (1993) *Geophysical Research Letters*, **20**, 2063.

Acknowledgements

We would like to take this opportunity to thank some of the many people who have made this book possible. Firstly, to all the formal and informal reviewers of the text including Jeremy Frey, Nick Green, John Griffiths, Mat Heal, Dwayne Heard, Stephen Leone, Ken McKendrick, Struan Robertson, Alison Tomlin, Stephen Scott, and Ben Whitaker. Their comments have all been very constructive and provide the basis for a significant portion of the text. To Robert Pilling for producing a number of questions and diagrams. To all the authors and publishers of the many papers from which we have used or adapted diagrams. To the staff of Oxford University Press who have kept faith in a project whose duration would have tried editors of less saintly dispositions. Their help and advice has been greatly appreciated. Finally, to our families who have received much less of our time and attention than is their due and to whom this book is dedicated.

In memoriam
Prof. David Gutman

Contents

0 Introduction—Why study reaction kinetics?

In classical terms reaction kinetics refers simply to the measurements of rates of reactions. Chemistry is all about reactions, so a knowledge of the rates at which they occur is of central and practical importance. For example the reaction of the hydroxyl radical with methane:

$$OH + CH_4 \rightarrow \text{products} \tag{R 1}$$

is the major route for scavenging methane, an important 'greenhouse' gas, from the Earth's atmosphere. The rate at which this reaction occurs determines the lifetime of methane in the atmosphere and hence its global warming potential.

The subject has developed considerably from just determining reaction rates. Kinetics can also tell us about the mechanism of complex reactions. These overall reactions take place via a series of *elementary reactions* (the simplest type of reactions) which combine together to give the overall reaction. Kinetic studies of the hydrogen/bromine reaction (Chapter 9)

$$H_2 + Br_2 \rightarrow 2HBr \tag{R 2}$$

show that the reaction occurs by the following mechanism

$$Br_2 \rightarrow 2Br \tag{R 3}$$

$$Br + H_2 \rightarrow HBr + H \tag{R 4}$$

$$H + Br_2 \rightarrow HBr + Br \tag{R 5}$$

$$Br + Br \rightarrow Br_2 \tag{R 6}$$

where reactions (4) and (5) can form a '*chain*' (the Br radical consumed in reaction (4) is regenerated in reaction (5)) which occurs many times before the sequence is terminated by reaction (6).

Reaction kineticists are also interested in more fundamental information about the elementary reactions themselves and seek to answer some of the questions listed below:

- why are some reactions fast and others slow?
- why do the rates of some reactions show a temperature dependence?
- how can we most efficiently promote a particular reaction?
- can we study the transition state, the fleeting intermediate between reactants and products, the structure of which would tell us so much about the molecular mechanism by which the reaction occurs?
- how is the energy liberated in a reaction distributed among products?

The same questions are posed somewhat differently in Fig. 0.1, a schematic plot of the potential energy of a chemical system as it proceeds from reactants to pro-

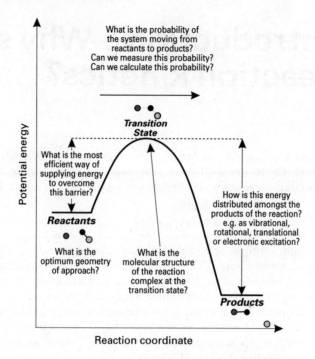

Fig. 0.1 A schematic diagram of the potential energy of a chemical system as it moves from reactants to products. The questions posed at various points along the reaction coordinate will be answered in subsequent chapters.

ducts and we can see the strong links between kinetic (e.g. rates of reaction and their temperature dependence) and dynamic (e.g. orientations of reacting molecules, energy distribution amongst products) and the potential energy surface of the reaction.

Major developments in experimental techniques, especially laser based methods, have allowed us to take significant steps forward in our understanding of the detailed mechanisms of elementary reactions. We shall investigate both the principles behind these experiments and discuss the fascinating new insights into reaction mechanisms which they have yielded.

As would be expected from such a core subject in chemistry, reaction kinetics overlaps with many other subjects. Strong links exist with thermodynamics, statistical mechanics and spectroscopy. Reaction kinetics is crucial in determining organic and inorganic reaction mechanisms and is of central importance for atmospheric chemistry. Recent interest in atmospheric problems (ozone depletion, acid rain, smog formation, and global warming) has provided a stimulus for research in reaction kinetics and provides many topical examples to illustrate the subjects discussed in this book.

We see, then, that there are two major components to reaction kinetics: a search for fundamental information about the molecular mechanisms of elementary reactions and the application of kinetic data to complex chemical systems e.g the Earth's atmosphere or industrial processes such as plastics manufacture. This division is broadly reflected in the layout of this book.

After an initial introduction and resume of basic kinetics, emphasizing the importance of elementary reactions, the next six chapters concentrate on the detailed study of elementary reactions. Firstly, in Chapter 2, we look at methods for measuring the rates of elementary reactions highlighting the improvements brought about by laser based experimental techniques. In Chapter 3 we begin to investigate theories of bimolecular reactions, such as reaction (1), which involve two reagent molecules. At present the predictive power of such theories is somewhat limited but developments in computing power, coupled with ever more searching experiments, hold out hope for quantitative validation of bimolecular theories in the near future.

Chapter 4 is concerned with the field of reaction dynamics—the molecular mechanisms by which reactions occur and how the reaction exothermicity is distributed amongst the products. Historically, reaction kinetics (measurement of the rates of reaction) and reaction dynamics (study of the molecular mechanism for a reaction) have been treated as almost separate subjects, but fundamentally kineticists and dynamicists are interested in looking at the same processes; i.e. what factors control the chemical changes from reactants to products, although they may approach the problem in differing ways.

Bimolecular reactions are not the only type of elementary processes. Several reactions, typified by the first step in the thermal decomposition of ozone:

$$O_3 \rightarrow O_2 + O \qquad \qquad (R\ 7)$$

appear to occur in a unimolecular step, i.e. no other molecule is involved in the elementary reaction. Further analysis shows that these reactions and their associated reverse, recombination reactions such as:

$$O + O_2 \rightarrow O_3 \qquad \qquad (R\ 8)$$

are much more complex in nature, involving collisional activation or stabilization, and the mechanisms of these reactions are discussed in Chapter 5.

Chapters 3–5 have concentrated entirely on reactions in the gas phase, where the structure of the reactants and products are well characterized. Chapters 6 and 7 discuss elementary reactions in different phases, namely in liquids and on metal surfaces. We shall see that we have to modify our theories, developed for the gas phase, to account for the reactant–solvent or reactant–surface interactions that occur in these environments.

The remainder of the book concentrates mainly on the more applied aspects of reaction kinetics. We introduce the subject in Chapter 8 where we look at how individual elementary reactions couple together to give the overall kinetics of a complex reaction and this theme of combining elementary reactions continues in succeeding chapters. In Chapter 9 we examine a class of complex reactions known as chain reactions, typified by the hydrogen/bromine system introduced above. Commercially important examples considered are hydrocarbon cracking (the first step in making chemical feedstocks from crude oil) and polymerization. In Chapters 10 and 11 we look at two cases where elementary reactions can combine to give spectacular and unusual results, namely explosions and oscillating reactions.

The subject of photochemistry, reactions initiated by the absorption of a photon of light, provides a good mix of the applied and the fundamental. The lifetime of a photolytically excited state is controlled by the coupling of its elementary removal processes (e.g. fluorescence, phosphorescence, chemical reaction), in ways similar to those described in Chapter 8. An alternative fate for a molecule after absorbing a photon is dissociation, e.g.

$$CH_3I + hv \rightarrow CH_3 + I. \tag{R 9}$$

Photodissociation can be thought of as a 'half reaction', absorption of a photon instantaneously produces a system at the transition state for the production of products and the study of the energy distribution of the photofragments yields similar information to reaction dynamical studies, i.e. how energy is distributed in moving from the transition state to products.

Throughout the book we include a number of worked examples illustrating the points raised in the text. Problems to test the understanding of the subject matter may be found at the end of each chapter. Equally important are the study notes which give access to, and some explanatory notes on, the original literature or more detailed texts. The student is strongly recommended to consult the references given in the study notes wherever possible.

We have decided to put some of the material in boxes. Boxed items are generally of a more tangential nature to the main thrust of the text and could be left out during the initial study of the chapter. However, we do recommend that you return to and read the material which may contain important background information, applications or an insight into current research in a particular field.

Whilst we believe that we have included all of the relevant ideas and principles for a second or third year undergraduate course on kinetics, we would not pretend that this book represents a comprehensive coverage of the whole field of reaction kinetics. In particular we recognize that there is a bias towards gas phase rather than solution or surface kinetics. Partially this reflects our backgrounds, which predominantly lie in gas phase kinetics, but from a more fundamental viewpoint we feel that gas phase studies, free from the complications of solvent or surface effects, provide a greater insight into the molecular processes governing the course of a reaction. An understanding of these processes is one of the important goals of reaction kinetics.

We hope you enjoy your study of the exciting field of reaction kinetics!

1 Time, concentration, and temperature

1.1 Introduction

The study of chemical kinetics is a fundamental part of chemistry. Chemistry is the study of reactions, therefore determinations of the rate at which these reactions occur (*kinetics*) are of central importance. The results of kinetic studies (i.e. *rates of reactions and rate coefficients*) give information which can be applied in different ways.

1.   Many important phenomena, such as combustion or stratospheric ozone depletion, involve many reaction steps. Determination of the rate of each step is necessary if we are to understand these processes fully. The practical advantages of such knowledge would be significant, e.g. developing more efficient combustion processes or reducing ozone depletion.

2.   Recent developments in experimental techniques (see Chapter 2) have allowed detailed studies of elementary reaction rates. The variation of such rates with temperature or pressure can give microscopic insights into the molecular mechanisms of these reactions.

Chemical kinetics therefore provides a wealth of information which is of both practical and fundamental interest. These are just two of the reasons why the field is so interesting and the subject of so much recent study.

The bread and butter of chemical kinetics is the *measurement of the rates of chemical reactions*. The subject started, as a quantitative field of study, in 1850 when Wilhelmy showed that the rate of hydrolysis of sucrose to form glucose and fructose:

$$C_{12}H_{22}O_{11} + H_2O \rightarrow C_6H_{12}O_6 + C_6H_{12}O_6 \tag{R 1}$$

depends on the first power of the sucrose concentration. He was able to follow the conversion by observing, with a polarimeter, the change in rotation of the plane of polarization of light as it passed through the reaction solution, sucrose being dextrorotatory. The hydrolysis, under the conditions he employed, takes the best part of a working day and, not having a thermostatted bath, he had to note the variation in temperature throughout the day and make small corrections to the observed rate.

Rate measurements were restricted to comparatively slow reactions, although not so slow as reaction (1), until the middle of this century, when a variety of experimental techniques were developed which enabled reactions of increasingly labile species to be studied on ever decreasing timescales. More recently kineticists have been able to take advantage of advances in fast electronics, lasers, and sensitive detection techniques, while microcomputer controlled experiments have greatly facilitated accurate data acquisition.

Figure 1.1 shows the decrease in concentration with time of methyl radicals, resulting from the combination reaction:

$$CH_3 + CH_3 \rightarrow C_2H_6 \tag{R 2}$$

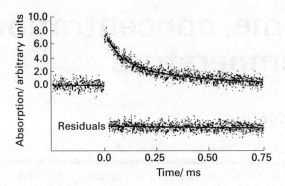

Fig. 1.1 Experimental trace for methyl radical recombination via reaction (2) following the instantaneous generation of methyl radicals at time $t = 0$ by a laser pulse (from a figure by J. Davies).

which is of importance in flame, combustion and pyrolysis chemistry. The methyl radicals were produced by photolysing propanone with a pulsed laser and their concentration measured by observing the absorption spectrum of CH_3 as a function of time. Analysis of the data demonstrates that, in this case, the rate of reaction depends on the square of the methyl radical concentration. The timescale of this reaction (several milliseconds) is short, but modestly so—it is now possible to study reactions on timescales of less than 10^{-12} s!

In this chapter we shall discuss the more phenomenological side of reaction kinetics, i.e. the way in which reaction rates depend on concentration and temperature, and how rate coefficients are defined and measured. Finally, we shall examine how rate coefficients for forward and reverse reactions are related via the equilibrium constant for the reaction and the general relationship between kinetic and thermodynamic parameters such as enthalpy and entropy. In later chapters we shall discuss the theoretical interpretation of rate coefficients, but it is as well to remember that, at least for the present, reaction kinetics is firmly based on experimental measurement.

1.2 Reaction order

The rate of reaction (2) may be defined as $-d[CH_3]/dt$, i.e. as the rate of decrease of the methyl radical concentration with time, where the square brackets denote concentration and t is time. Experiments show that the rate is proportional to the square of the methyl radical concentration, or

$$-d[CH_3]/dt = k_2[CH_3]^2 \tag{E 1}$$

where the constant of proportionality, k_2, is the *rate coefficient*. This latter term is also known as the rate constant, but as we shall see this so-called constant varies with temperature and sometimes pressure, and hence we shall use the former term. In Wilhelmy's experiment the rate was directly proportional to the sucrose concentration so:

$$-d[C_{12}H_{22}O_{11}]/dt = k_1[C_{12}H_{22}O_{11}] \tag{E 2}$$

These equations can be generalized, so that, for a reaction A + B → Products, we may write a *rate law* in the form:

$$-da/dt = ka^{\alpha}b^{\beta} \qquad \text{(E 3)}$$

where a and b are the concentrations of the reactants A and B. The exponents, α and β, *which are determined experimentally*, are termed the *orders of the reaction* with respect to A and B. The overall order of the reaction is equal to the sum of α and β. For example, for reaction (3)

$$H_2 + I_2 \rightarrow 2HI \qquad \text{(R 3)}$$

it is found that the reaction is first order in hydrogen and in iodine and hence second order overall. It must be stressed at this stage that the order of a reaction is not necessarily related to the stoichiometry of the chemical reaction. It is possible for a species which is not consumed in the reaction to occur in the rate law (see Box 1.2) and conversely a reaction may display zeroth-order kinetics in one of the reactants (i.e. it may have a rate which is independent of the concentration of that reactant). A familiar example is the iodination of propanone, which is first order in propanone and hydrogen ions, but zeroth in iodine.

$$I_2 + H^+ + CH_3COCH_3 \rightarrow CH_2ICOCH_3 + HI + H^+ \qquad \text{(R 4)}$$

$$-d[I_2]/dt = k_4[H^+][CH_3COCH_3]. \qquad \text{(E 4)}$$

Neither is it necessary for the order of a reaction to be integral. For example, the $H_2 + Br_2$ reaction is first order in hydrogen but of order one-half in Br_2. Experimental observations of this type provide important clues to the mechanisms of such reactions.

1.3 Molecularity

We ought, at this stage, to distinguish between elementary and multi-step reactions. Hydrogen atoms react with bromine molecules

$$H + Br_2 \rightarrow HBr + Br \qquad \text{(R 5)}$$

by forming a collision complex H–Br–Br; the two reactants collide and stay together for a very short period of time, before rearranging to form the products. The *molecularity* of the reaction is determined by the number of species involved in forming the collision complex. For reaction (5) this is two and hence the reaction is termed *bimolecular*. As we shall see in Chapter 5, the thermal dissociation of azomethane,

$$CH_3N_2CH_3 \rightarrow 2CH_3 + N_2 \qquad \text{(R 6)}$$

entails the acquisition by a reactant molecule of sufficient thermal energy to enable it to fall apart. The process of molecular rearrangement starts to occur in a single molecule and the reaction is termed *unimolecular*.

The distinction between order and molecularity is important. The order of a reaction is based directly on *experimental observations* of the dependence of the reaction rate on concentration. It is an *empirical* quantity and in itself says nothing about the mechanism of the reaction, although it may give some clues. The molecularity is deduced from experimental determinations of reaction order, but depends on the proposal of a consistent mechanism. Thus, experiments show that the $CH_3 + CH_3$ reaction is second order in methyl radicals. Provided the experiments have been conducted accurately, this assertion is incontrovertible. When the reaction is overall second order, we might infer that it involves the coming together of two reactant molecules. This remains an inference, and may be questioned. For example, the decomposition of dinitrogen pentoxide is first order in N_2O_5; from this we might presume that the reaction is unimolecular and proceeds via a mechanism similar to that described for the decomposition of azomethane. However, we would be wrong! Additional experiments have shown that the reaction involves several steps which combine to give a rate that depends linearly on $[N_2O_5]$.

A great many important reactions have multi-step mechanisms. We have already alluded to the iodination of propanone, in which the enol is formed in a slow step involving the protonation of the ketone,

$$\underset{CH_3-\overset{\overset{\displaystyle O}{\|}}{C}-CH_3}{\overset{\displaystyle H^+ \searrow}{}} \xrightarrow{\text{slow}} \underset{H}{\overset{H}{\diagdown}}C=C\underset{CH_3}{\overset{OH}{\diagup}} + H^+ \qquad (R\ 7)$$

followed by rapid attack by I_2 on the enol;

$$\underset{H}{\overset{H}{\diagdown}}C=C\underset{CH_3}{\overset{OH}{\diagup}} + I_2 \xrightarrow{\text{fast}} CH_2I-\underset{\overset{\|}{O}}{C}-CH_3 \qquad (R\ 8)$$
$$+ HI$$

The overall rate of the reaction is governed by the slow step, which explains why the rate equation contains the first power of the propanone and H^+ concentrations, but is independent of $[I_2]$. In Chapter 8 we shall discuss multi-step reactions examining the idea of a *rate-determining step*, such as the first step discussed above and also spend some time on the important class of *chain reactions* (Chapter 9). The latter contain several repeating steps, an example being the $H_2 + Br_2$ reaction. All of these complex reactions are made up of component elementary reactions (such as (R 5)), which may be ascribed both an order and a molecularity. The overall complex reaction has an experimentally determinable order but strictly speaking no molecularity. Sometimes it is assigned a molecularity corresponding to that of its rate-determining step as with the famous S_N1 reaction.

It should be apparent by now that the rate equation of a reaction may be totally unrelated to its stoichiometric equation. An example is the iodination of propanone discussed above. The $H_2 + Br_2$ reaction looks particularly simple on the basis of its stoichiometry,

$$H_2 + Br_2 \rightarrow 2HBr \qquad (R\ 9)$$

but its order is non-integral (rate $\propto [H_2][Br_2]^{\frac{1}{2}}$) which tells us that appearances can be deceptive. Non-integral orders are a sure sign that a reaction has a multi-step mechanism, but as we have noted in the case of N_2O_5 an integral order is no guarantee that a reaction is elementary.

Many multi-step reactions involve the reaction of labile species such as atoms, radicals or energetically excited species. The advent of laser based experiments has allowed the isolation and study of many of these elementary reactions. Once the component elementary rates have been determined the overall reaction can be simulated using a computer model (Chapter 8) and the results compared with experiments. Such studies have taken much of the guess work out of elucidating complex mechanisms and, because the elementary steps involve only a small number of species (generally 1–3), the experimental results often give great insight as to exactly how the molecules interact on a molecular level.

1.4 Determination of reaction orders and rate coefficients

In Chapter 2 we shall discuss experimental methods for studying the rates of chemical reactions. It will be helpful, however, if we first examine how the order and rate coefficient can be determined from the data such experiments provide. The data are normally in the form of something that is related to concentration of reactants or products (for example, conductivity) versus time. Several alternative methods may be used to analyse the data and we shall first discuss them individually before making comparisons between them.

1.4.1 Integral method

For a first-order reaction, such as the thermal decomposition of azomethane, the rate law can be written in the following way:

$$A \rightarrow \text{Products}$$

$$- da/dt = ka \qquad \text{(E 5)}$$

integrating, we find (see question 1.3):

$$\ln(a_0/a) = kt \qquad \text{(E 6)}$$

where a_0 is the concentration of A at time $t = 0$. Similar equations may be derived for reactions of other orders and some of these are shown in Table 1.1. These equations are valid for like reactants (e.g. A + A), or for stoichiometrically equivalent reactants (e.g. A + B) with equal initial concentrations. When the initial molar concentrations are not the same, or when the reactants are not stoichiometrically equivalent (e.g. A + 2B), more complex expressions are required, and two of the more common ones are shown in Table 1.2.

When we measure the reactant concentration as a function of time, the data should fit an expression of the type shown in the third column of Table 1.1., provided the order is integral. If we think the reaction may be first order, we plot $\ln(a)$ vs. t, and expect to get a straight line of gradient $-k$. If we think it is second order we plot $(1/a)$ against time and so on.

Table 1.1 Differential and integral forms of rate equations

Order	da/dt	kt	Units of k
0	$-k$	$(a_0 - a)$	mol dm^{-3} s^{-1}
1	$-ka$	$\ln(a_0/a)$	s^{-1}
2	$-ka^2$	$(1/a) - (1/a_0)$	dm^3 mol^{-1} s^{-1}
3	$-ka^3$	$(1/2a^2) - (1/2a_0^2)$	dm^6 mol^{-2} s^{-1}

Table 1.2 Integrated rate laws for more complex situations

Reaction	Order	da/dt	kt
$A + B \rightarrow P$	2	$-kab$	$\dfrac{1}{(b_0 - a_0)} \ln\left(\dfrac{a_0 b}{b_0 a}\right)$
$A + 2B \rightarrow P$	3	$-kab^2$	$\dfrac{a_0 - b_0}{a_0 b_0 (2a_0 + b_0)} + \left(\dfrac{1}{2a_0 - b_0}\right)^2 \ln\left\{\dfrac{a_0(b_0 - 2b)}{(a_0 - a)b_0}\right\}$

Box 1.1 Dimensions and units of the rate coefficient

The dimensions of the rate coefficient will depend on the order or molecularity of the reaction. The rate of the reaction always has a unit of concentration time^{-1}. Apart from first-order reactions they will always be a function of concentration and time (first-order rate coefficients have the dimensions time^{-1}). The dimensions can always be determined from a dimensional analysis of the rate equation. For example for a third-order reaction:

$$-da/dt \propto a^3$$

Dimensions: concentrations $\cdot$ time$^{-1} \propto$ concentration^3

To turn the above relationship into an equation the dimensions on both sides of the equation must be equal and therefore the dimensions of k for a third-order reaction are: concentration$^{-2} \cdot$ time^{-1}.

$$-da/dt = ka^3$$

Dimensions: concentration $\cdot$ time^{-1} = concentration$^{-2} \cdot$ time$^{-1} \cdot$ concentration^3

concentration $\cdot$ time^{-1} = concentration $\cdot$ time^{-1}

The *units* of k will depend on the units of concentration and time (usually seconds or minutes). In this book we shall usually use mol dm^{-3} and seconds respectively for these quantities giving units of dm^6 mol^{-2}s^{-1} for our third-order rate coefficient, but often you will come across different units for concentration in the literature. Gas phase kineticists often use units of molecule cm^{-3} for concentration and the units of torr, mtorr, or mbar, which strictly speaking are units of pressure (remember pressure and concentration are directly linked for gases), can also be encountered. Appendix I contains conversion factors for a number of common units.

Example 1.1

In the decomposition of azomethane (R 6) at a pressure of 2.18×10^4 Pa and a temperature of 576 K, the following time dependent azomethane concentrations were recorded:

t/minutes	0	30	60	90	120	150	180
[azomethane]/ 10^{-3} mol dm^{-3}	8.70	6.52	4.89	3.67	2.75	2.06	1.55

Show that the reaction is first order in azomethane and determine the rate coefficient at this temperature.

We use the integrated expression given in Table 1.1 and plot ln[azomethane] vs. time.

t/minutes	0	30	60	90	120	150	180
ln ([azomethane]/ 10^{-3} mol dm^{-3})	2.16	1.90	1.55	1.29	1.07	0.64	0.46

The plot, shown in Fig. 1.2, is a good straight line confirming the first-order kinetics. The slope ($= -9.6 \times 10^{-3}$ min^{-1}) is equal to $-k$, so that the rate coefficient is; $k_6 = 1.58 \times 10^{-4}$ s^{-1}. We may show that zeroth- or second-order kinetics do not apply by plotting [A] and 1/[A] vs. time, where [A] = [azomethane]. These plots are shown in Figs 1.3 and 1.4 and clearly demonstrate substantial curvature.

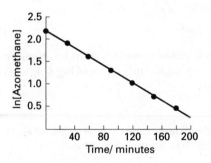

Fig. 1.2 Plot of ln[azomethane] vs. time. The straight line indicates that the reaction is first order with respect to [azomethane] and the gradient of the line is $-k$.

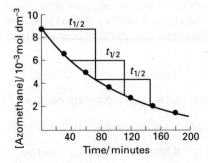

Fig. 1.3 Plot of [azomethane] vs. time. Significant curvature is observed showing that, as expected, the reaction is *not* zeroth order with respect to [azomethane]. Three half lives for the reaction are shown. For a first order decay $t_{\frac{1}{2}}$ should be independent of initial concentration on the decay curve.

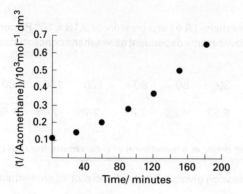

Fig. 1.4 Plot of $1/$[azomethane] vs. time. Once again the graph is curved, showing that second order kinetics do not apply.

1.4.2 Isolation method

When more than one reactant is involved the integrated rate equation becomes more complex. A common way of simplifying the situation is to arrange for all the reagents except one, say A, to be present in such great excess that their concentrations can be regarded as effectively constant during the course of the reaction. In practice this generally means that their initial concentrations should be at least $10a_0$ (and hence would have fallen by 10 per cent to $9a_0$ for a 1:1 stoichiometric reaction by the end of the reaction) and preferably $100a_0$, although this may be difficult for technical reasons, such as solubility.

Under these conditions the order with respect to A may be determined by applying the integral equations of Table 1.1. For example:

$$A + B \rightarrow \text{Products}$$

$$-\mathrm{d}a/\mathrm{d}t = ka^\alpha b^\beta = k'a^\alpha \qquad (E\,7)$$

where

$$k' = kb_0{}^\beta \qquad (E\,8)$$

and b_0 is the concentration of B which is assumed to remain constant. β, the order with respect to B may then be found by varying b_0, since a plot of $\log k'$ against $\log b_0$ should give a straight line of gradient β.

Example 1.2.

The flash photolysis (see Chapter 2) of iodine vapour in a large excess of argon produces iodine atoms, which then recombine according to the reaction:

$$I + I + Ar \rightarrow I_2 + Ar. \qquad (R\,10)$$

The role played by Ar is discussed briefly in Box 1.2 and more fully in Chapter 5. For the present we examine simply its effect on the rate equation. The rate of reaction may be determined by measuring $[I_2]$ as a function of time at several different argon pressures. In each case the iodine concentration before the flash was $1 \times 10^{-5}\ \mathrm{mol\ dm^{-3}}$.

$t/10^{-3}\,\mathrm{s}$	2	4	6	8	10	
$[I_2]/10^{-6}\,\mathrm{mol\,dm^{-3}}$	8.45	8.70	8.90	9.05	9.15	$[\mathrm{Ar}] = 2.0 \times 10^{-3}\,\mathrm{mol\,dm^{-3}}$
$[I_2]/10^{-6}\,\mathrm{mol\,dm^{-3}}$	8.53	8.96	9.19	9.34	9.44	$[\mathrm{Ar}] = 4.0 \times 10^{-3}\,\mathrm{mol\,dm^{-3}}$
$[I_2]/10^{-6}\,\mathrm{mol\,dm^{-3}}$	8.77	9.19	9.39	9.52	9.60	$[\mathrm{Ar}] = 6.0 \times 10^{-3}\,\mathrm{mol\,dm^{-3}}$

Note that the timescale for this reaction is in milliseconds, flash photolysis allows for the study of reactions at these and at much shorter timescales. The concentration of argon atoms is very much greater than that of iodine and, in any case, Ar is not consumed in the reaction so that we are justified in using the isolation method, since [Ar] is not disturbed by reaction.

From the stoichiometry of the reaction, $[I] = 2([I_2]_0 - [I_2])$, where $[I_2]_0$ is the molecular iodine concentration before the flash and also, since all the iodine atoms must eventually recombine, the molecular iodine concentration at long times after the flash. Since [Ar] is constant, we evaluate [I] and use the integral expressions shown in Table 1.1. for the second-order case. Figure 1.5 shows second-order plots, whose linearity demonstrates that the reaction is second order in I;

$$-d[I]/dt = k'[I]^2. \tag{E 9}$$

The slope of the plots give k', which clearly depend on [Ar], as expected from the relationship $k' = k_{10} [Ar]^\alpha$;

$[Ar]/10^{-3}$ mol dm^{-3}	2	4	6
$k'/10^7$ dm^3 mol^{-1} s^{-1}	3.5	6.9	10.4
$(k'/[Ar])/10^6$ dm^6 mol^{-1} s^{-1}	1.75	1.73	1.73

Figure 1.6 shows a plot of log k' vs. log [Ar]; its unit slope demonstrates that $\alpha = 1$, i.e. that the order with respect to argon is one, and that overall the reaction is third order. The table confirms this conclusion, since $k'/[Ar]$ is constant, and also gives us a mean value for the third-order rate coefficient k_{10}:

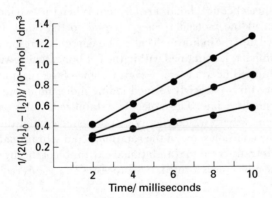

Fig. 1.5 Second-order plots for reaction (10) for varying initial argon pressure. The straight lines confirm second-order kinetics with respect to iodine.

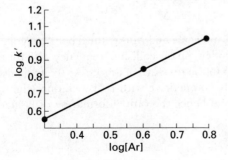

Fig. 1.6 Plot of log k' vs. log [Ar]. initial methyl concentration). The straight line graph with unit slope shows that the reaction is first order with respect to argon.

$$k_{10} = 1.74 \times 10^{10} \, \text{dm}^6 \, \text{mol}^{-2} \, \text{s}^{-1}.$$

There are two general points arising from this example, which warrant further comment. Firstly, the isolation method is not limited to use in conjunction with the integral method. It can also be used with the half-life and differential methods which are discussed below. Secondly, some ambiguity can arise in the definition of rate coefficients for elementary reactions which are second order in one of the reactants. Thus we may write:

$$-\text{d}[I]/\text{d}t = k_{10}[I]^2[\text{Ar}] \tag{E 10}$$

and

$$\text{d}[I_2]/\text{d}t = k'_{10}[I]^2[\text{Ar}]. \tag{E 11}$$

Since, by stoichiometry, $-\text{d}[I]/\text{d}t = 2\text{d}[I_2]/\text{d}t$ (two iodine atoms are removed for every iodine molecule produced), it follows that $k_{10} = 2k'_{10}$. It is therefore necessary when quoting the rate coefficient for iodine atom recombination to state clearly whether one is referring to k_{10} or k'_{10}. Kineticists generally (but not always!) use the convention associated with k'_{10}, so that

$$-\text{d}[I]/\text{d}t = 2k'_{10}[I]^2[\text{Ar}] \tag{E 12}$$

and this is the form that we shall adopt throughout the rest of this book. To be consistent, therefore, we should revise our rate coefficient in Example 1.2 to $8.7 \times 10^9 \, \text{dm}^6 \, \text{mol}^{-2} \, \text{s}^{-1}$.

Box 1.2 Pressure dependent reactions

Energy is released when chemical bonds are formed. When the two iodine atoms combine to form an iodine molecule, energy equal to the bond dissociation energy of I_2 is released into the iodine molecule. This energy, corresponding to all the reaction exothermicity, is contained within the I–I bond, which will rupture, reforming the reactant iodine atoms. However, if some of the reaction exothermicity can be removed from a newly formed iodine molecule it will no longer have enough energy to dissociate. The molecule is stabilized. Collisions with the argon bath gas are one way in which energy can be removed and the molecule stabilized. During the collision some of the rotational and vibrational energy of the iodine molecule is transferred to translational excitation of the argon atom.

Therefore once the molecule is formed, two processes can occur, redissociation or stabilization.

$$I + I \rightleftharpoons I_2^* \xrightarrow{\text{M}} I_2$$

where I_2^* represents the nascent I_2 molecule and M the 'third body', argon.

The rate of stabilization is dependent on the collision frequency with the argon bath gas, which is in turn related to the pressure of argon. As the pressure rises stabilization becomes increasingly competitive with redissociation (the rate of which is pressure independent) and hence the rate of formation of I_2 increases with the pressure of bath gas.

1.4.3 Half-life method

The half-life of a reaction $(t_{\frac{1}{2}})$ is the time taken for 50 per cent of the reactant(s) to be consumed and hence $t_{\frac{1}{2}}$ may be found by substituting $a = a_0/2$ in column three of Table 1.1 and equating this to $kt_{1/2}$. For a first-order reaction:

$$\ln\left[(a_0/2)/a_0\right] = -kt_{\frac{1}{2}} \tag{E 13}$$

therefore,

$$t_{\frac{1}{2}} = \ln 2/k \tag{E 14}.$$

The resulting expressions for integral orders and for a general order α are shown in Table 1.3. Note that the half-life is always inversely proportional to k and that its dependence on a_0 is determined by the order of the reaction. These relationships may therefore be used to determine both the order of a reaction and its rate coefficient. There are two possible methods.

1. We can follow a reaction over several half-lives and examine the dependence of the sequence of half-lives on the concentration at the start of that half-life period.

2. Several experiments can be performed, each with a different initial concentration and the first half-life for each experiment determined.

These two methods are demonstrated in Examples 1.3 and 1.4.

Table 1.3 Half-life expressions

Reaction Order	$t_{\frac{1}{2}}$
0	$a_0/(2k)$
1	$(\ln 2)/k$
2	$1/(ka_0)$
3	$3/(2ka_0^2)$
α	$\dfrac{2^{\alpha-1}-1}{(\alpha-1)ka_0^{\alpha-1}}$

Example 1.3

We apply the first method to the decomposition of azomethane (R 6). Figure 1.3 (which we used to demonstrate that the reaction is not zeroth order in azomethane) shows a plot of [azomethane] vs. t and three half-lives are indicated. The following data can be obtained on three half-lives of the reaction:

Initial concentration/ 10^{-3} mol dm^{-3}	8.70	6.00	4.35
Half-life/minutes	72	70	72

The half-lives are clearly independent of the concentration at the start of the relevant time period, as is required for a first-order reaction (Table 1.3). The mean half-life may be used to calculate k_6:

$$k_6 = \ln 2/t_{\frac{1}{2}} = \ln 2/(71 \times 60s)$$
$$k_6 = 1.63 \times 10^{-4}\,s^{-1}$$

in good agreement with the value obtained using the integral method.

Example 1.4
We showed in Fig. 1.1 the decay of methyl radicals as determined in a flash photolysis experiment. In this example, we determine the order and rate coefficient for the reaction from the measurement of half-lives.

$[CH_3]_0/10^{-8}\,mol\,dm^{-3}$	2.0	5.0	10.0	20.0
$t_{\frac{1}{2}}/10^{-3}\,s$	80	35	16	7.5
$10^{10}\,dm^3\,mol^{-1}/(2[CH_3]_0)\cdot t_{\frac{1}{2}}$	3.1	2.9	3.1	3.3

Table 1.3 demonstrates that, except for a first-order reaction, $t_{\frac{1}{2}}$ is proportional to $a_0^{-(\alpha-1)}$ so that a plot of log $t_{\frac{1}{2}}$ vs. log a_0 has a slope of $-(\alpha-1)$. Figure 1.7 shows the half-lives plotted in this way and the slope has a value of -0.95 or approximately -1. Hence the reaction order, α, is 2. The rate coefficient may now be determined from the relevant expression in Table 1.3. We must remember though our convention since two radicals are removed in each elementary reaction. Thus,

$$k_2 = 1/(2.[CH_3]_0.t_{\frac{1}{2}})$$

The values of k_2 calculated for each datum point are shown in the above table; they agree reasonably well and give a mean rate coefficient of $k_2 = 3.1 \times 10^{10}$ $dm^3\,mol^{-1}\,s^{-1}$.

Finally we note that the general relationship shown in Table 1.3 between $t_{\frac{1}{2}}$ and the order, α, applies even if α is non-integral.

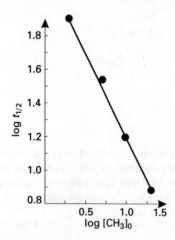

Fig. 1.7 Plot of log $t_{1/2}$ vs. log (initial methyl concentration). The gradient of the graph $(-0.95) = -(\alpha - 1)$ confirms the expected order of two with respect to methyl radical concentration.

1.4.4 The differential method

Taking logs of the generalized rate law at time $t = 0$,

$$da_0/dt = ka_0^\alpha b_0^\beta \ldots \tag{E 15}$$

gives:

$$\log(\mathrm{d}a_0/\mathrm{d}t) = \log k + \alpha \log a_0 + \beta \log b_0 + \ldots \qquad \text{(E 16)}$$

By varying the values of a_0, (holding b_0 etc. constant) and measuring the *initial* rates, we can find the order, α, with respect to A and by a similar method, this time varying b_0 only, β can be obtained.

So called initial rate methods can be useful in elucidating the mechanism of complex reactions, where the rate of reaction may be dependent on the concentration of the products (e.g. in the H_2/Br_2 system described in Chapter 9 it is found that increasing concentrations of the product, HBr, slows down the rate of reaction and alters the order with respect to bromine). At the start of the reaction the concentration of products will be extremely small and the effects of reactions involving the products will be negligible.

Alternatively, for a reaction rate given by

$$-\mathrm{d}a/\mathrm{d}t = ka^{\alpha} \qquad \text{(E 17)}$$

$$\log(-\mathrm{d}a/\mathrm{d}t) = \log k + \alpha \log a. \qquad \text{(E 18)}$$

Tangents to the concentration vs. time plot at various points along the graph give $-\log(\mathrm{d}a/\mathrm{d}t)$. If these are plotted against the corresponding values of $\log a$ then k and α can be calculated from the intercept and gradient respectively. This procedure is illustrated schematically in Fig. 1.8.

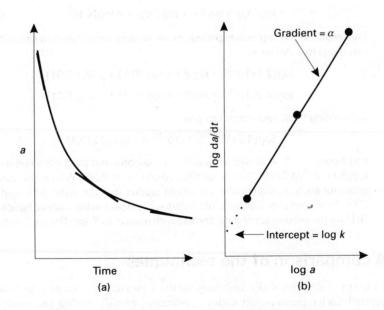

Fig. 1.8 (a) Estimates of the rate of reaction can be obtained by drawing slopes to the concentration vs. time graph. (b) A plot of the rates vs. the corresponding value of $\log a$ gives a straight line plot with gradient = α. Note that the intercept with the y-axis corresponds to $\log k$ (see (E 18)).

Example 1.5

The initial rates of reaction between two organic bromides, X and Y, and hydroxide ions, were measured as a function of initial concentrations:

$[X]_0$ /mol dm^{-3}	10^{-1}	2×10^{-1}	2×10^{-1}
$[OH^-]_0$ /mol dm^{-3}	10^{-2}	10^{-2}	2×10^{-2}
Initial rate /dm^3 mol^{-1} s^{-1}	2.1×10^{-5}	4.2×10^{-5}	8.1×10^{-5}
$[Y]_0$ /mol dm^{-3}	2×10^{-3}	4×10^{-3}	2×10^{-3}
$[OH^-]_0$ /mol dm^{-3}	10^{-3}	10^{-3}	2×10^{-3}
Initial rate /dm^3 mol^{-1} s^{-1}	1.5×10^{-2}	2.8×10^{-2}	1.4×10^{-2}

For the first set of experimental data we shall calculate the orders with respect to the organic halide X and OH$^-$. The reaction can be written as

$$X + OH^- \rightarrow products$$

therefore the generalized rate equation is:

$$-d[X]/dt = R = k[X]^\alpha[OH^-]^\beta \qquad (E\,19)$$

and the initial rate will be given by:

$$-d[X]/dt = R_0 = k[X]_0^\alpha[OH^-]_0^\beta \qquad (E\,20)$$

taking logs:

$$\log(R_0) = \log k + \alpha \log [X]_0 + \beta \log[OH]_0 \qquad (E\,21)$$

The results of the first two experiments allow us to formulate two simultaneous equations to solve for α.

$$\log(2.1 \times 10^{-5}) = \log k + \alpha \log (0.1) + \beta \log(0.01)$$
$$\log(4.2 \times 10^{-5}) = \log k + \alpha \log (0.2) + \beta \log(0.01)$$

Subtracting these two equations gives:

$$\log(4.2 \times 10^{-5}/2.1 \times 10^{-5}) = \alpha \log(0.2/0.1)$$

and hence $\alpha = 1$. A similar analysis for the second and third experiments gives $\log(8.1 \times 10^{-5}/4.2 \times 10^{-5}) = \beta \log(0.02/0.01)$, $\beta = 0.95$. Within the expected errors for such an experiment we would predict that the order with respect to [OH$^-$] was also one. Calculate the orders with respect to the organic halide Y and OH$^-$ for the second set of data. (orders with respect to Y and OH are 1 and 0).

1.5 A comparison of the techniques

In many cases use of the isolation method means that many reactions can be studied under pseudo-first order conditions, greatly easing the analysis. Two important points should be noted about first order kinetics:

1. With the integral method *we do not need to know the absolute concentration of A, but rather its relative value (or some quantity directly proportional to it)*. For many kinetic determinations, especially those occurring on short timescales, it is much more

convenient to measure some property (conductivity, light absorption or fluorescence) that is proportional to concentration. Because we only require relative values we do not need to calibrate the apparatus to the absolute concentration, thus simplifying the experiment.

2. Once again, using the half-life technique we do not require absolute concentrations, as the half-life is independent of initial concentration for a first-order reaction. We shall return to, and illustrate these points in Chapter 2.

Generally we are interested in the magnitude of the rate coefficient and this is best extracted from an integrated plot. It is always worthwhile calculating k over a number of half-lives to check that there are no secondary reactions systematically affecting the data.

If, on the other hand, we are interested in the order of the reaction, this can be obtained in one plot using the differential method. However, a few notes of caution about the differential method:

1. The rate coefficient can only be found by extrapolation to an intercept, increasing the error in k.

2. Determining tangents by eye from a concentration vs. time plot is generally subject to considerable uncertainty.

3. Merely following the initial rates can be an oversimplification if the rate of reaction is affected by the formation of products. While the rate coefficient and order obtained will be correct at times close to $t = 0$, they may not be valid over the whole course of the reaction. A good example is the reaction of hydrogen and bromine to give hydrogen bromide (p. 223), a *chain reaction*.

$$H_2 + Br_2 \rightarrow 2HBr \tag{R 9}$$

$$d[HBr]/dt \propto [H_2][Br_2]^{\frac{1}{2}}. \tag{E 22}$$

However, owing to the complex interaction of reactants and products the overall expression is (E 23):

$$\frac{d[HBr]}{dt} \propto \frac{[H_2][Br_2]^{\frac{3}{2}}}{[Br_2] + C[HBr]} \tag{E 23}$$

where C is a constant and the term in [HBr] becomes more significant as reaction proceeds. Note the non integer order, typical of a chain mechanism (Chapter 9).

1.6 Dependence on temperature

Thus far, we have discussed the rates of chemical reactions and their dependence on concentration only. It is also found experimentally that the rate and, hence the rate coefficient, both depend on the temperature (T) and that this dependence is often very strong. For the present we shall content ourselves with a phenomenological discussion.

It is found *experimentally* that the vast majority of reactions have rate coefficients that follow the relationship:

$$k = A \exp(-E_{expt}/RT) \tag{E 24}$$

where A is called the pre-exponential or A factor and E_{expt} is the activation energy. We have added the subscript expt to indicate that the latter is experimentally determined and defined. We see from this relationship that, provided

E_{expt} is positive (which is generally the case), the rate coefficient increases with temperature, and that a straight line relationship can be obtained, and hence A and E_{expt} determined by plotting $\ln k$ vs. $1/T$:

$$\ln k = \ln A - E_{\text{expt}}/RT. \tag{E 25}$$

Table 1.4 gives activation energies for a few representative reactions and demonstrates the wide range of temperature dependencies.

Table 1.4 Typical activation energies for a number of reactions

Reaction	Activation Energy/kJ mol^{-1}
$2HI \rightarrow H_2 + I_2$	183.0
$H_2 + I_2 \rightarrow 2HI$	157.0
$C_4H_9Cl + H_2O \rightarrow C_4H_9OH + HCl$	180.0
$OH + CH_4 \rightarrow H_2O + CH_3$	19.5
$Cl + O_3 \rightarrow ClO + O_2$	2.1
$CH + CH_4 \rightarrow C_2H_4 + H$	−1.7

Example 1.6

In Example 1.1 we determined the rate coefficient for azomethane decomposition at 576 K. Similar determinations over a range of temperatures allow us to calculate A and E_{expt}

T/K	523	541	560	576	593
k/s^{-1}	1.8×10^{-6}	1.5×10^{-5}	6.0×10^{-5}	1.6×10^{-4}	9.5×10^{-4}

We plot $\ln k$ vs. $1/T$ (Figure 1.9):

$\ln (k/10^{-6}\,s^{-1})$	0.59	2.71	4.09	5.08	6.86
$(1000\ \text{K})/T$	1.912	1.848	1.786	1.736	1.686

A linear least squares fit to the data gives a value of $2.6 \times 10^4\ \text{K}^{-1}$ for E_{expt}/R and hence an activation energy of $220\ \text{kJ mol}^{-1}$, and returns a value of $2.0 \times 10^{16}\ \text{s}^{-1}$ for A. If we have drawn the graph by eye, then instead of making the long extrapolation to $1/T = 0$, it is best to take a value from the line and use the

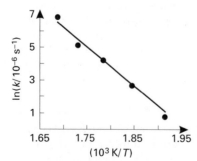

Fig. 1.9 Arrhenius plot for the decomposition of azomethane.

calculated value E_{expt} and eqn (24) to calculate A. e.g. at 1000 K/T = 1.75, ln k = −8.67 (remember we have plotted ln $(k/10^{-6}\,s^{-1})$, hence:

$$\ln A = -8.67 + 1.75 \times 26.0 = 36.83$$

$$A = 1.0 \times 10^{16}\,s^{-1}.$$

The temperature-dependent form of the rate coefficient, eqn (E 24) is called the Arrhenius equation after Svante Arrhenius, who published a paper in 1889 demonstrating that a wide range of reactions have temperature dependencies that conform to this equation. It arises because the reactants have to overcome an energy barrier (Fig. 1.10) when their valence electrons are rearranged as the products are formed. In Chapter 3 we shall examine, in some detail, theories of chemical kinetics which attempt to rationalize the Arrhenius equation. Arrhenius's paper followed one by van't Hoff, in which an equation compatible with (E 24) was proposed on the basis of the relationships between the equilibrium constants and the rate constants for the forward and reverse reactions. An interesting account of these developments may be found in articles by Logan and Laidler.

It is important to emphasize that the Arrhenius equation is an *experimental observation* that is only followed *approximately* over a finite temperature range. In the paragraph above we introduced some ideas as to what physical property at least one of the parameters might represent and we shall investigate the relationship between experimental observations of $k(T)$ (the rate coefficient as a function of temperature) and theories of bimolecular reactions in Chapter 3.

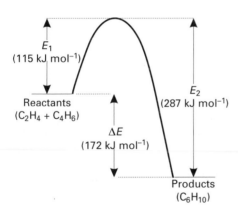

Fig. 1.10 Reaction profile for the reaction of ethene with butadiene, showing the relationship between forward and reverse activation energies and the overall reaction exothermicity.

1.7 Thermodynamics and its relationship to kinetics

Equilibrium is dynamic with reaction proceeding in both forward and reverse directions, but with equal rates so that the equilibrium concentrations are maintained. Thus, for the isomerization equilibrium between cis and trans 3-phenyl-propenonitrile;

$$\text{cis-}C_6H_5CH{=}CHCN \rightleftharpoons \text{trans-}C_6H_5CH{=}CHCN \qquad (R\ 11)$$

we have at equilibrium

$$\frac{\rho_f}{\rho_r} = \frac{k_f[\textit{cis-}C_6H_5CH{=}CHCN]_{eq}}{k_r[\textit{trans-}C_6H_5CH{=}CHCN]_{eq}} = 1 \qquad \text{(E 26)}$$

where ρ_f and ρ_r are the rates of the forward and reverse reactions and the subscript eq refers to equilibrium. Therefore

$$\frac{k_f}{k_r} = \frac{[\textit{trans-}C_6H_5CH{=}CHCN]_{eq}}{[\textit{cis-}C_6H_5CH{=}CHCN]_{eq}} \qquad \text{(E 27)}$$

or

$$k_f/k_r = K_{11} \qquad \text{(E 28)}$$

where K_{11} is the equilibrium constant.

Returning to reactive systems far from equilibrium, it sometimes turns out that a reaction rate is easier to measure in one direction than another. For example, the recombination of methyl radicals to form ethane (R 2)

$$CH_3 + CH_3 \rightarrow C_2H_6 \qquad \text{(R 2)}$$

has been studied by flash photolysis over a wide range of temperatures and pressures. The reverse decomposition reaction is important in a variety of high temperature processes, such as combustion and there is considerable interest in the value of k_{-2}. The decomposition rate has been measured, but it is difficult to make such measurements over a wide range of temperatures. We discussed above the relationship between the rate coefficients and the equilibrium constant at equilibrium; the rate coefficients are constant (at a given T and P) and do not change as equilibrium is approached. Therefore it is possible to use the measured value of k_2 and the equilibrium constant K_2 to calculate the value of the dissociation rate coefficient k_{-2}.

$$k_{-2} = k_2/K_2. \qquad \text{(E 29)}$$

However, we need to know K_2. Firstly we recall that rate coefficients generally have concentration units, so that K in equation (29) is K_c. Converting to $K_p/p°$ ($K_c = K_p/p°(RT)^{-1}$) and invoking the relationship between $K_p/p°$ and $\Delta G°$ ($\Delta G° = -RT \ln K_p/p°$ and $\Delta G° = \Delta H° - T\Delta S°$):

$$k_{-2} = k_2/(K_p/p°(RT)^{-1}) \qquad \text{(E 30)}$$

therefore

$$k_{-2} = (k_2 RT)\exp(\Delta H°_T/RT)\exp(-\Delta S°_T/R) \qquad \text{(E 31)}$$

where $\Delta H°_T$ and $\Delta S°_T$ are the standard enthalpy and entropy of reaction at temperature T:

$$\Delta H°_T = \Delta H°_{fT}(C_2H_6) - 2\Delta H°_{fT}(CH_3) \qquad \text{(E 32)}$$

$$\Delta S°_T = S°_T(C_2H_6) - 2S°_T(CH_3) \qquad \text{(E 33)}$$

where $\Delta H°_{fT}$, $S°_T$ are the molar standard enthalpy of formation and the molar third law entropy of the respective species at temperature T. These quantities are well known for stable compounds such as ethane, but what about radicals, for which we cannot use standard experimental thermochemical techniques? CH_3 is spectroscopically well characterized, so that its geometry and vibrational frequencies are known and the techniques of statistical mechanics may be used to calculate S_T° (CH$_3$). For ΔH_{fT}°(CH$_3$), we simply turn the problem on its head and measure forward and backward rate coefficients, choosing a system where both measurements are feasible. The approach is illustrated in the following example for determining the heat of formation of the t-butyl (methyl-2-propyl) radical.

Example 1.7
At 298 K the forward and reverse rate coefficients for reaction (12)

$$Br + i\text{-}C_4H_{10} \rightarrow HBr + t\text{-}C_4H_9 \qquad\qquad (R\,12)$$

$\Delta H_{f,298}^\circ$/kJmol^{-1}	112	−134.2	−36.4	??
S_{298}°/Jmol^{-1} K^{-1}	175	295.3	198.7	313

were determined to be $k_f = 1.02 \times 10^6$ dm^3 mol^{-1} s^{-1} and $k_r = 1.67 \times 10^{10}$ dm^3 mol^{-1}s^{-1}. The equilibrium constant is therefore equal to 1.02×10^6 dm^3 mol^{-1} s^{-1}/ 1.67×10^{10} dm^3 mol^{-1} s^{-1} = 6.10×10^{-5}. In $K_p = -9.70$ and hence $\Delta G_{r,298}^\circ = 24.0$ kJ mol^{-1} (using $\Delta G^\circ = -RT$ In K_p/p°). $\Delta S_{r,298}^\circ = +41.4$ J mol^{-1} K^{-1}. Using the relationship $\Delta G^\circ = \Delta H^\circ - T\Delta S^\circ$ the reaction enthalpy $\Delta H_{r,298}^\circ = +36.3$ kJ mol^{-1}. Using the measured values for the enthalpy of formation of the other three components, a value for $\Delta H_{f,298}^\circ$ (t-C$_4$H$_9$) of + 50.4 kJ mol^{-1} is calculated.

1.8 Parallel and consecutive reactions

We conclude this introductory chapter with a brief look at how elementary reactions can couple together. We shall return to the subject in much greater detail in Chapter 8.

1.8.1 Parallel reactions

In some circumstances a reactant may be removed by two or more reactions. We are interested in knowing what is the overall rate of removal of the reactant and what fraction of it is removed by the various different channels. A relevant example involves chlorine atoms generated by the ultraviolet (UV) photolysis of chlorofluorocarbons (CFCs) in the stratosphere. Chlorine can react with ozone or methane, the former reaction being part of a catalytic cycle removing ozone, the latter forming a relatively harmless 'reservoir' species holding the chlorine in an unreactive form.

$$Cl + O_3 \rightarrow ClO + O_2 \qquad\qquad (R\,13)$$

$$Cl + CH_4 \rightarrow HCl + CH_3. \qquad\qquad (R\,14)$$

As both of these are elementary reactions we can immediately write down the rate equations for each reaction:

$$-d[Cl]/dt = k_{13}[Cl][O_3] \qquad\qquad (E\,34)$$

$$-d[Cl]/dt = k_{14}[Cl][CH_4]. \qquad\qquad (E\,35)$$

Chlorine atoms will be lost by both processes and hence the overall rate of loss will be the sum of the two individual rates:

$$-d[\text{Cl}]/dt = k_{13}[\text{Cl}][\text{O}_3] + k_{14}[\text{Cl}][\text{CH}_4]. \qquad \text{(E 36)}$$

The concentrations of ozone and methane will be much higher than that of the chlorine atoms and hence we can make a pseudo-first-order approximation

$$-d[\text{Cl}]/dt = k'[\text{Cl}] \qquad \text{(E 37)}$$

where $k' = k_{13}[\text{O}_3] + k_{14}[\text{CH}_4]$. The fraction of chlorine atoms reacting with either ozone or methane will depend on the ratio of the pseudo-first-order rate coefficients for the particular reaction to the overall rate of loss:

$$f(\text{ClO}) = k_{13}[\text{O}_3]/k' \qquad \text{(E 38)}$$

where $f(\text{ClO})$ is the fraction of ClO formed.

1.8.2 Consecutive reactions

An elementary reaction may form part of a chain of reactions, the product of the first reaction going on to form a second product e.g. for two consecutive first order reactions we have:

$$A \rightarrow B \rightarrow C. \qquad \text{(R 15,16)}$$

The overall reaction involves the conversion of A to C via the *intermediate* B. The concentration profiles of the three species will depend on the relative values of the rate coefficients k_{15} and k_{16}. The differential equations controlling the concentrations of A, B and C can be solved and Fig. 1.11 shows two possible extremes. In the first case, where k_{16} is much smaller than k_{15}, we get substantial conversion of A into B before much C is formed. The growth of C approximately matches the decay of B, the rate-determining step of the reaction is the conversion of B to C. In the second case, B is a very reactive intermediate, as k_{16} is very large. The concentration of B never builds up to any substantial level; as soon as a B molecule is formed it rapidly reacts to form C. The growth of C now approximately matches the decay of A. After a short initiation period we note that the concentration of B remains approximately constant and this value is known as the

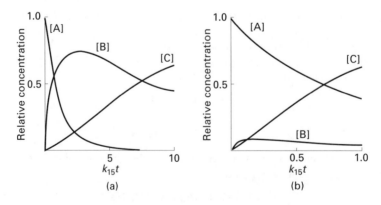

Fig. 1.11 Concentrations of A, B, and C during the conversion of A to C via the consecutive reactions, (R 15) and (R 16). (a) $k_{15} \gg k_{16}$, (b) $k_{15} \ll k_{16}$.

steady-state concentration. During this period of constant [B], written as $[B]_{ss}$, *the rate of formation of B must equal its rate of decay.* We shall make use of the so-called steady-state approximation in Chapters 5 and 6 and discuss in more detail its derivation and applicability in Chapter 8.

1.9 Questions

1.1   Write a concise definition of order and molecularity.

1.2   When concentrations of reactants are measured in mol dm^{-3}, what are appropriate units for (i) the rate, (ii) the rate coefficient, for
(a) a reaction of zero order;
(b) a reaction of order 1.5?

1.3   Derive equation $\ln(a_0/a) = kt$ from the rate law for a first order reaction.
   The following data give the pseudo-first-order rate coefficient for (R 17) under conditions where $[CH] \ll [O_2]$. :

$$CH + O_2 \rightarrow \text{Products} \qquad\qquad (R\ 17)$$

$[O_2]/10^{-7}$ mol dm^{-3}	2.56	6.41	9.19	13.12	17.28	21.43	24.09
k'/s^{-1}	19 600	29 850	34 590	45 700	59 700	64 700	80 700

Determine the bimolecular rate coefficient.
   CH radicals are known to play important roles in combustion. For example the reaction of CH with N_2 is thought to lead to the formation of nitrogen oxides in car exhausts. These compounds play an important role in urban smog formation.

1.4   The thermal decomposition of dimethyl ether has been studied by measuring the increase in pressure with time.

$$(CH_3)_2O \rightarrow CH_4 + H_2 + CO \qquad\qquad (R\ 18)$$

The data in the table were obtained at 777 K with an initial pressure of ether equal to 312 mmHg (torr).

Time/s	390	777	1195	2000	3155
Pressure increase/mmHg	96	179	250	363	467

Show that the reaction is first order and find the rate coefficient.

1.5   Methane is an important trace constituent of our lower atmosphere (the Troposphere). It is a significant greenhouse gas absorbing infrared radiation from the Earth's surface leading to potential global warming. The main mechanism for methane removal is reaction with the hydroxyl radical:

$$OH + CH_4 \rightarrow H_2O + CH_3. \qquad\qquad (R\ 19)$$

The temperature of the troposphere declines with altitude. Calculate the relative change in the rate coefficient for reaction (19) at the Earth's surface ($T = 295$ K) and at the top of the troposphere ($T = 220$ K). Reaction (19) has an activation energy of 19.5 kJ mol^{-1}. .

1.6   The rate of a certain reaction doubles on increasing the temperature from 290 K to 300 K. Evaluate the activation energy.

1.7   Experiments have established that the rate coefficient for the reaction

$$H + H_2O_2 \rightarrow OH + H_2O \qquad\qquad (R\ 20)$$

is given by the expression $k_{20} = 1.0 \times 10^{10} \exp(-1800/T)$ dm^3 mol^{-1} s^{-1}. If ΔH° and ΔS° for the reaction are -285 kJ mol^{-1} and 24.5 J K^{-1} mol^{-1} respectively, obtain an Arrhenius expression for the rate coefficient of the reverse reaction.

References

Laidler, K. J., and King, M. C. (1984). The development of the Arrhenius equation. *Journal of Chemical Education*, **61**, 494–8.

Logan, S. R. (1982). The origin and status of the Arrhenius equation. *Journal of Chemical Education*, **59**, 279–81.

2 Experimental techniques

2.1 Introduction

We have seen from Chapter 1 that in order to carry out kinetic investigations we need to measure the time-dependence of the concentrations of reactants (or products). This generalization applies whether we are determining the magnitude of the rate coefficient or investigating the order of a particular reaction. A kinetic experiment must therefore fulfil the following criteria:

- bring the reactants together, mix them and initiate the reaction on a timescale that is negligible compared to that of the reaction
- measure the concentration of reactants or products as a function of time after initiation
- accurately measure and control the temperature (and for some reactions, the pressure) at which the reaction occurs.

These criteria are relatively easily fulfilled for reactions occurring on timescales of minutes or hours and are illustrated in Section 2.2, for the study of the hydrolysis of 2-chloro-2-methylpropane. However, much of the thrust of this book involves the study of *elementary reactions* and the timescales of these reactions are often many orders of magnitude smaller. The rates of these reactions are of both practical interest (for example in modelling atmospheric chemistry or surface catalysis) but also provide information about the mechanisms of the molecular interactions (*molecular dynamics*).

After a review of classical techniques, the bulk of this chapter deals with techniques to measure elementary reactions. Four basic techniques are described for the mixing and initiation of such reactions (discharge flow, flash photolysis, relaxation methods, and shock tubes). For each technique a number of different methods can be used to monitor the reactant (or product) concentrations and some of the more common detection techniques are described for each of the basic methodologies. In Chapter 4 we shall return to a discussion of experimental techniques to describe the specialized methods used to study reaction dynamics. Finally a number of methods of attaining high (e.g. for flame modelling) or low temperatures (e.g. for interstellar chemistry or stratospheric ozone depletion) are briefly outlined.

2.2 Classical techniques

As an example of a kinetic analysis that might be encountered in the early stages of an undergraduate kinetics course, we shall look at the hydrolysis of 2-chloro-2-methylpropane, followed using conductivity measurements. Reaction (1) proceeds with the evolution of ionic products and hence the conductivity of the reaction solution will increase.

$$C_4H_9Cl + H_2O \rightarrow C_4H_9OH + H^+ + Cl^-. \tag{R 1}$$

The rate equation for the formation of products (P) is given in eqn (1):

$$dP/dt \propto [C_4H_9Cl]^{\alpha}[H_2O]^{\beta} \qquad \text{(E 1)}$$

Reactant and product concentrations are directly linked via the stoichiometry of reaction (1). At any time t the concentration of C_4H_9Cl is given by:

$$[C_4H_9Cl]_t = [C_4H_9Cl]_0 - \tfrac{1}{2}([H^+]_t + [Cl^-]_t) \qquad \text{(E 2)}$$

and

$$[C_4H_9Cl]_0 = \tfrac{1}{2}([H^+]_\infty + [Cl^-]_\infty) \qquad \text{(E 3)}$$

where $[H^+]_\infty$ and $[Cl^-]_\infty$ represent the concentrations of the ionic products at 'infinite' time, i.e. at the end of the reaction. The reaction is carried out in an 80:20 (v/v) solution of water and propanone. As the concentration of C_4H_9Cl is kept low the concentration of water is essentially constant and the reaction is carried out under isolation conditions. Therefore:

$$dP/dt \propto \{([H^+]_\infty + [Cl^+]_\infty) - ([H^+]_t + [Cl^+]_t)\}^{\alpha}. \qquad \text{(E 4)}$$

The conductivity (C_t) of the solution will be directly proportional to the concentration of ionic products and hence we can rewrite eqn (4) as:

$$dC_t/dt = k_1(C_\infty - C_t)^{\alpha}. \qquad \text{(E 5)}$$

We shall *assume* that the reaction is first order in C_4H_9Cl and therefore $\alpha = 1$. Equation (5) can now be integrated to give:

$$\ln\{(C_\infty - C_t)/C_\infty\} = -k_1 t. \qquad \text{(E 6)}$$

If we are correct in our assumption then a plot of $\ln(C - C_t)$ vs. t should give a straight line of gradient $-k_1$.

Figure 2.1 shows the apparatus. The reaction is carried out in a thermostatted and well-stirred reaction vessel. Conductivity is measured using a pair of platinum black electrodes and the conductivity meter is connected directly to a chart recorder. The solvent (100 cm^3 of 80/20 water/propanone) is added to the vessel and the temperature allowed to stabilize. At this point 1 cm^3 of C_4H_9Cl solution (in dry propanone) is added to the mixture and half-way through the injection the chart recorder is started. Figure 2.2 illustrates a typical conductivity plot as a function of time. Once the reaction has finished (typically two minutes at 310 K) the data can be analysed according to eqn (6). Figure 2.3 shows a typical set of results confirming our assumption of first-order kinetics with respect to C_4H_9Cl and giving a value for k_1 of 0.0132 s^{-1} at 303 K. The temperature of the thermostatted bath can now be altered and the experiment repeated to determine an activation energy and A factor for the reaction.

A potential source of error in this experiment is the time taken for mixing. Reaction (1) has a high activation energy ($\approx 80 \text{ kJ mol}^{-1}$) and k_1 increases rapidly with temperature. Above 323 K the half-life of the reaction becomes comparable with the time taken for mixing to occur and the experiment is then subject to considerable error i.e. where exactly is $t = 0$ on the x-axis and under what conditions of concentration was most of the reaction occurring?

Note that for kinetic studies of a first-order reaction we do not need to know the *absolute* concentration of the reactant. In order to obtain our rate coefficient we can plot the relative values of the infinite and time dependent conductivities. However, for any other order of reaction we do need to know the absolute concentration, i.e. we can either measure the concentration directly e.g. by removing

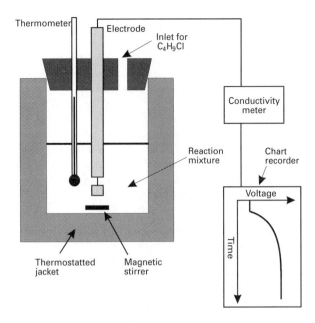

Fig. 2.1 Schematic diagram of conductivity apparatus.

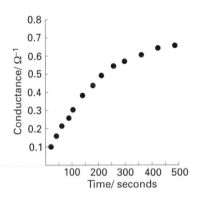

Fig. 2.2 Variation of conductance with time following initiation of reaction (1).

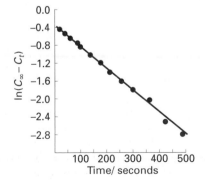

Fig. 2.3 Plot of $\ln(C_\infty - C_t)$ vs. time. The linear nature of the graph confirms first order kinetics and the gradient yields a value of $0.0132\,\text{s}^{-1}$ as the first order rate coefficient.

aliquots of reaction mixture and titrating for very slow reactions, or by directly calibrating our measure of concentration (e.g. conductivity) against absolute concentrations. Other properties of a reaction system which are proportional to concentration and might be used to follow reactions include: volume or pressure changes, light transmission, or fluorescence intensity.

Classical methods of this type are limited by the time taken for mixing and initiating the reaction and in some cases for the analysis technique (e.g. titrations). For example, a second-order reaction with a pseudo-first-order rate coefficient of 1 s^{-1} has a half-life of 0.7 s. Given a mixing time of 5 s this reaction would be 127/128 complete before any observation could be made! Nevertheless, a multitude of important reactions do occur on slow enough timescales to be measured by these so called 'classical methods'.

2.3 Discharge flow

The first 'fast' method we shall investigate is the *discharge flow technique*. A schematic diagram of the apparatus is shown in Fig. 2.4. The time between reaction initiation (point A) and reactant (or product) detection (point D) can be calculated if the velocity of the gas mixture in the flow tube is known. As an example we shall consider the reaction of OH radicals with cyclohexane.

$$OH + C_6H_{12} \rightarrow \text{products.} \qquad \text{(R 2)}$$

OH radicals are generated by an indirect process. Initially H atoms are produced by the microwave discharge of a mixture of molecular hydrogen in helium (~ 1

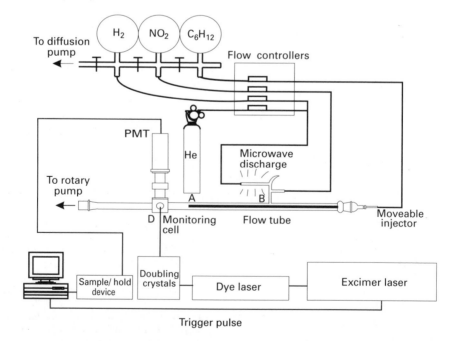

Fig. 2.4 Schematic diagram of a discharge flow apparatus for the study of OH reactions with detection by laser induced fluorescence (from an original drawing by P. Smurthwaite).

per cent H_2 at a total pressure of $\approx 1 \times 10^2$ Pa) in one of the side arms of the flow tube. The H atoms are then injected into the main body of the flow tube where they react, in a well characterized reaction, with an excess concentration of NO_2.

$$H + NO_2 \rightarrow OH + NO. \qquad (R\ 3)$$

Reaction (3) is very fast and by point B all the H atoms have been converted to OH radicals. Subsequently, cyclohexane is introduced from a *moveable* injector and starts to react with the OH radicals. The reaction is performed under pseudo-first-order conditions, i.e. $[OH] \ll [C_6H_{12}]$ and hence the OH radicals will decay exponentially with time. The *relative* concentration of OH radicals is detected at point D, in this case, by the technique of laser induced fluorescence (LIF). Detection techniques for flow systems are discussed in the following section. The distance (X) between reaction initiation and detection is known, as is the flow velocity of the gas mixture (V). The time (t) between initiation and detection can then be calculated by eqn (7):

$$t = X/V. \qquad (E\ 7)$$

The relative concentration of OH is then plotted as the appropriate point on a graph of relative [OH] vs. time (or distance). The cyclohexane inlet is moved to a new position and the experiment is repeated. Gradually a decay profile of the relative OH concentration as a function of injector position (and hence reaction time) is built up (Fig. 2.5). The pseudo-first-order rate coefficient ($k_2' = k_2[C_6H_{12}]$) can be calculated from a plot of $\ln[OH]$ vs. time (Section 1.4) and the second-order rate coefficient from a plot of k_2' vs. $[C_6H_{12}]$ (Fig. 2.6).

We saw that for classical techniques a major handicap was the time taken for mixing to occur and once again this is a major limitation in discharge flow studies. Ideally we require a uniform concentration of reactants at point A, where the OH and cyclohexane first meet. However, even in gases, mixing (either by diffusion or turbulent mixing) takes a finite time (of the order of a fraction of a millisecond at 1×10^2 Pa) and hence a uniform concentration will not be obtained until some point downstream of the moveable injector. This does not matter if this distance (or time) is small compared to that for the reaction but for very fast reactions a significant portion of the reaction will be complete before a uniform concentration is achieved. Timescales of reactions studied by discharge flow are therefore limited to the millisecond range.

A further restriction on experimental conditions is the need to retain a uniform flow velocity along the entire cross-section of the tube. So called plug flow con-

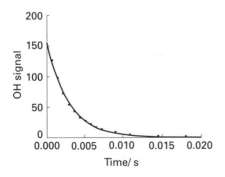

Fig. 2.5 Typical first order decay of OH radicals in the presence of an excess of cyclohexane obtained with the apparatus shown in Fig 2.4. The conversion from injector position to reaction time has already been performed. The solid line is an exponential fit to the data points to give the pseudo-first-order rate coefficient for the reaction conditions (i.e. the equivalent of plotting $\ln$ [OH] *vs. t*).

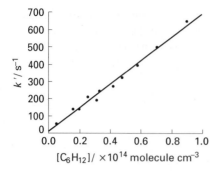

Fig. 2.6 Plot of pseudo-first-order rate coefficient vs. [cyclohexane]. The gradient of the graph corresponds to the bimolecular rate coefficient for reaction (2). The positive intercept on the y-axis represents the reaction of OH radicals with the flow tube wall coating.

ditions can only be attained at low pressures and hence discharge flow experiments are generally restricted to pressures below 10^3 Pa. Diffusion and mixing will also be slower at higher pressures providing a further limitation on high pressure studies. A number of workers have been investigating the possibility of high pressure flow reactors in which the effects of non plug flow are accounted for (see for example, Abbatt *et al.* 1990). Whilst the simplicity of the basic flow system is lost, the high pressure systems do enhance the range of experimental conditions that can be investigated.

Potential complications can arise from heterogeneous reactions. The reactant gases are in contact with the walls of the flow tube which may catalyse the reaction under investigation, promote secondary reactions or contribute to unwanted loss of radical reactants. Contributions from heterogeneous reaction can readily be distinguished by two experiments. Firstly the walls of the flow tube can be coated with Teflon or halocarbon wax. These materials have very few active sites for the molecules to 'stick' to and hence lower the probability of surface catalysed reactions. Alternatively the diameter of the flow tube can be varied altering the surface to volume ratio and hence the relative importance of heterogeneous reactions. If surface reactions are unimportant the measured bimolecular rate coefficient will be independent of either wall coating or reactor diameter.

Despite these limitations discharge flow techniques have been a major and successful tool in gas phase kinetics for many years. They are relatively cheap to construct and can be coupled to a number of sensitive detection techniques. A variety of atomic species can readily be generated from the microwave discharge of an appropriate molecular precursor e.g.

$$Br_2 \rightarrow 2Br \qquad (R\ 4)$$

$$CF_4 \rightarrow F + CF_3 \qquad (R\ 5)$$

and a number of molecular radicals can be made by indirect reactions such as those described above for the generation of OH.

2.4 Detection techniques for discharge flow apparatus

2.4.1 Resonance fluorescence

Resonance fluorescence is a well-used technique for monitoring the concentration of atomic species (e.g. H, N, O, Br, Cl, or F) and can also be applied to a limited number of molecular species but here the technique has largely been replaced by laser induced fluorescence (p. 34). Helium gas, containing a trace of the atomic precursor (e.g. H_2, O_2, Br_2 etc.), is passed through a microwave discharge in which some of the molecular species is dissociated (cf. the discussion of the source of atomic species in the discharge flow method). A fraction of the resulting atoms are collisionally excited (by electrons or energetic helium atoms) to form electronically excited atoms. These atoms subsequently return to the ground electronic state with the emission of a photon of light, a process known as fluorescence. The frequency of this light is characteristic of the atoms in the discharge lamp. The light is then filtered and directed into the flow tube. Atoms *of the same species* in the flow tube can absorb the light (the transitions are resonant) and are themselves raised to an exited state; they then return to the ground state by emitting resonant fluorescence. The fluorescence is proportional to the atomic concentration, provided the latter is small. A photomultiplier tube (a sensitive light detector), placed at right angles to the axis of the lamp, is employed to monitor the fluorescence and hence obtain a relative measure of the atomic concentration. Figure 2.7 schematically illustrates the processes occurring in both the resonance lamp and the flow cell.

Major advantages of the technique are its relatively low cost, simplicity and its great specificity. Atomic lines are so sharp that the chance of two atomic transi-

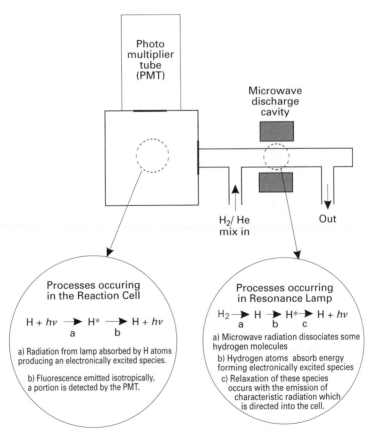

Fig. 2.7 Schematic diagram illustrating the various processes occurring during resonance fluorescence detection of H atoms.

tions of different species coinciding is very small. For instance bromine atoms can be determined in the presence of chlorine atoms or even D atoms in the presence of H. Molecular transitions are much wider and care must be taken that light from the lamp or fluorescence signal is not absorbed by molecular species in the flow tube before reaching the photomultiplier tube.

2.4.2 Laser induced fluorescence

Similar in principle to resonance fluorescence, laser induced fluorescence (LIF) offers greater sensitivity and applicability, especially for molecular radicals, at the cost of greater experimental complexity. Figure 2.8 illustrates the technique for the detection of OH radicals. Once again the species to be detected is excited to an upper electronic state. Dye lasers are used as the excitation source as their output frequency can be precisely tuned to match a particular rovibronic transition, thus giving the technique great selectivity. Subsequently the molecules relax to the ground electronic state with the emission of fluorescence. Relaxation can occur to several vibrational levels in the ground state and hence the fluorescence can be detected at a different wavelength from excitation. The timescale of this relaxation (nanoseconds) is typically very much less than that of the reaction (micro–milliseconds). At high pressures vibrational relaxation will occur in the upper electronic state and fluorescence will occur from the vibrational ground state. Fluorescence is detected by a photomultiplier tube (PMT).

The greater sensitivity of the technique arises from the power of the laser sources. For low levels of absorption the fluorescence intensity rises linearly with the number of photons absorbed and hence with excitation intensity. The number of molecules raised to the upper state, n', and hence the intensity of the fluorescence, are proportional to the number of molecules in the ground state, n'':

$$I_f \propto n''. \tag{E 8}$$

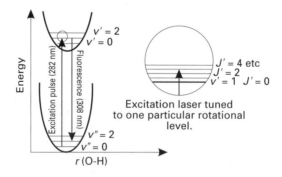

Fig. 2.8 Schematic diagram of OH radical detection by laser induced fluorescence (LIF). The initial laser induced excitation occurs to one particular rovibrational level in the upper electronic state. Fluorescence to vibrational levels in the ground electronic state can occur at longer wavelengths from either $v' = 1$, or following vibrational relaxation at higher pressures, $v' = 0$.

It must be stressed that LIF only provides a relative measure of concentration and not an absolute concentration. This is the major disadvantage of the technique in comparison to less sensitive, but absolute, methods such as absorption (see p. 43).

There are two practical points to notice. Figure 2.4 shows that the excitation (or probe) laser and PMT are mounted at right angles. This is to minimize the amount of laser light that the PMT sees. The intensity of the laser light is very much greater than the fluorescence and in a poorly designed experiment would swamp the real signal. Secondly, the wavelength of the fluorescence can be different from the excitation laser (note the difference from resonance fluorescence). An example of this is shown in Fig. 2.8. Laser excitation is tuned to excite ground state $v'' = 0$ molecules to the $v' = 1$ level of the first excited state. Fluorescence will be dominated by non-resonant transitions to $v'' = 1$. A suitable filter can be placed in front of the PMT to pass the fluorescence signal but further cut down on the scattered laser light.

Laser induced fluorescence is limited to species with a bound and accessible upper electronic state and some important radical species (e.g. CH_3) cannot be detected using this method. Table 2.1 lists some of the radicals which have been detected by LIF.

Table 2.1 Examples of excitation and fluorescence wavelengths used for LIF studies

Species	Excitation wavelength/ nm	Fluorescence wavelength/ nm
OH	282	308
CN	388	422
CH	431	431
CH_3O	303.9	375
NH	336	336
H	121.6	121.6
SO	266.5	283.4

2.4.3 Mass spectrometry

Mass spectrometers act as sensitive universal detectors and are hence widely used in gas phase kinetic experiments. We illustrate their use in discharge flow experiments by reference to work on the reaction

$$C_2H_5 + O_2 \rightarrow C_2H_5O_2 \qquad \text{(R 6a)}$$

$$\rightarrow C_2H_4 + HO_2 \qquad \text{(R 6b)}$$

by Plumb and Ryan (1981). Reaction (6) is of significance in the atmospheric oxidation of ethane and in higher temperature combustion processes. Figure 2.9 shows a schematic diagram of the apparatus. Ethyl radicals are generated by the reaction of Cl atoms (from the microwave discharge of Cl_2) with ethane.

$$Cl_2 \rightarrow 2Cl \qquad \text{(R 7)}$$

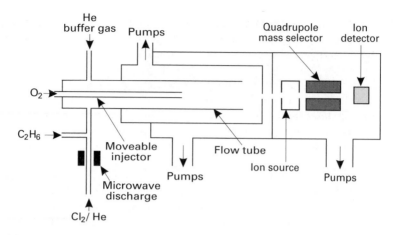

Fig. 2.9 Schematic diagram showing the combination of flow tube with mass spectrometric detection. High capacity pumps are required to remove most of the gas flow with only a small fraction entering the very low pressure ($\sim 10^{-6}$ torr, or 1.33×10^{-4} Pa) mass spectrometer.

$$Cl + C_2H_6 \rightarrow HCl + C_2H_5. \tag{R 8}$$

Once all the Cl atoms have been converted to ethyl radicals, O_2 is introduced from a moveable injector. A 0.5 mm pinhole extracts a small portion of the reacting gas mixture and a beam of this sample is generated by a second pinhole and the pumps. Ethyl radicals in the molecular beam are ionized by electron impact and then mass selected by a quadrupole so that only mass 29 ($C_2H_5^+$) is detected by the electron multiplier. The relative concentration of ethyl radicals is detected by the mass spectrometer at various injector positions, giving the pseudo-first-order rate coefficient for the reaction. The oxygen concentrations are then varied and for the conditions of the experiment (295 K and 263 Pa) a value of 2.2×10^{-12} cm^3 molecule^{-1} s^{-1} is determined for the rate coefficient for removal of the ethyl radical.

The great advantage of the mass spectrometric technique is illustrated by Plumb and Ryan's ability to determine the rates of the two channels (R 6a) and (R 6b). C_2H_4 can also be detected by the mass spectrometer and by measuring the absolute final yield of C_2H_4 for a given absolute initial concentration of ethyl radicals, they were able to show that, under their conditions, 11 per cent of the reaction proceeds via channel (b), the rest occurring via channel (a). Plumb and Ryan's paper gives a detailed discussion of their experimental system and of the difficulties associated with measurements of this sort. In certain cases ions can be produced by photoionization rather than electron impact ionization. Although photoionization may produce lower concentrations of ions, less ion fragmentation occurs, simplifying product identification.

Example 2.1
The following data were obtained by Plumb and Ryan in their study of reaction (6). The reaction was carried out under pseudo-first-order conditions and therefore a plot of ln[C_2H_5] vs. t should be a straight line of gradient, $-k'$, the pseudo-first-order rate coefficient. The [C_2H_5] will be proportional to the mass

spectrometer ion count and as we have seen, the time is proportional to the injector distance and therefore a plot of ln(ion count) vs. injector position should also be a straight line.

O_2 injector position (cm)	3	5	7	10	12	15
$C_2H_5^+$ signal (arbitrary units)	6.14	3.95	2.53	1.25	0.70	0.40
$\ln(C_2H_5^+$ signal)	1.84	1.37	0.93	0.22	-0.36	-0.92

Figure 2.10 shows, as expected, that the graph is a straight line of gradient -0.23 cm^{-1}. To convert this to the pseudo-first-order rate coefficient we simply multiply by the linear flow velocity (1080 cm s^{-1}) to give a value of 248 s^{-1} for k'. Plumb and Ryan then repeated their experiments at different oxygen concentrations to obtain the bimolecular rate coefficient for the reaction of ethyl radicals with molecular oxygen.

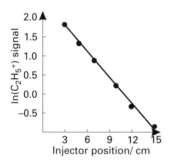

Fig. 2.10 Typical first-order plots of the $\ln(C_2H_5^+$ signal) with injector position.

2.4.4 Laser magnetic resonance

Laser magnetic resonance (LMR) has been applied to the detection of radicals in discharge flow systems. A CO_2 laser is used to probe for radical concentrations. These lasers produce a variety of lines in the infrared region of the spectrum (corresponding to rovibrational transitions). These rarely coincide with a radical vibration–rotation absorption line, however, provided that the radical has a magnetic moment (the rule rather than the exception), its absorption may be 'tuned' to a laser frequency by applying a magnetic field (the Zeeman effect [see Atkins 1994]). Laser absorption may then be used to monitor the concentration of the radical as at low absorbances concentration and absorbance are related via the simplified Beer–Lambert Law

$$I_{abs}/I_0 = \varepsilon cl \qquad \text{(E 9)}$$

where I_{abs} is the change in measured light intensity, I_0 the intensity before the photolysis pulse, ε is the absorption coefficient, c the reactant concentration and, l is the path length of the absorption.

An example of the application of this technique is provided by the biradical triplet methylene. The ground electronic state of methylene has two unpaired electrons with parallel spins and so exists as a triplet species, which we shall label as $^3CH_2$. At slightly higher energies there is another electronic state in which the

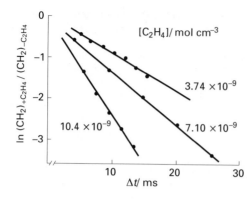

Fig. 2.11 Typical pseudo-first-order decay plots for reaction (10) at T = 540 K.

spins are paired; singlet methylene can be detected by LIF, but only the triplet state can be detected by LMR. This illustrates the fact that many of the detection techniques that we have discussed are complementary in their nature and there is no one ideal detection system for all radicals. $^3CH_2$ may be generated by the reaction of oxygen atoms with ketene (CH_2CO) in a discharge flow tube (the oxygen atoms being produced from the microwave dissociation of molecular oxygen).

$$O + CH_2CO \rightarrow {}^3CH_2 + CO_2. \tag{R 9}$$

The reaction of $^3CH_2$ with ethene

$$^3CH_2 + C_2H_4 \rightarrow C_3H_6 \tag{R 10}$$

was studied by monitoring the LMR signal as a function of the distance between the mixing point and the laser detector, in the presence of a range of ethene concentrations. Figure 2.11 shows a first-order decay plot for the data at a temperature of 540 K for various ethene concentrations; the slopes show a linear dependence on [C_2H_4] and give a bimolecular rate coefficient of 3.8×10^{-14} cm^3 molecule^{-1} s^{-1}.

2.5 Liquid and stopped-flow systems

Flow systems are not limited to the study of gaseous reactions but, with certain alterations can be applied to the study of liquid phase reactions (Fig. 2.12). Mixing is significantly slower in the liquid phase and careful attention has to be paid to the design of the mixing chamber. The simple injector used in gas phase studies is certainly not adequate, and liquid flow systems often have a fixed mixing position and a moveable detector as shown in Fig. 2.12. The most common detection technique is absorption spectrophotometry which is relatively compact and can be moved to various positions along the flow tube. Although the design of the apparatus has changed from the discharge flow systems, the same relationship between distance from mixing to detector and flow rate gives the time dependence of the reactant concentrations.

High flow rates are required, both to promote mixing and, for fast reactions, to ensure that the reaction is not complete in the initial portion of the flow tube. This can involve large quantities of reagents which may not be available to the experimenter. The stopped-flow technique provides an alternative method of

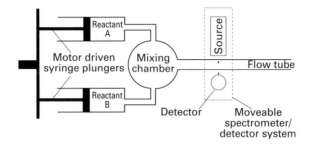

Fig. 2.12 Schematic diagram of a liquid flow tube with reactant detection by absorption spectroscopy. The mixing chamber will contain an elaborate set of baffles to promote turbulent flow and mixing of the two reactant solutions.

study for liquid phase reactions and, from a monitoring viewpoint, has some similarities with the flash photolysis technique described below. The initial part of the apparatus is similar to the liquid flow system with the reactants being injected from two syringes into a mixing cell. However, instead of the effluent going to waste it fills a third syringe. When filled to a certain volume the plunger hits a backstop stopping the flow in the whole system (including the reactant syringes). The *real time* evolution of the reactant (or product) concentrations is then monitored in one section of the experimental system. The stopped-flow system is essentially a method for rapidly mixing two reactant solutions.

An illustration of this technique is provided by the study of the adsorption of manganese ions on to a surface site, in this case MnO_2 suspended in solution.

$$Mn^{2+} + 2S^- - H \rightleftarrows 2S^- - Mn + H^+ \qquad \text{(R 11)}$$

where $S^- - H$ represents a negatively charged surface site occupied by a hydrogen atom. The apparatus is shown schematically in Fig. 2.13. The two reactants are injected from the syringes and into a mixing/reaction cell. As this process occurs a third syringe is filled. Contact between the syringe plunger and an activation switch triggers the data acquisition. $[Mn^{2+}]$ is monitored in real time by electron paramagnetic resonance (see Atkins 1994). Because the detection position is stationary in the stopped-flow system a greater variety of detection systems can be utilized as detector mobility is no longer an experimental criterion. Figure 2.14 shows a typical decay trace recorded in these experiments. Note that the timescale of these reactions is similar to that of other flow experiments. The reactions of metal ions with colloidal suspensions are of great interest in a variety of chemical fields including catalysis, electrochemistry and soil chemistry. Further details of the experimental technique and applications to soil science can be found in a paper by Fendorf *et al.*, 1993.

2.6 Flash photolysis

Mixing times and pressure limitations are the major disadvantages of flow systems. They are overcome in the kinetic technique of flash photolysis developed by Norrish and Porter in Cambridge in the late 1940s and for which they received the Nobel prize in 1967. The basis of the technique is very simple: reactants and precursors are premixed and flowed into the photolysis cell at the required pres-

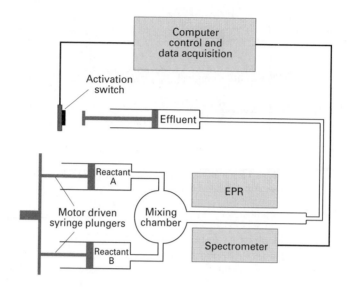

Fig. 2.13 Schematic diagram of a stopped-flow system.

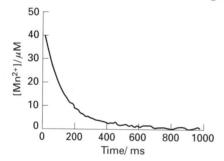

Fig. 2.14 A typical rate curve of Mn^{2+} sorption on δ-MnO_2.

sure $(1 \times 10^2 - 1 \times 10^6$ Pa, 0.001–10 atmospheres pressure). A pulse of light is used to produce a transient species, an atom, radical or excited state, whose concentration is then monitored as a *function of time*. The precursors and reactants are premixed and if the photolysis beam is of uniform intensity a homogeneous concentration of reactants is obtained. There is no mixing time. The limitation on the timescales of reactions which can be studied is the duration of the light pulse. For the flash lamps that Norrish and Porter used this could be several milliseconds, however, these light sources have been replaced by high powered lasers with pulse durations of nanoseconds $(10^{-9}$ s) or less. We saw an example of the use of flash photolysis in Fig. 1.1, which showed the second order decay of methyl radicals following their generation from propanone by pulsed laser photolysis.

$$CH_3COCH_3 + h\nu \rightarrow 2CH_3 + CO \qquad (R\ 12)$$

where $h\nu$ represents a photon.

Another advantage of flash photolysis over flow methods is that because the reactant species are generated and monitored in the centre of the reaction cell there are no possible complications from wall catalysed reactions.

Flash photolysis is not limited to the study of reactions in the gas phase. Diffusion is even slower in liquids and hence flash photolysis techniques, where the reactants and precursors are uniformly premixed, is of great importance.

Because of the real time nature of flash photolysis experiments detection techniques must be able to respond to rapidly changing concentrations. Many of the same techniques discussed in Sections 2.4.1–4 are applicable and Section 2.7 describes their real time implementation.

2.6.1 Light sources for flash photolysis

Early experiments used flash lamps, not unlike photoflash lamps, to generate the transient species. Whilst lamps are still employed, the majority of flash photolysis experiments are now conducted with lasers, which have the advantage of short pulse duration, high repetition rate, a narrow, precisely defined wavelength range and a well defined spatial profile. The excimer (or exciplex) laser is particularly widely employed. Such lasers have a pulse duration of 10–20 ns (cf. microseconds for flashlamps), repetition rates of up to 500 Hz and high pulse energies. The latter property is of great importance in that a significant quantity of radicals or transient species can be generated from a very low precursor concentration. Low precursor concentrations simplify the overall kinetics of the system. Even so some thought needs to be given to the precursor used so that secondary reactions of the transient with the precursor are negligible.

The laser radiation is produced by passing a pulsed electric discharge through a mixture of helium, a noble gas (Ar, Kr, or Xe), and either F_2 or HCl. Ions are generated which can combine to give electronically excited species such as KrF* from a mixture of $He/Kr/F_2$. Lasing action occurs between this upper electronic state and the dissociative ground state. The wavelength depends on the exciplex formed and hence on the gas mixtures involved. The major wavelengths are given in Table 2.2.

Table 2.2 Excimer wavelengths and energies.

Mix	Exciplex	λ/nm	Photon energy/ $kJ\ mol^{-1}$	Photon energy/eV
$Ar/F_2/He$	ArF	193	621	6.44
Kr/HCl/He	KrCl	222	540	5.60
$Kr/F_2/He$	KrF	248	483	5.01
Xe/HCl/He	XeCl	308	389	4.04
$Xe/F_2/He$	XeF	351	342	3.54

These wavelengths are in a particularly active photochemical region as we note that the photon energies are comparable to bond dissociation energies (300–500 $kJ\ mol^{-1}$) and hence absorption of such a UV photon often results in one or more bonds being broken. Some examples of transient species which may be generated from these lasers are listed below:

$$CH_2CO + h\nu \rightarrow {}^3CH_2 + CO \ (351\ nm) \tag{R 13}$$

$$RCHO + h\nu \rightarrow R + HCO \ (308\ nm) \tag{R 14}$$

$$O_3 + h\nu \rightarrow O_2 + O(^1D) \text{ (248 nm)} \qquad\qquad \text{(R 15)}$$

$$HBr + h\nu \rightarrow H + Br \text{ (193 nm)}. \qquad\qquad \text{(R 16)}$$

A brief discussion of the mechanism of excimer lasers can be found in Box 2.1 and a full description of their operation, and that of many other laser systems, may be found in books by Hollas (1992) or Andrews (1992).

By means of a process called mode locking it is possible to generate laser pulses which are only a few picoseconds (10^{-12} s) or less in duration. Such lasers can be used to study a variety of very fast kinetic phenomena, such as the isomerization of an electronically excited olefin, the rotational motion of a large molecule in solution or some of the fast processes occurring in photosynthesis. Whilst these are not 'reactions' in the sense that we have discussed them so far, they are important kinetic processes that contribute to overall reactions and illustrate the detail which may be gleaned from the study of so-called ultrafast processes. We shall encounter a further example in Chapter 12, however, a detailed discussion is beyond the scope of this book, but an excellent account may be found in Fleming (1986).

An alternative technique for generating transient species is that of pulse radiolysis, in which a short (10^{-9}–10^{-6} s) pulse of high energy electrons (1–5 MeV) is produced from a linear accelerator or van de Graaff machine. The technique has been primarily used for studies of liquid phase kinetics and especially of the solvated electron, although it is a very flexible method and has also been used for gas phase studies.

A comprehensive review of radiation chemistry, including applications of pulse radiolysis, is provided by Farhataziz and Rodgers (1987).

Box 2.1 The Excimer Laser as an example of Lasers

Laser is an acronym for *L*ight *A*mplification by *S*timulated *E*mission of *R*adiation. To understand how lasers work we need to look at the three basic radiative processes that can link any two energy levels.

1. Absorption—a photon can be absorbed by the lower state.

2. Spontaneous Emission—the upper state spontaneously emits a photon (fluorescence) as it returns to the lower state.

3. Stimulated Emission—a photon stimulates the transition from the upper state to the ground state i.e. two photons are produced. The photon which is released is coherent with the stimulating photon (same direction and phase).

A *population inversion*, in which the population of an excited state exceeds that of a lower state, is needed to obtain laser action. When light of the correct wavelength passes through the *laser cavity*, where the inversion is maintained, it is amplified because the rate of stimulated emission exceeds that of absorption. The laser cavity consists of the *active medium*, where the population inversion is generated and is bounded by mirrors which reflect the light back and forth through the active medium. One of the mirrors is made partially reflecting so that some of the light is allowed out.

Exciplex compounds are bound in an upper electronic state but dissociative in the ground state. Figure 2.15 shows a schematic diagram of the energy levels for KrF. The potential for creating a population inversion is obvious since the ground state population is immediately removed by dissociation. A pulsed electrical discharge creates excited Kr^+ and F^- ions from a mixture of $Kr/F_2/He$. The ions combine to create the exciplex. Helium acts as a third body in the same way as argon does in the iodine recombination reaction (see Box 1.2). The energy separation between the two levels is such that the stimulated photon is in the UV region of the spectrum. Different exciplexes will have different energy separations giving the laser a range of wavelengths depending on the chosen gas mixture. Short wavelengths and high power make excimer lasers ideal tools for generating significant quantities of radicals or atoms. The same properties have led to uses in eye surgery, microchip manufacture and materials processing applications.

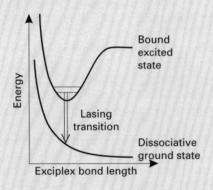

Fig. 2.15 The potential energy curves for the excited and ground states of the KrF exciplex.

2.7 Detection techniques for flash photolysis experiments

Once the transient species has been produced it must be monitored. A brief resume of a few of the major techniques are listed below. Where the technique has already been introduced for flow experiments, we simply note how it has been adapted for real time measurements.

2.7.1 Absorption spectroscopy

Kinetic absorption spectroscopy is an ideal method of following fast reactions. Light of a frequency corresponding to an absorption (vibrational or electronic in nature) of the radical reactant is passed through the reaction cell and is monitored by a suitable detector. The absorption signal of the reactant radical will increase instantaneously (on the timescale of the reaction) following its generation in the photolysis pulse, and then decay away as the radical reacts. Concentration and absorption are related via the Beer–Lambert Law:

$$I = I_0 \exp(-\varepsilon cl) \tag{E 9a}$$

if εcl is small then the exponential term can be expanded giving:

$$I_{abs}/I_0 = \varepsilon cl \tag{E 9b}$$

where I_{abs} is the measured change in light intensity, I_0 the intensity before the photolysis pulse (see Fig. 2.16(b)), ε is the absorption coefficient, c the reactant concentration, and l is the path length of the absorption. For studies under first-order conditions ε need not be known. It is enough for us to know that the absorption is directly proportional to radical concentration (*for small $\varepsilon c l$*), but for reactions with an order different from one, ε and the cell path length must be known.

A variety of light sources have been used in absorption experiments ranging from the infrared to the vacuum ultraviolet regions of the spectrum. Broadband xenon arc lamps are still very popular light sources providing intense UV light over a range of frequencies. A monochromator is used to monitor the frequency corresponding to the radical absorption. For example in the apparatus used to record the data presented in Fig. 1.1 on the methyl radical recombination reaction, the temporal decay of the methyl radical concentration was followed by monitoring the methyl radical absorption at 216 nm. Further details of the experiment can be found in Box 2.2. Broadband absorption spectroscopy is a common technique for monitoring concentrations in liquid solvents (question 2.1).

There are two major advantages of the absorption technique. Firstly, one can obtain a direct measurement of the concentration, provided that the absorption coefficient and pathlength are known. Secondly, a complete decay trace is recorded each time the photolysis laser fires. Although a number of traces are co-added to improve the signal-to-noise ratio the technique is still considerably more efficient than LIF (see section below).

Potential complications can arise when more than one reactant or product absorbs at the monitoring wavelength. For example in a flash photolysis study of the ethyl peroxy recombination reaction, absorption by the $C_2H_5O_2$ radical was used to follow the reaction. However, HO_2, a secondary product of the reaction also absorbs light at the monitoring wavelength and sophisticated analysis techniques are required to deconvolute the measured absorption signal. The development of tunable UV, visible and IR lasers has greatly enhanced the selectivity and sensitivity of kinetic absorption spectroscopy. The very narrow line widths of laser probes allow the selective isolation of one particular transition in the reactant radical. The stability of the laser sources means that random fluctuations in I_0, and hence in the overall signal-to-noise ratio, are minimized. Another important property of laser beams is their low beam divergence and therefore the laser beam may be passed through the reaction cell many times using a mirror arrangement known as a White cell. This increases the path length l and so enhances the sensitivity of the technique. Examples of laser absorption studies may be found in Opansky *et al.* (1993) for IR probing and Baggott *et al.* (1986) or Biggs *et al.* (1991). For UV/visible probing Hollas (1992) and Andrews (1992) provide excellent accounts of the principles and operation of the lasers used to monitor radical concentrations.

Box 2.2 Signal Averaging

Figure 2.16(a) shows the experimental design of an apparatus used to monitor the methyl radical recombination reaction. Light from a 'white light' source, a xenon arc lamp, is passed through the cell, dispersed by a monochromator, whose grating is set so that light of a wavelength absorbed by the radical falls on the outlet slit and so on to a photomultiplier (PMT). Some of the light of this wave-length is absorbed by the radicals generated in the flash, so that the signal generated by the PMT falls sharply following the flash, returning to the original level as the radicals react (Fig 2.16 (b)). The voltage change ΔV from the PMT, which is proportional to the change in light intensity (and hence via eqn (9) is related to the radical concentration), is passed to a device known as a digitizer which records the signal and stores it, in digital form, in a series (typically 1024) of channels which correspond to a series of time intervals, chosen by the experimentalist to suit the radical decay time. This set of digital data is then passed to a microcomputer for storage. The laser fires again and another set of data is recorded and transferred to the computer. The process is repeated several hundred times and finally the data at each time point (or channel) is averaged. The idea behind this process of signal averaging is the improvement of the signal-to-

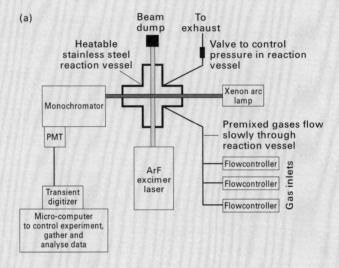

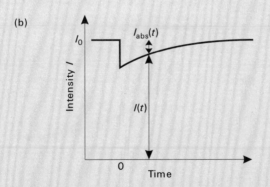

Fig. 2.16 (a) Schematic diagram of the apparatus used to follow the methyl radical recombination reaction. (b) Variation of absorbed light during the course of the methyl radical recombination reaction.

noise ratio (S/N). The voltage signal generated by the PMT contains not only the contribution from the radical absorption, but also noise from a variety of sources. The absorption signal is reproducible from shot to shot, whilst the noise is random. Adding together a whole series of data points averages out the noise leaving behind the absorption profile with a much enhanced S/N ratio. To a good approximation the S/N ratio improves with the square root of the number of decay traces averaged, i.e. to double the S/N ratio one averages four times the number of decay traces. It should be noted that in a multishot experiment it is necessary to flow the reactants slowly through the cell to avoid extensive depletion of the radical precursor and reactants and to prevent the build up of interfering products.

2.7.2 Fluorescence techniques

Absorption techniques require accurate measurement of two intensities, I and I_0, which in most cases will be of similar magnitude. Absorptions of 1 part in 10 000 are not uncommon and an analogy would be to tell the difference between two

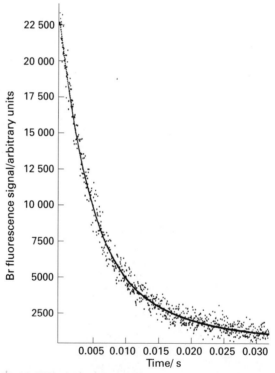

Fig. 2.17 Typical resonance fluorescence trace for the decay of bromine atoms in the presence of an excess of silane. Owing to the rapid nature of the reverse reaction ($HBr + SiH_3$) the decay is not a straightforward exponential process. Analysis of the decay profiles gives both the forward and reverse rate coefficients (Section 1.7) and was used to calculate ΔH_f (SiH_3).

piles of pennies containing 9999 and 10 000 coins separated by several metres! In all fluorescence techniques the signal is detected relative to a zero background and, up until saturation of the absorption transition, the signal increases linearly with excitation intensity. Both of these effects increase the sensitivity of fluorescence (after all it is much easier to see one penny rather than to tell the difference between two similar piles of pennies), which in favourable circumstances can detect concentrations as low as 10^8 molecules cm^{-3} (cf. absorption 10^{12}–10^{13} molecules cm^{-3}). Lower concentrations of transient species reduce the effects of secondary reactions.

Resonance fluorescence

The technique is very similar to that described on p. 32. The resonance lamp is a continuous source and hence the fluorescence signal will rise rapidly following photolysis and then decay as the reactant is removed. Signal averaging is used to enhance the signal-to-noise ratio and a typical decay trace is shown in Fig. 2.17

Laser induced fluorescence

Once again the principles of the technique are similar to those described above, however, implementation in a real time experiment can be more complex. For cases when the probe laser can be run in a continuous mode the fluorescence profile will be very similar to that shown in Fig. 2.17. However, in a majority of cases dye lasers operate in a pulsed mode and careful laser control is required to build up a decay trace. Figure 2.18 shows schematically how this is achieved. Firstly the photolysis laser fires generating the radical species and setting time $t = 0$. At a known time later the dye laser fires, exciting the radical species (p. 34) and the fluorescence is collected. The magnitude of the fluorescence signal is proportional to the concentration of the radical species *at the time at which the dye laser was*

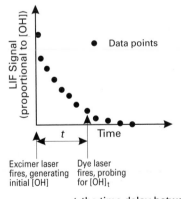

Fig. 2.18 Schematic diagram showing the timing sequence required to build up a decay curve in a pulsed flash photolysis, pulsed LIF experiment.

triggered. The experiment is repeated many times but with different delay times between the photolysis laser ($t = 0$) and the probe laser pulses. In this way a complete decay profile of the radical species is built up.

Example 2.2

Figure 2.19 shows a typical decay trace obtained from a study of the reaction

$$CH + O_2 \rightarrow \text{products.} \qquad \text{(R 17)}$$

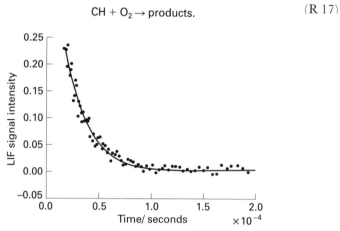

Fig. 2.19 Decay of CH radicals in the presence of an excess of oxygen (reaction (17)), $[O_2]=8.8 \times 10^{14}$ molecules cm^{-3}.

The relative concentration of CH radicals is monitored using LIF in the presence of excess oxygen. The discrete nature of the data points obtained is clearly seen in this figure. To determine the pseudo-first-order rate coefficient graphically we take natural logarithms of the LIF signal (I_F) at various times. A plot of $\ln(I_F)$ vs. t should be a straight line of gradient k'. (In order to make the calculation simpler, averaged values of the data have been taken at 5 μs intervals. In a real analysis, all of the data points would be analysed by a computer package.)

time/μs	20	25	30	35	40	50	60	70	80
I_F	0.230	0.188	0.144	0.119	0.088	0.050	0.033	0.020	0.010
$\ln(I_F)$	−1.47	−1.67	−1.94	−2.13	−2.44	−2.99	−3.41	−3.92	−4.56

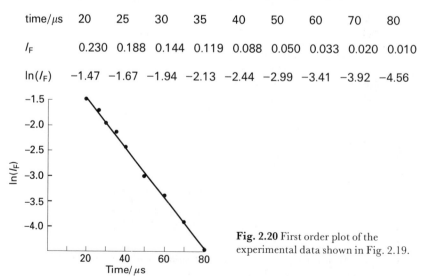

Fig. 2.20 First order plot of the experimental data shown in Fig. 2.19.

Figure 2.20 shows that as expected the graph is a straight line and the gradient, the pseudo-first-order rate coefficient is 51 500 s^{-1}. The experimental data are gathered on a microcomputer which can automatically fit the data with the expected first-order decay profile. The computer program varies the value of k' until the best fit to the data is obtained, shown as the solid line in Fig. 2.19.

2.8 Shock tubes

At very high temperatures the equilibria of many dissociation reactions are shifted towards the dissociated products:

$$\text{e.g. } C_3O_2 \rightleftharpoons C + 2CO. \tag{R 18}$$

In a shock tube well mixed reactants and precursors (e.g. C_3O_2 for the study of C atom reactions) are subjected to a very rapid increase in pressure (the shock) which causes rapid heating of the mixture to several thousand kelvin and dissociation of the precursor species. The reaction of the transient species is then followed by detection techniques such as absorption discussed above. Figure 2.21 shows a schematic diagram of the apparatus. An inert gas (the driver) is maintained at a very high pressure on one side of the diaphragm. On the other side a suitable mixture of reactants is made up and allowed to mix. The diaphragm is then pierced and a high pressure shock wave moves through the reaction mixture, causing rapid heating and reaction initiation. The temperature rise can be controlled by the pressure and composition of the driver gas. The technique is ideal for the study of reactions relevant to combustion.

There are two major drawbacks to the technique. Firstly the method of initiation is rather indiscriminate! Reactants as well as the precursor may decompose and it can be very easy to end up with a 'soup' of reactive species. Under these conditions a complex reaction scheme may be required to model the decay of the transient species under study as it will be removed in a number of secondary reactions as well as by the target reaction. Secondly, only one decay trace is obtained from each experiment, significant signal averaging to produce very

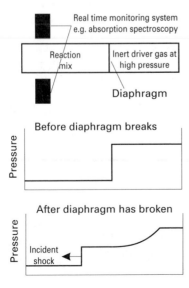

Fig. 2.21 Schematic diagram of shock tube apparatus.

high S/N ratios is not realistic. In addition the signal traces may be affected by subsequent reflections of the initial shock wave. Advances in detection techniques such as laser absorption spectroscopy have greatly improved the sensitivity and accuracy of the shock tube method. An article by Dean *et al.* (1991) on the study of CH and C atom reactions with NO, illustrates the excellent S/N that can be obtained in a well-designed experiment, but, even with this relatively simple system, 56 reactions are required to explain the detailed time dependence of the transient concentrations in the shock tube. Suitable choice of conditions are, therefore, needed in order to ensure that one or two reactions only dominate the time profile of the transient species. Sensitivity analysis (see Chapter 8) can be used to determine the optimal conditions.

2.9 Relative rate determinations

Under certain circumstances direct isolation of an elementary reaction for real time analysis may not be possible. Relative rate methods with gas chromatographic (GC) analysis of the end reaction mixture provide an alternative analysis technique. We illustrate the method by a relative rate determination of the reactions of OH radicals with two alkanes, one a reference compound (RH) for which the rate coefficient is known; the rate coefficient for the other alkane (SH) is to be determined.

$$OH + RH \rightarrow H_2O + R \qquad\qquad (R\ 19)$$

$$OH + SH \rightarrow H_2O + S. \qquad\qquad (R\ 20)$$

Known concentrations of the two reactants are admitted to the reaction vessel and are allowed to mix with the OH precursors. Photolysis lamps are then switched on generating a small but essentially constant concentration of OH radicals. At certain times samples are withdrawn from the reaction vessel and analysed by gas chromatography to determine $[RH]_t$ and $[SH]_t$. Gas chromatography is a very flexible detection technique and may be used in addition to determine the eventual fates of the radicals R and S, but for the present we shall concentrate on the analysis to yield k_{20}. Note that the concentration of OH does *not* need to be known. The rate laws for reactions (19) and (20) can be written down as below and because the concentration of OH is constant, designated $[OH]_{ss}$ (steady-state) can be straightforwardly integrated (see Section 1.4)

$d[SH]/dt = k_s[OH][SH]$ As $[OH]$ is constant: $\ln([SH]_t/[SH]_0) = k_s[OH]_{ss}t$

$d[RH]/dt = k_r[OH][RH]$ As $[OH]$ is constant: $\ln([RH]_t/[RH]_0) = k_r[OH]_{ss}t$.

Combining the above two equations gives:

$$\ln([SH]_t/[SH]_0) = k_s/k_r \ln([RH]_t/[RH]_0). \qquad\qquad (E\ 10)$$

Concentrations of substrate and reference alkanes are measured as a function of time. By using low concentrations the reaction can take tens of minutes, allowing for sampling by GC. A plot of $\ln([SH]_t/[SH]_0)$ vs. $\ln([RH]_t/[RH]_0)$ will be a straight line of gradient k_s/k_r (Fig. 2.22). Obviously k_r must be accurately known and this will have to be determined by other techniques.

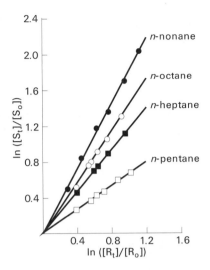

Fig. 2.22 Relative rate determinations for the reaction of OH with a number of hydrocarbons.

2.10 Relaxation techniques

Another way of overcoming the problem of mixing is to allow the reactants to come to equilibrium and then in some way, to perturb that equilibrium. The system will then relax back to the new equilibrium position and, as we shall see, the rate of relaxation gives information about the forward and reverse rate coefficients. Obviously the technique is only applicable to systems that are relatively close to equilibrium.

As an example of this method we shall consider the equilibrium between N_2O_3, NO and NO_2

$$N_2O_3 + Ar \rightleftarrows NO + NO_2 + Ar \qquad (R\ 21, -21)$$

Argon plays the same role as it did in reaction (12) of Chapter 1, stabilizing the initial complex that is formed (Box 1.2 and Chapter 5). The reaction mixture contains NO, NO_2, N_2O_3, Ar, and a small amount of a strong infrared absorber SiF_4. The reactants are allowed to mix and equilibrium is established. A CO_2 (infrared) laser is then fired and the pulse of light is absorbed by the SiF_4 molecules. Collisions with the argon bath gas and the reactant molecules rapidly turn this energy into heat and the temperature of the reaction mixture rises by several degrees kelvin. Increasing the temperature of the mixture tends to shift the equilibrium towards NO and NO_2. An analysis of the kinetics (Chapter 8) shows that

$$-dz/dt = \{k_{-21}([NO] + [NO_2]) + k_{21}\}z + k_{-21}\,z^2 \qquad (E\ 11)$$

where z is the difference in the concentration of N_2O_3 from its equilibrium value. If the perturbation is kept small then the quadratic term in z can be ignored simplifying the analysis. The equation is now in first-order form with a relaxation time of $\{k_{-21}([NO][NO_2]) + k_{21}\}$. z is measured by monitoring the absorption of N_2O_3 at 253 nm. [NO] and [NO_2] before the perturbation are known and the final values can be calculated from the known stoichiometry of the reaction. Figure 2.23 shows a typical relaxation trace and further experimental details can be found in an article by Markwalder *et al.* (1993). The forward and reverse rate

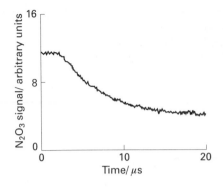

Fig. 2.23 Typical experimental results for the temperature jump study on the equilibrium (R 21, −21). The trace is the result of signal averaging over 100 laser shots.

coefficients can be found by measuring the relaxation time as a function of [NO] and [NO$_2$]. The reaction is pressure dependent and was studied over a wide range of argon pressures to investigate the applicability of theories of such reactions to this system.

Relaxation techniques are widely employed for solution phase reactions, where a number of different types of perturbation (e.g. pressure, temperature, electric field) can be employed. Eigen received the Nobel prize in 1967, along with Norrish and Porter, for work in this field. In the temperature jump method, a small increase in temperature is achieved by discharging the electrical energy stored in a high voltage capacitor through the solution. Alternatively, absorption of UV radiation, from a pulsed laser, or pulsed microwave radiation can be used to achieve the same effect. The pressure jump technique relies on the dependence of the equilibrium constant on pressure, via the change in volume of the reaction, while the electric field jump method relies on changes in dipole moment.

The advantages of relaxation techniques is that they allow both forward and reverse rate coefficients to be measured. Proton transfer reactions have been widely studied in this way. The prototype reaction:

$$H^+ + OH^- \rightleftarrows H_2O \qquad \text{(R 22, −22)}$$

has a relaxation time for neutral water at 298 K of 3.5×10^{-5} s, corresponding to $k_{22} = 1.4 \times 10^{11}$ dm^3 mol^{-1} s^{-1} and to $k_{-22} = 2.5 \times 10^5$ s^{-1}. The large value for k_{22} requires a high mobility of the reactant ions and a large encounter distance (see Chapter 6). Relaxation methods have also been employed to study enzyme kinetics.

A thorough treatment of the subject may be found in the book *Relaxation Kinetics* by Bernasconi (1976) however, some of the examples used are now somewhat dated.

2.11 Temperature control and measurement

In order to determine A factors, activation energies or to investigate possible deviations from the Arrhenius equation, rate coefficients must be measured over a range of temperatures. The wider the range of temperature (or more strictly the wider the range of $1/T$) covered the more accurate the determinations and the more obvious any deviations from the Arrhenius equation will be. Such variations provide information useful to developing theories of elementary reactions (Chapter 3). The measurement of rate coefficients at other than ambient temperature is of obvious practical importance. Rate data are required to model such diverse systems as interstellar clouds (10–100 K), stratospheric chemistry (200 K), urban smog (295 K), combustion (700–3000 K) and plasmas (>3000 K). Many of these temperatures are difficult to achieve in the laboratory and extrapolation of experimental rate data to these conditions is required. Such estimates can be made with much greater confidence if a wide temperature range has been covered in the laboratory experiments.

2.11.1 Temperature control

High temperatures (300–1300 K) can be routinely attained by a wide range of conventional resistive heaters. The maximum temperature that can be reached will depend on a number of factors including heater power, degree of insulation, geometry and complexity of the apparatus, and the construction materials (pyrex <700 K, quartz <1000 K, stainless steel, <1300 K or alumina, <2000 K). Fontijn (1979) and coworkers have specialized in measuring elementary reactions at high temperatures (<1900 K) using flash photolysis techniques. Shock tubes, especially when combined with flash photolysis and real time detection provide a method of studying elementary reactions up to ∼3000 K. In this case the shock is merely used to raise the temperature of the reactants, no thermal decomposition occurs. The reaction is then initiated by a laser pulse providing a more selective initiation. An example of this technique is given by the study of Davidson and Hanson (1990) who investigated the reaction of N atoms with NO, an important reaction in combustion. They shock heated the reactants to 1500–3500 K, nitrogen atoms were then produced by the photolysis of a small portion of the NO reactant and their concentration followed by absorption spectroscopy.

$$NO + h\nu \rightarrow N + O \qquad\qquad (R\ 23)$$

$$N + NO \rightarrow N_2 + O. \qquad\qquad (R\ 24)$$

In other cases the power of a CO_2 laser can also be used to provide instantaneous heating. The reaction mixture is made up with a known concentration of a bath gas such as SF_6 which has a strong absorption cross-section at CO_2 laser wavelengths. As the CO_2 laser fires, its energy is absorbed by the SF_6. Subsequent collisions distribute the energy as heat throughout the reaction vessel. The technique is similar to that described in Section 2.10 on relaxation methods. In that case low intensity pulses were used to slightly increase the reactor temperature. With high power laser pulses temperature rises of several hundred kelvin are possible. Alternatively, many combustion systems can be analysed *in situ*, i.e. in the flame itself. For example, relative concentrations of OH in a flame may be monitored by LIF.

Low temperatures, relevant for stratospheric modelling, can be achieved by flowing liquid nitrogen (77 K) or any other suitable coolant (e.g. pentane/liquid N_2 slush) through an outer jacket which surrounds the apparatus. The extremely low temperatures found in some interstellar clouds may only be attained by supersonic expansion of the reactants (see Chapter 4). Rowe and co-workers (Sims *et al.* 1992) have recently developed such an apparatus which is presently yielding interesting kinetic data at temperatures as low as 10 K.

2.11.2 Temperature measurement

Arrhenius plots will only be meaningful if the 'independent axis' $1/T$ is accurately known, i.e. there is little uncertainty in the temperature measurement. Thermo-couples are the commonest way of conveniently and accurately measuring temperature, although platinum resistance thermometers are also used at high temperatures and standard mercury in glass thermometers may be used closer to ambient conditions.

In all experimental systems there will be some degree of temperature variation across the apparatus. In flow tube experiments it is vital to minimize these variations otherwise the reaction will be occurring at different rates in different portions of the flow tube! The criteria are less stringent in flash photolysis experiments, at least when 'point' measurements such as laser induced fluorescence are employed, because the interaction zone where the reaction is initiated and monitored is generally small and well defined and hence will have virtually no temperature variation. It is however important that the temperature sensor is placed as close to the observation region as possible.

In flames and other complex systems where an intrusive method of temperature measurement, such as a thermocouple, would affect the transport properties of the flame, temperature measurements can be achieved using LIF. The relative populations of rotational states (e.g. of OH) in a Boltzmann distribution is a function of temperature. Relative populations can be determined by tuning the dye laser to various rotational states and thus obtaining a non-intrusive temperature measurement. Coherent anti-stokes Raman spectroscopy (CARS) on N_2 is another spectroscopic method which is frequently used.

Example 2.3

The following example illustrates the importance of measuring rate coefficients over as wide a range as possible and also shows that significant deviations can occur from the Arrhenius equation at high temperatures. We shall develop theories to explain both Arrhenius behaviour and deviations from Arrhenius behaviour in subsequent chapters.

From the following data determine the pre-exponential factor and the activation energy for the reaction of hydroxyl radicals with ethane. Using the determined parameters estimate the rate coefficient for the reaction at 1700 K.

Temperature/K	222	250	286	333	400	500
$k/10^{-13}$ (cm^3 molecule^{-1} s^{-1})	0.786	1.349	2.512	4.266	7.413	14.130

To evaluate A and E_{expt} we plot $\ln(k)$ vs. $1/T$. From eqn (25) of Chapter 1 we see that the slope of the plot is equal to the activation energy and the intercept with the y-axis corresponds to the pre-exponential factor.

Temperature/K	222	250	286	333	400	500
1000 K/T	4.5	4.0	3.5	3.0	2.5	2.0
$k/10^{-13}$ (cm^3 molecule^{-1} s^{-1})	0.786	1.349	2.512	4.266	7.413	14.130
$\ln k$	−13.12	−12.87	−12.60	−12.37	−12.13	−11.85

A linear least squares fit to the data shown in Fig. 2.24(a) returns values of 1.393×10^{-11} cm^3 molecule^{-1} s^{-1} and 9.62 kJ mol^{-1} for the pre-exponential and activation energy respectively. Using these data we can extrapolate a value of 7.5×10^{-12} cm^3 molecule^{-1} s^{-1} at 1700 K. However, experimental data have been obtained for the temperature region 500 – 2000 K and the complete data set is shown in Fig. 2.24(b). It can be readily be seen that our simple extrapolation of the lower temperature data does not predict the high temperature measurements and the rate coefficient at 1700 K is underestimated by a factor of 7.5. In the next chapter we shall look at some of the theories which seek to explain these so called curved Arrhenius plots.

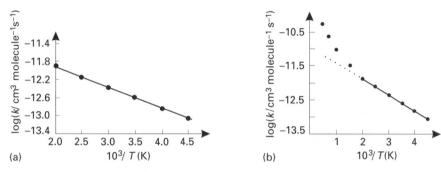

(a) (b)

Fig. 2.24 (a) Linear Arrhenius plot for the reaction of OH radicals with ethane over a limited temperature range (222–500 K). (b) Full Arrhenius plot showing the substantial curvature at high temperatures.

2.12 Questions

2.1   The following data were taken for a liquid phase flash photolysis study of the reaction of SO_4^- ions with an excess of $HCOO^-$. The concentration of SO_4^- was followed as a function of time after reaction initiation by absorption spectroscopy. By making an appropriate plot confirm the first-order kinetics and determine the pseudo-first-order rate coefficient for the reaction.

Time/μs	10	20	30	40	50	60	70	80
I_{abs}/I_o	0.0218	0.0191	0.0163	0.0143	0.0121	0.0107	0.0094	0.0078

2.2   The following data were taken from a shock tube study of the reaction of C atoms with NO. Following the shock C_3O_2 is rapidly pyrolysed

$$C_3O_2 \rightarrow 2CO + C \qquad \text{(R 18)}$$

and C then reacts with an excess of NO. From the table of data and a suitable plot calculate the initial three half-lives for the disappearance of C atoms. Do your results confirm the expected first order kinetics? What is the pseudo-first-order rate coefficient for this reaction?

Time/μs	10	20	30	40	50	60	80	100	140	180
[C]/arbitrary units	5.74	4.33	3.13	2.33	1.83	1.40	0.80	0.47	0.10	0.0

2.3   The following data are taken from an aqueous phase flash photolysis experiment on the oxidation of the nitrate ion by the sulphate ion SO_4^-. The sulphate ion is generated by the flash photolysis of a low concentration of $S_2O_8^{2-}$

by an excimer laser pulse. The $[SO_4^-]$ is monitored as a function of time following the reaction initiation, by absorption spectroscopy. The $[SO_4^-]$ is always much less than the nitrate ion concentration ensuring first-order conditions. From the data below calculate the bimolecular rate for reaction (R 25)

$$SO_4^- + NO_3^- \rightarrow SO_4^{2-} + NO_3 \qquad \qquad (R\ 25)$$

$[NO_3^-]/10^{-2}$ M	1.0	1.5	2.0	2.5	3.0	3.5	4.0
Pseudo-first-order rate coefficient/10^3 s^{-1}	1.5	2.0	2.63	3.13	3.73	4.23	4.83

What possible reasons are there for the positive intercept on the y-axis?

Reaction (25) and a number of other similar reactions are thought to take place in cloud droplets and be involved in the formation of acid rain.

2.4 From the following data calculate the activation energy for the reaction of hydroxyl radicals with formaldehyde (methanal). What are the likely products of the reaction?

Temperature/K	250	286	333	400	500	667
$k\,/10^7$ (mol dm^{-3} s^{-1})	4.02	8.49	18.0	39.9	85.1	179.5

2.5 Neodymium yttrium aluminium garnet (YAG) lasers are alternative photolysis sources for the generation of transient species. Lasing transitions occur between various electronic levels of neodymium ions in a YAG lattice. The output of the Nd–YAG laser is in the near infrared (at 1062 nm) but by processes known as frequency doubling and frequency mixing, the fundamental wavelength can be converted to output at 532, 355 and 266 nm. Calculate the maximum energies of the bonds that can be broken by each of these photons. Why do you think that the output at 532 nm is rarely used in the generation of radical species?

2.6 Lasers are excellent photolysis sources. In a series of experiments to study bromine atom reactions, CF_2Br_2 was photolysed by an argon fluoride excimer laser at 193 nm.

$$CF_2Br_2 + h\nu \rightarrow CF_3Br + Br.$$

If the laser pulse is 50 mJ, calculate the number of photons produced per pulse.

What other information would we require in order to calculate the number of Br atoms produced by the laser pulse?

References

Abbatt, J. P. D., Demerjian, K. L., and Anderson, J. G. (1990). A new approach to free radical kinetics. *Journal of Physical Chemistry*, **94**, 4566–75.

Andrews, D. L. (1992). *Lasers in Chemistry*, (2nd edn). Springer Verlag, Berlin.

Atkins, P. W. (1994). *Physical Chemistry*, (5th edn), p. 457. Oxford University Press.

Atkins, P. W. (1994). *Physical Chemistry*, (5th edn), p. 654. Oxford University Press.

Baggott, J. E., Frey, H. M., Lightfoot, P. D., and Walsh, R. (1986). The absorption cross-section of the HCO radical @614.59 nm and the rate constant for the HCO+HCO reaction. *Chemical Physics Letters*, **132**, 225–30.

Bernasconi, C. F. (1976). *Relaxation kinetics*. Academic Press, New York.

Biggs, P., Hancock, G., Heal, M. R., McGarvey, D. J., and Parr, A. D. (1991). Temperature dependencies of CH_2 removal rates by Ar, NO and H_2.

Davidson, D. F., and Hanson, R. K. (1990). High temperature rate coefficients. *International Journal of Chemical Kinetics*, **22,** 843–61.

Dean, A. J., Hanson, R. K., and Bowman, C. T. (1991). A shocktube study of reactions of C and CH with NO including product channel measurements. *Journal of Physical Chemistry*, **95,** 3180–9.

Farhataziz and Rodgers, M. A. J. (1987), *Radiation Chemistry—principles and applications*. VCH, New York.

Felder, W., and Fontijn, A. (1979). High temperature photochemistry, a new technique. *Chemical Physics Letters*, **67,** 53–6.

Fendorf, S. E., Sparks, P. L., Franz, J. A., and Camaiori, D. M. (1993). E.P.R. stopped-flow kinetic study of Mn (II) sorption–desorption on birnessite. *Journal of American Soil Science*, **57,** 57–62.

Fleming, G. (1986). *Chemical applications of ultrafast spectroscopy*. Oxford University Press.

Hollas, J. M. (1992). *Modern spectroscopy*, (2nd edn), Chapter 9. Wiley, Chichester.

Markwalder, B., Guzel, P., and van den Bergh, H. (1993). Laser induced temperature jump measurements. *Journal of Physical Chemistry*, **97,** 5260–5.

Opansky, B. J., Seakins, P. W., Pedersen, J. P., and Leone, S. R. (1993). Kinetics of the C_2H+O_2 reaction from 193–350 K using laser photolysis/kinetic IR absorption spectroscopy. *Journal of Physical Chemistry*, **97,** 8583–9.

Plumb, I. C., and Ryan, K. (1981). Kinetic studies of the $C_2H_5+O_2$ reaction. *International Journal of Chemical Kinetics*, **13,** 1011–28.

Sims, I. R., Quefiec, J. L., Defrance, A., Rebrion–Rowe, C., Travers, D., Rowe, B. R., and Smith, I. W. M. (1992). Ultra low temperature kinetics of neutral–neutral reactions. *Journal of Chemical Physics*, **97,** 8798–800.

3 An introduction to theories of bimolecular reactions

3.1 Introduction

One of the basic aims of theoretical reaction kinetics is to understand why some reactions are fast and others slow, why some reactions have strong positive temperature dependencies and others virtually none. We wish to be able to rationalize, *quantitatively* if possible, the magnitude of a given rate coefficient and of its temperature dependence. We might even envisage the possibility of predicting rate coefficients in the absence of experimental data.

Obviously we need to compare our theoretical expressions with experimental data or empirical expressions for the temperature dependence of the rate coefficient. Initially we shall limit ourselves to trying to match the Arrhenius expression $(k_{expt} = A \exp(-E_{expt}/RT))$ for the *experimental* variation of k with temperature. Although as we saw in Example 2.4 this expression is not perfect, it is a good starting point for our discussion. In addition, we shall limit ourselves primarily to a discussion of gas phase reactions. The kinetics of reactions in liquids, with all the extra complexity involved with solvent interactions, are discussed in Chapter 6.

We commence our discussion of theories of chemical reactions with an examination of bimolecular reactions, recognizing first of all that collisional processes are involved and restricting ourselves initially to a model in which the molecules are thought of as hard, structureless spheres. We then describe a more detailed theory which is statistical, rather than collisional, in its approach. This is the famous transition state theory (TST) which was developed in the 1930s and has since formed a framework for much of the discussion of rate processes. Both collision theory and transition state theory are currently enjoying detailed investigation and in Chapter 4 we shall briefly discuss current thinking on both theories. For the moment however, we shall limit ourselves to a discussion of their more traditional and elementary aspects.

3.2 Collision theory

We begin by assuming that the molecules are hard, structureless spheres like billiard balls; by this we mean that (1) there is no interaction between them until they come into contact and (2) they are impenetrable, i.e. they maintain their size and shape on collision and no matter how hard they strike one another, their centres cannot come closer than the sum of their radii. These are similar assumptions to those made in deriving the kinetic theory of gases.

When the molecules react there is a rearrangement of their valance electrons and this *generally* means that some energy has to be expended in overcoming an energy barrier—energy is required to disturb the electrons from their initial arrangement, even if the final arrangement in the products is more stable. This seems to be a reasonable assumption to make and as we shall see, it is necessary if we are to explain the temperature dependence of most bimolecular rate coefficients.

We consider first a collision between two spheres A and B with radii r_A and r_B and velocities v_A and v_B (Fig. 3.1). Our problem is easier to follow if we keep one of them (say B) fixed and allow A to move with *relative velocity* v, equal to $v_A - v_B$. We define a quantity b, the impact parameter, as the closest perpendicular distance of approach of the centres of the two molecules. On collision A is deflected through an angle θ, which depends on b. For a head on collision, $b = 0$ and $\theta = \pi$, while $\theta = 0$ for $b > d$, where $d = r_A + r_B$. We consider a collision to have occurred if a deflection takes place. We can define a total cross-section, S, as the effective target area presented to A by B, so that $S = \pi d^2$.

An A molecule passing with velocity v through N_B stationary B molecules, contained in unit volume, will suffer a collision with a B molecule whose centre lies within an area πd^2 about the trajectory of A. Thus, in unit time, an A molecule sweeps out a collision volume $\pi d^2 v$ (Fig. 3.2), and so undergoes collisions with $N_B \pi d^2 v$ B molecules. If there are N_A A molecules per unit volume, the total number of collisions per unit time and per unit volume, $Z' = N_A N_B \pi d^2 v$.

In a gas at temperature T, the molecules do not all have the same speed. There is instead a Maxwell distribution of speeds and we must substitute the mean value

$$\bar{v} = (8k_B T/\pi\mu)^{\frac{1}{2}} \tag{E 1}$$

in our expression. μ is the reduced mass, $m_A m_B/(m_A + m_B)$, and occurs in the equation because we are interested in the relative speed of approach. Thus

$$Z' = N_A N_B \pi d^2 (8k_B T/\pi\mu)^{\frac{1}{2}} \tag{E 2}$$

So far we have considered only elastic collisions, and we must now see how our approach can be modified for reactive collisions. As we discussed above, there is an interaction energy between the molecules, which we have so far neglected. In other words, before they can get close enough to react, the molecules must surmount an energy barrier ε^0. Only those molecules which have sufficient kinetic energy along the line of centres (AB) to overcome this barrier will react. This

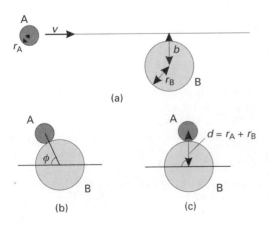

(a)

(b)

(c)

Fig. 3.1 (a) Approach of molecule A, radius r_A to molecule B, radius r_B, with relative velocity v. The impact parameter, b, is the closest perpendicular distance between the centres of the two molecules. (b) Collision of A and B at a general angle ϕ. (c) Glancing collision of A and B. This represents the maximum value of the impact parameter b, $(= r_A + r_B)$.

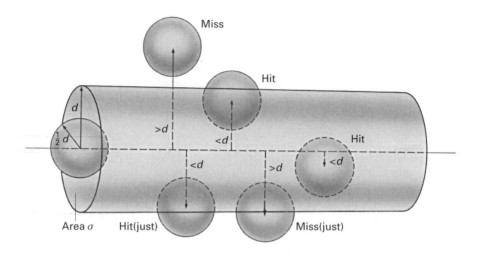

Fig. 3.2 Collision volume swept out by an A molecule passing through stationary B molecules.

stipulation about the line of centres is necessary since energy perpendicular to this line will be of no use in overcoming the barrier, which lies between A and B. For a Maxwell distribution, this fraction is $\exp(-E^0/RT)$, where $E^0 = L\varepsilon^0$ and L is the Avogadro number. Our predicted rate of reaction ρ becomes

$$\rho = N_A N_B \pi d^2 (8k_B T/\pi\mu)^{\frac{1}{2}}\exp(-E^0/RT) \qquad\qquad\text{(E 3)}$$

and since $k = \rho/N_A N_B$ we have:

*molecules H $h\nu$ suff en-
to rxt only.*

$$k = \pi d^2 (8k_B T/\pi\mu)^{\frac{1}{2}}\exp(-E^0/RT) \qquad\qquad\text{(E 4)}$$

or

$$k = Z\exp(-E^0/RT) \qquad\qquad\text{(E 5)}$$

where $Z = Z'/N_A N_B$, known as the *collision frequency factor*. We can immediately see that our expression for the rate coefficient is made up of two components. Firstly, Z, the frequency with which the reactants collide and secondly, an expression which determines the fraction of collisions which have enough energy to overcome the barrier to reaction. Further information on the velocity distributions may be found in Chapters 1 and 24 of Atkins (1994).

Box 3.1 An alternative derivation

We may derive eqn (5) more explicitly and with only a slight increase in complexity by considering each range of relative velocities separately rather than averaging at the outset. Our approach reveals more clearly the ability of the relative translational energy to overcome the energy barrier. In addition, it gives us a flexible approach which we shall refer to later.

Once again we consider A approaching B with a relative velocity v. As the particles collide, only that component of the velocity along the line of centres may be used to overcome the energy barrier. The magnitude of this velocity, v_{lc} is given by (Fig 3.3)

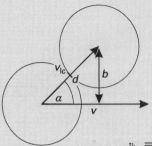

Fig. 3.3 Orientation of molecules in a general collision. Only the kinetic energy directed along the line of centre $\frac{1}{2}\mu v_{lc}^2$ can be used to overcome the reaction barrier. $v_{lc} = \cos \alpha$. ($\sin \alpha = b/d$, $\cos^2\alpha + \sin^2\alpha = 1$, therefore $v_{lc} = v\{d^2 - b^2/d^2\}^{\frac{1}{2}}$.

$$v_{lc} = v\cos\alpha = v\{d^2 - b^2/d^2\}^{\frac{1}{2}} \tag{E 6}$$

and if we define the total translational energy as $\varepsilon (=\frac{1}{2}\mu v^2)$ then μ_{lc} the energy along the line of centres is:

$$\varepsilon_{lc} = \varepsilon\{(d^2 - b^2)/d^2\}. \tag{E 7}$$

Thus, as b increases, ε_{lc} decreases; for a given ε, ε_{lc} has its maximum value for $b = 0$, i.e. for a head-on collision. If ε^0 is the critical or threshold energy then reaction only occurs for a given ε provided that b is less then b_{max}, defined by:

$$\varepsilon^0 = \varepsilon\{(d^2 - b_{max}^2)/d^2\} \tag{E 8}$$

or

$$b_{max}^2 = d^2(1 - \varepsilon^0/\varepsilon) \tag{E 9}$$

For collisions occurring with total energy ε and with impact parameters greater than b_{max}, the energy along the line of centres is less than ε^0. b_{max} increases with ε, since larger values of α can be accommodated within the restriction that $\varepsilon_{lc} > \varepsilon^0$.

We can now begin to evaluate the rate coefficient for reaction. For a given collision energy, ε, the collision cross-section is given by:

$$S(\varepsilon) = \pi b_{max}^2 = \pi d^2 (1 - \varepsilon^0/\varepsilon). \tag{E 10}$$

An A molecule, passing with relative velocity v (and relative kinetic energy ε) through $\mathcal{N}_B$ B molecules per unit volume strikes $vS(\varepsilon)\mathcal{N}_B$ B molecules with sufficient directed energy to react. The overall rate of reaction may then be obtained by integrating over all the A molecules, whose velocities relative to B conform to a Maxwell–Boltzmann distribution:

$$\text{Rate} = k(T)\mathcal{N}_A\mathcal{N}_B = \int_\varepsilon^\infty (2\varepsilon/\mu)^{\frac{1}{2}} f(\varepsilon,T)S(\varepsilon)\, d\varepsilon \mathcal{N}_A\mathcal{N}_B \tag{E 11}$$

where the substitution $v = (2\varepsilon/\mu)^{\frac{1}{2}}$ has been made and where $f(\varepsilon,T)d\varepsilon$ is the fraction of collisions taking place with relative kinetic energy between ε and $\varepsilon+d\varepsilon$ at temperature T. Substituting

$$f(\varepsilon,T) = 2\varepsilon^{\frac{1}{2}} \exp(-\varepsilon/kT)/\sqrt{\pi}(k_B T)^{\frac{3}{2}} \tag{E 12}$$

and integrating gives

$$k(T) = (8k_B T/\pi\mu)^{\frac{1}{2}} \pi d^2 \exp(-\varepsilon^0/k_B T) \tag{E 13}$$

in agreement with eqn (5) with ε^0/k_B replacing E^0/R.

   Collision theory has led to an expression of the correct form, i.e. $A\exp(-E/RT)$, and gives us some insight into the physical significance of A (collision frequency) and E (activation energy). The increase in the rate coefficient with increasing temperature may be appreciated by considering Fig. 3.4, which shows a plot of the number of collisions with relative kinetic energy between ε and $\varepsilon + d\varepsilon$, vs. energy. An increase in temperature shifts the distribution to the right, and substantially increases the number of collisions with a high relative kinetic energy. The shaded portions of Fig. 3.4 indicate those collisions with relative energies in excess of $10\,\text{kJ mol}^{-1}$. Although this diagram is not strictly applicable in our case, since no stipulation has been made about the direction of the relative velocity, it does demonstrate the striking effect of increasing temperature. Figure 3.5 shows a plot of the fraction of collisions with collision energy along the line of centres greater than E, $\exp(-E/RT)$, for various temperatures for an activation energy, E, of $10\,\text{kJ mol}^{-1}$.

   Calculations of activation energies are extremely difficult. Only recently has it been possible to calculate with any accuracy the energies of stable molecules and atoms; the calculation of molecular energies and geometries corresponding to the peak of the energy profile is increasingly feasible but expensive and time consuming. We shall return to such calculations in Chapter 4. At present all we shall ask of our theory is to estimate A. Collision theory gives for the pre-exponential factor A_{th}

$$A_{\text{th}} = (8k_{\text{B}}T/\pi\mu)^{\frac{1}{2}}\pi d^2 \qquad\qquad (\text{E }14)$$

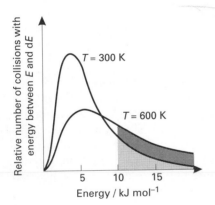

Fig. 3.4 The increase with temperature of the number of collisions with relative kinetic energy greater than $10\,\text{kJ mol}^{-1}$.

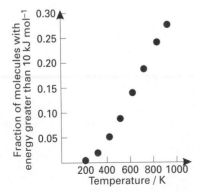

Fig. 3.5 A graphical representation of the increase with temperature of the fraction of molecules with relative translational energy along the line of centres greater than $10\,\text{kJ mol}^{-1}$.

A_{th} is temperature dependent, but for most reactions this dependence is masked by the exponential term. For example, A increases by less than 2 per cent on raising the temperature from 300 to 310 K, while, for a reaction with an activation energy of 100 kJ mol^{-1}, the exponential term increases by a factor of 320 per cent over the same temperature range. For many reactions, the small variation in A is swamped by the large variation in the exponential term, and by the random experimental error involved in measuring the rate coefficient. Therefore the predicted temperature dependence of A and the observed temperature independence are not incompatible.

We can now use eqn (14) to calculate the magnitude of A_{th} and compare it to the experimental values. Molecular diameters may be calculated from gas viscosity data and used to calculate d. A selection of values are given in Table 3.1 and are typically in the region of 0.3 nm. Table 3.1 also lists the collision frequency, calculated from the molecular parameters, at one atmosphere of pressure (1×10^5 Pa) and 300 K. The mean time between collisions, τ, is the reciprocal of the collision frequency. Under the conditions of Table 3.1 τ is less than 10^{-9} s.

Table 3.1 Collision parameters

Species	Molecular diameter/nm	Collision frequency/s^{-1} ($P = 10^5$ Pa)
NH$_3$	0.30	9.3×10^9
CO	0.32	5.2×10^9
H$_2$	0.24	10.2×10^9

Table 3.2 compares calculated and experimental A factors for a range of reactions. Typically, the calculated values are of the order of 10^{11} dm^3 mol^{-1} s^{-1}. One of the estimates is smaller than the experimental value, but the others are much too large, in two cases by factors of more than 10^5. How can the calculation be so far out? We can go some way towards understanding the over-estimated values if we accept that orientation is important in a reactive collision. Some collisions will be ineffective because the colliding molecules are not pointing their reactive 'ends' at each other.

Table 3.2 Comparison of collision theory and experiment

Reaction	$T/$(K)	$E/$(kJ mol^{-1})	$10^{-11} \times A_{exp}/$ (dm^{3}mol^{-1}s^{-1})	$10^{-11} \times A_{th}/$ (dm^{3}mol^{-1}s^{-1})	P
K + Br$_2$ → KBr + Br	600	0	10	2.1	4.8
CH$_3$ + CH$_3$ → C$_2$H$_6$	300	0	0.24	1.1	0.22
2NOCl → 2NO + Cl$_2$	470	102	0.094	0.59	0.16
[CHO + CHO → (cyclohexene CHO)]	500	83	1.5×10^{-5}	3.0	5×10^{-6}
H$_2$ + C$_2$H$_4$ → C$_2$H$_6$	800	180	1.24×10^{-5}	7.3	1.7×10^{-6}

hookline

An example might be the reaction of hydroxyl radicals with $CHBr_3$

$$OH + CHBr_3 \rightarrow H_2O + CBr_3 \tag{R 1}$$

For most directions of OH attack the small hydrogen atom will be shielded from the hydroxyl group by the bulky bromine atoms and only in certain very restricted approaches will the attacking radical be able to access the hydrogen atom. To allow for this effect we insert into eqn (5) a probability factor P. P will be very small ($\ll 1$) for reactions which require very specific orientation of the reactant molecules on collision, e.g. the Diels-Alder reaction, and will attain a maximum value of unity for reaction in which orientation is unimportant. Although orientation must be of some importance in a great many reactions, it cannot wholly account for the wide range of P values shown in Table 3.2.

Box 3.2 Further developments in collision theory

A modified form of collision theory has been described by Smith, I. W. M. (1982) (A new collision theory for bimolecular reactions. *Journal of Chemical Education* **59**, 9–14) which may be thought of as an extension of the simple collision theory to reactions between more structured molecules, in this case an atom, A and a diatomic molecule, B. For the purposes of the collisions, B is still thought of as a sphere, radius r_B, but we can now define the direction of the molecular axis of B within this sphere. It is assumed that the activation barrier ε_{eff} depends on the angle ϕ between this molecular axis and the line XY joining centres of A and B on collision (Fig. 3.6)

$$\varepsilon_{eff} = \varepsilon^0 + 2\varepsilon(1 - \cos \phi). \tag{E 15}$$

Thus the activation barrier is minimized if ϕ is zero and rises to a maximum when XY is perpendicular to the molecular axis. Effectively, the model is predicting that the collision is likely to be more efficient if A strikes the diatomic molecule end on. Thus the collision cross-section is now both angle and energy dependent and, as is described in Smith's paper, both of these dependencies must be accommodated in the derivation of the rate coefficient.

A second, simpler, change is also introduced; it is assumed that the molecules must attain a separation D ($D < d$) before reaction can take place, in recognition of small reaction distances (significantly less than gas kinetic collision diameters) found in detailed calculations of reaction cross-sections.

The combined result of these changes is that the rate coefficient now becomes:

$$k = (k_B T/2\varepsilon') (8k_B T/\pi\mu)^{\frac{1}{2}} \pi D^2 \exp(-\varepsilon^0/k_B T) \tag{E 16}$$

which is less than that described by eqn (4) if $2\varepsilon' > k_B T$ and $D < d$.

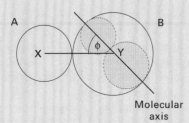

Molecular axis

Fig. 3.6 Schematic diagram showing the impact of A relative to a molecular axis for B.

3.3 Summary of collision theory

Collision theory is a pictorially simple model which provides a good initial visualization of bimolecular reaction. It emphasizes the importance of collision events providing the energy for reaction, and predicts qualitatively the form of the temperature dependence of the rate coefficient. The predicted values of A_{th} are far from the experimental results. When we consider the sweeping assumptions made at the beginning of the section we can begin to appreciate how these discrepancies arise.

Firstly, in making the 'hard spheres' assumption we have completely ignored the structure of the molecules. By introducing the steric factor, P, we can go some way in allowing for conformational effects. However, the assigned values do not always correlate with molecular complexity, nor is there any *a priori* method of calculating them.

Secondly, we have assumed that molecules react instantaneously. In practice the changes in structure take place over a finite period. The structure of the reaction complex will evolve and this must be considered.

Finally, all molecules interact with each other over distances greater than the sum of their hard sphere radii. In many cases these interactions will be considerable (e.g. between oppositely charged ions). Long range attractive interactions are important in explaining reactions where $P > 1$.

However, great difficulties arise in incorporating these considerations into collision theory and we must turn to transition state theory (TST) to make further progress. While studying Section 3.4 note how the points raised above are considered. The effect of molecular interactions is further considered in Chapter 4.

3.4 Transition state theory

Once simple collision theory was found to be inadequate, the problem of predicting reaction rates was examined using a different type of model, developed in the 1930s , initially by Wigner and Pilzer, later extended by Eyring and known as *activated complex theory* or now more commonly as *transition state theory*. As an introduction let us examine Fig. 3.7, which is the familiar energy profile of a reaction, a plot of energy versus reaction coordinate. The *transition state* is the highest energy point on the graph labelled $X^{\ddagger}$, and the reactants at this configuration are known as the *activated complex*. We must first study this profile in more detail, and also decide what we mean by the term reaction coordinate.

Consider a schematic atom + diatomic molecule reaction:

$$A + BC \rightarrow X^{\ddagger} \rightarrow AB + C. \tag{R 2}$$

The potential energy, V, of the reacting system depends on the relative positions of A, B, and C, which are fully specified by the three variables r_{AB}, r_{BC}, and ϕ (see Fig. 3.8). Four variables (V, r_{AB}, r_{BC}, and ϕ) are difficult to handle, but if we keep ϕ constant, i.e. specify the direction of approach of A, we may plot V as a function of r_{AB} and r_{BC}, when we obtain a three-dimensional potential energy surface. This is shown in Fig. 3.9 for $\phi = \pi$, i.e. for a collinear collision. We need a whole family of such curves to characterize fully the angular dependence of the potential energy. It is also rather difficult to draw and interpret 3-D diagrams, and to overcome this problem we generally use contour diagrams, with r_{AB}, and r_{BC} as

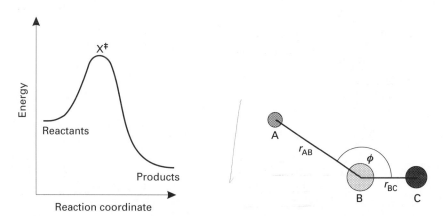

Fig. 3.7 Energy profile for a typical reaction.

Fig. 3.8 General orientation of the A + BC reaction.

the x- and y-axes, and the potential energy plotted as contours. Figure 3.10 shows the surface from Fig. 3.9 plotted in this way.

For large values of r_{AB}, there is no interaction between A and BC, and a section through DE corresponds to the normal Morse-type potential energy curve for the ground electronic state of BC. As A approaches, the potential energy rises, until the point $X^{\ddagger}$ is reached, after which the energy falls as C moves away. $X^{\ddagger}$ represents the maximum in the minimum energy path between reactants and products, i.e. the *saddle point* between the valleys $RX^{\ddagger}$ and $X^{\ddagger}P$. The dotted line $RX^{\ddagger}P$ is the reaction coordinate. Motion along this coordinate implies that as A approaches and the A–B distance decreases, B–C increases, at first slowly, from the equilibrium bond length. The contraction of r_{AB} and expansion of r_{BC} occur in such a way that the energy traces the minimal energy path through the saddle point and into reaction. The activated complex is formed at $X^{\ddagger}$ and although it is unstable with respect to motion along $RX^{\ddagger}P$, it is stable, i.e. it lies at the potential energy minimum, with respect to movement along $OX^{\ddagger}E$. Close to O, the atoms are compressed on to one another, their electron clouds will tend to repel each other

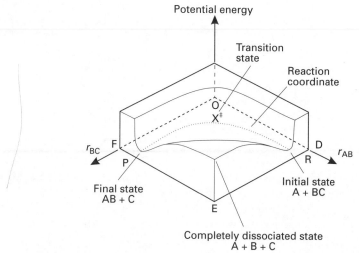

Fig. 3.9 3-D potential energy surface for the colinear reaction A + BC → AB + C.

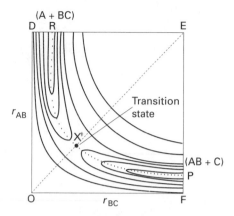

Fig. 3.10 Contour diagram for the surface shown in Fig. 3.9.

and hence this is a high energy state, as is E, which corresponds to dissociation into three separate atoms. We know that E must be a high energy state because we have to supply energy to break either the A–B or B–C bond.

Transition state theory examines a reaction from an equilibrium, or really a statistical aspect, in contrast to the dynamic, collision orientated point of view we encountered in the last section. At this point it is worth briefly reviewing the link between equilibrium and kinetics as this relationship underpins the whole of transition state theory.

The old adage 'thermodynamics can predict the position of equilibrium but not its rate of attainment' whilst true, implies that there is little relationship between thermodynamics and kinetics which as we saw in Section 1.7 is certainly not the case. Equilibrium is a dynamic state. In the system shown below reaction is constantly occurring in both directions.

$$X \rightleftarrows Y. \tag{R 3, R −3}$$

At equilibrium there will be an equal number of events (reactions) occurring in each direction in unit time. Hence:

$$k_3[X] = k_{-3}[Y] \tag{E 17}$$

(NB only in the special case that $k_3 = k_{-3}$ will there be equal concentrations at equilibrium.) Therefore:

$$k_3/k_{-3} = [Y]/[X] = K_c \tag{E 18}$$

and k_3 and k_{-3} reflect the relative stability of each component, and depend on the two activation energies E_3 and E_{-3}.

Transition state theory starts by assuming that the transition state is in equilibrium with the reactants.

$$[X^{\ddagger}]_f = K^{\ddagger}[A][BC] \tag{E 19}$$

where $[X^{\ddagger}]_f$ represents the concentration of activated complexes moving forward to products, and $K^{\ddagger}$ is the equilibrium constant for the formation of $X^{\ddagger}$ from A + BC. The overall rate of forming products, ρ, is $k^{\ddagger}[X^{\ddagger}]_f$, where $k^{\ddagger}$ is the first order rate coefficient for the decomposition of $X^{\ddagger}$ into products. Thus, since $\rho = k_2[A][BC]$ we can substitute from eqn (19) to give:

$$k_2 = k^{\ddagger}K^{\ddagger} \tag{E 20}$$

and, assuming that the removal of the transition state to form products does not affect the equilibrium between $X^{\ddagger}$ and reactants, our problem reduces to one of evaluating $k^{\ddagger}$ and $K^{\ddagger}$.

3.4.1 Equilibria and energy levels

Equilibrium constants can be calculated from molecular parameters, such as bond lengths, bond frequencies and molecular weights, from partition functions using the results of statistical mechanics. The distribution of molecules throughout a set of energy levels (ε_i) is given by the Boltzmann Law:

$$n_i \propto g_i \exp(-\varepsilon_i/k_B T) \qquad \text{(E 21)}$$

where n_i is the number of molecules in the ith level and g_i is the degeneracy i.e. the number of states which occur with energy ε_i. The *partition function* is defined as:

$$q = \sum_{i=0}^{\infty} g_i \exp(-\varepsilon_i/k_B T). \qquad \text{(E 22)}$$

It may be thought of as a sum of all the energy levels weighted according to the probability of their occupation at a given temperature. The larger is q, the more the population is distributed among the available levels: when $q = 1$, only one state is occupied (e.g. at absolute zero). It may be shown that the equilibrium constant, K, expressed in terms of molecules per unit volume, for the reaction:

$$H + J \rightleftarrows F + G \qquad \text{(R 4)}$$

is given by

$$K_4 = (Q_F Q_G/Q_H Q_J) \exp(-\Delta\varepsilon_0/k_B T) \qquad \text{(E 23)}$$

where Q is the partition function per unit volume (q/V) and $\Delta\varepsilon_0$ is the difference in energy between the lowest energy levels of the reactants and products. For a fuller discussion on partition functions and other aspects of statistical thermodynamics see *Entropy and Energy Levels* by Gasser and Richards (1979), Chapters 19 and 20 in *Physical Chemistry* by Atkins (1994), or *An Introduction to Statistical Thermodynamics* by Hill (1962).

By applying the Born–Oppenheimer approximation, we may subdivide the molecular energy into its component parts, translation, rotation, vibration, and electronic energy. Expressions for the energy levels ε_i can then be obtained by solving the Schrödinger equation for each form of energy. In this way we can obtain expressions for the partition functions and each form of motion can be determined; they are listed in Table 3.3 together with typical numerical values at 300 K.

The translational partition function is very large, as one would expect, since many states are occupied, but $Q_V \approx 1$ at 300 K, because the spacing between energy levels is large, and molecules occupy only the lowest state. The partition functions depend on molecular masses, moments of inertia and vibrational frequencies, which may be determined from molecular spectroscopy. The total partition function Q is the product of the individual partition functions, Q_T, Q_R, Q_V, Q_E. This follows from its definition in terms of an exponential function of energy.

Table 3.3 Partition functions

Motion	Degrees of freedom	Partition function (Q_i^n)	Order of magnitude
Translation	3	$Q_T^3 = (2\pi m k_B T)^{3/2}/h^3$	10^{33} m^3
Rotation linear molecule	2	$Q_R^2 = 8\pi^2 I k_B T/\sigma' h^2$	$10{-}10^2$
Rotation (non-linear molecule)	3	$Q_R^3 = 8\pi^2 (8\pi^3 I_A I_B I_C)^{\frac{1}{2}} (k_B T)^{3/2}/\sigma' h^3$	$10^2{-}10^3$
Internal rotation	n_1	$Q_{IR}^{n_1} = \{(8\pi^3 I' k_B T)^{1/2}/h\}^{n_1}$	$1{-}10^{n_1}$
Vibration	n_2	$Q_V^{n_2} = \{1 - \exp(-h\nu/k_B T)\}^{-n_2}$	$1{-}10^{n_2}$
Electronic	—	$Q_E = \Sigma_j g_j \exp(-\varepsilon_j/k_B T)$	1

m = mass of molecule, I = moment of inertia (linear molecule), I_A, I_B, I_C = principal moments of inertia (polyatomic molecule), σ' = symmetry factor, I' = effective moment of inertia for internal rotation, n_1 = number of internal rotational modes, ν = vibrational frequency, n_2 = number of vibrational modes, g_j = degeneracy of electronic level of energy ε_j. The electronic partition function is summed over all the electronic levels, but for most molecules, all but the lowest level are so high in energy that they can be neglected. The degeneracy of the ground electronic state is usually one, notable exceptions being radicals (two).

3.4.2 Application of statistical mechanics to transition state theory

Applying these results to transition state theory we obtain:

$$k = k^{\ddagger} (Q_X/Q_A Q_{BC}) \exp(-\Delta\varepsilon_0/k_B T) \qquad (E\ 24)$$

where $\Delta\varepsilon_0$ is the difference between the lowest energy levels of A and BC and of $X^{\ddagger}$, and corresponds to the activation energy at 0 K.

In assigning a partition function to the activated complex, we are presuming that its motion and associated energy levels may be treated in much the same way as those of stable molecules. There are some problems associated with vibrational motion, which warrant attention. A stable linear triatomic molecule has three vibrational modes, as shown in Fig. 3.11. What can we say about the corresponding modes of our unstable complex ABC? In the symmetric stretching mode (ν_1), A–B and B–C both extend at the same time, and they both contract at the same time. Thus ν_1 corresponds to motion along OXE in Fig. 3.10, perpendicular to the reaction coordinate. $X^{\ddagger}$ is the minimum on the potential energy surface for motion along OXE and any displacement from $X^{\ddagger}$ in this direction, i.e. any extension or contraction of both A–B and B–C, is consequently accompanied by a restoring force. The symmetric stretching mode is thus a true vibration. The same goes for ν_2, but we cannot discuss this in terms of Fig. 3.10, since the bending mode involves a deformation in the bond angle ϕ, while our potential surface restricts the atoms to $\phi = \pi$. In the asymmetric stretching mode (ν_3) A–B contracts as B–C extends. This is equivalent to motion perpendicular to OXE, i.e. to motion along the reaction coordinate, and we might rename ν_3 the dissociation mode. Since $X^{\ddagger}$ lies at a maximum in the potential energy surface for motion in this direction, there is no restoring force. As A moves in to form the A–B bond, the B–C bond breaks and C moves out. The dissociation mode is therefore not a true vibration and we have to treat it differently.

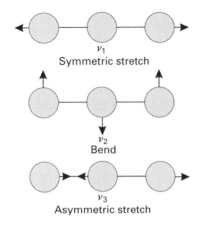

ν_1
Symmetric stretch

ν_2
Bend

ν_3
Asymmetric stretch

Fig. 3.11 Vibrational modes of a triatomic species.

It can be shown that, with a minor modification to the potential energy surface the reaction may be simulated by the motion of a point of mass μ over the potential energy surface, where μ equals the reduced mass of the reaction system (p. 107). This means that rolling a ball bearing on a 3-D model of the surface reproduces the actual motion of the reacting system. Extending this analogy we see that oscillatory motion of the point mass along $OX^{\ddagger}E$ corresponds to vibration in the symmetric stretching mode, while translation of the point mass across the saddle in the direction $RX^{\ddagger}P$ corresponds to dissociation of the complex. If we specify that the complex exists when the reaction coordinate lies within a length δ, centred around $X^{\ddagger}$, the mean rate coefficient for dissociation of the complex is:

$$k^{\ddagger} = v/\delta \qquad \frac{\text{cm s}^{-1}}{\text{cm}} = \text{s}^{-1} \tag{E 25}$$

where v is the mean velocity of the point mass along the reaction coordinate (i.e. the mean velocity in one dimension and one direction only). Since we are assuming an equilibrium distribution for the transition state, we are justified in calculating v for a Maxwell distribution. The one dimensional average velocity is given by:

$$v = (k_B T/2\pi\mu)^{\frac{1}{2}}. \qquad (1D) \tag{E 26}$$

Before we leave the dissociation mode and move on in our calculation, we need to calculate the associated partition function for this mode, Q_d. The motion is aperiodic, and a one-dimensional translational partition function is more applicable than a vibrational one. The translational partition function for one dimensional motion depends on the size of the 'box' in which the motion takes place. In our case we can use the length δ as the dimension of the box.

$$Q_d = (2\pi\mu k_B T)^{\frac{1}{2}} \delta/h. \qquad (1D) \tag{E 27}$$

except $Q_{d,trans}$

The other partition functions contained in Q_X may be treated normally, and $Q_X^{\ddagger}$ is the product of all the partition functions of the activated complex except Q_d. A and BC are stable molecules, and the partition functions of Table 3.3 may be used without modification. Combining eqns (25–27) gives:

$$k = k^{\ddagger} (Q_X/Q_A Q_{BC}) \exp(-\Delta\varepsilon_0/k_B T) \tag{E 24}$$

substituting (E 25) and partitioning Q_X into Q_d and $Q_X^{\ddagger}$

$$k = \frac{v}{\delta}\left(\frac{Q_d Q_X^{\ddagger}}{Q_A Q_{BC}}\right)\exp(-\Delta\varepsilon_0/k_B T)$$

$$k = \left(\frac{k_B T}{2\pi\mu}\right)^{\frac{1}{2}}\frac{1}{\delta}\frac{(2\pi\mu k_B T)^{\frac{1}{2}}\delta}{h}\frac{Q_X^{\ddagger}}{Q_A Q_{BC}}\exp(-\Delta\varepsilon_0/k_B T)$$

substituting $v = (k_B T/2\pi\mu)^{\frac{1}{2}}$ (E 26) and $Q_d = (2\pi\mu k_B T)^{\frac{1}{2}}\delta/h$ (E 27)

$$k = \frac{\kappa k_B T}{h}\left(\frac{Q_X^{\ddagger}}{Q_A Q_{BC}}\right)\exp(-\Delta\varepsilon_0/k_B T). \qquad (E\ 28)$$

Note that parameter δ cancels out in the product of $k^{\ddagger}$ (v/δ) and Q_d $((2\pi\mu k_B T)^{\frac{1}{2}}$ $\delta/h)$ and therefore we do not need to define an actual value for this parameter. We have inserted a quantity κ, known as the transmission coefficient, which allows for the possibility that not all activated complexes lead to products, since some may be reflected back to the reactants. Frequently, however, κ is unity. The quantity $k_B T/h = 6 \times 10^{12}\ s^{-1}$ at 300 K, and has the units of frequency.

The partition function quotient represents the most interesting aspect of this theory, since it permits us to make allowance for molecular complexity. If we move to a reaction involving more than three atoms, then the simple surfaces which we have discussed above are no longer applicable (the reactions takes place on a multi-dimensional hypersurface), however the general conclusions embodied in eqn (28) are still valid. Note that because we are using molecular partition functions, our rate coefficient has units of volume molecule^{-1} s^{-1} and must be multiplied by Avogadro's number, L, to convert it to molar units.

3.4.3 Comparison of transition state theory and experiment

For a non-linear molecule with N atoms, the number of internal rotational plus vibrational degrees of freedom is $3N - 6$. In the following discussion, T refers to a translational, R to a rotational, IR to an internal rotational, and V to a vibrational degree of freedom. Let us first apply TST to a reaction between two structureless particles, A and B, radii r_A and r_B. They have no internal energy modes, and simply have 3T degrees of freedom each. All being well, we should obtain the same result as for collision theory as our basic model of structureless molecules is the same. The complex $X^{\ddagger}$ is like a diatomic molecule, and has $3T + 2R + V$ degrees of freedom; V may be associated with the hypothetical dissociation mode of $X^{\ddagger}$. Applying eqn (28) gives:

$$k = \frac{\kappa k_B T}{h}\frac{(Q_R^2)_X(Q_T^3)_X}{(Q_T^3)_A(Q_T^3)_B}\exp(-\Delta\varepsilon_0/k_B T). \qquad (E\ 29)$$

If we assume that in the complex the particles are in contact, the moment of inertia of $X^{\ddagger}$ is:

$$I_X = m_A m_B (r_A + r_B)^2/(m_A + m_B). \qquad (E\ 30)$$

On inserting the expressions from Table 3.3 into the pre-exponential term in eqn (29) we find:

$$A = \frac{\kappa k_B T}{h} \frac{(Q_R^2)_X (Q_T^3)_X}{(Q_T^3)_A (Q_T^3)_B}$$

$$A = \frac{\kappa k_B T}{h} \frac{\left(\frac{8\pi I k_B T}{h^2}\right) \left\{\frac{(2\pi(m_A + m_B) k_B T)^{\frac{3}{2}}}{h^3}\right\}}{(2\pi k_B T)^3 (m_A^{\frac{3}{2}} m_B^{\frac{3}{2}})/h^6}$$

$$A_{31} = \kappa (8 k_B T / \pi \mu)^{\frac{1}{2}} \pi (r_A + r_B)^2 \tag{E 31}$$

which is the result we obtained using collision theory, eqn (14), apart from the inclusion of κ. For structureless particles the two theories agree. Collision theory is a pictorially simple model and it is easy to visualize the processes occurring. In transition state theory we move one step away from just considering collisions and look at how the reaction system crosses the energy barrier to products. It is worth remembering that the gas laws and kinetic theory can all be derived from the application of statistical mechanics and therefore the ideas of collision theory are inherently bound up in transition state theory even if they are not as immediately obvious.

We now move on to a consideration of more complex reactions. The pre-exponential component of the rate coefficient in TST allows for molecular complexity and therefore we have high hopes of TST providing a better qualitative or even quantitative agreement with experiment. As an example we shall consider a reaction from Table 3.2

$$2NOCl \rightarrow 2NO + Cl_2. \tag{R 5}$$

The reactants each have 3T, 3R and 3V modes. The complex has 3T, 3R, 3IR and 8V modes plus the dissociation mode (see Table 3.4). Thus

$$A_5 = \frac{\kappa k_B T}{h} \frac{(Q_R^3)(Q_T^3)(Q_{IR}^3)(Q_V^8)_{(NOCl)_2}}{(Q_R^6)(Q_T^6)(Q_V^6)_{2NOCl}}. \tag{E 32}$$

As an initial approximation we shall assume that the corresponding partition functions in the complex and reactants are of equal magnitude, i.e. $(Q_T)_{NOCl} = (Q_T)_{(NOCl)_2}$. Equation (32) then simplifies to:

$$A_5 = \frac{\kappa k_B T}{h} \frac{(Q_{IR}^3)(Q_V^2)_{(NOCl)_2}}{(Q_R^3)(Q_T^3)_{2NOCl}}. \tag{E 33}$$

As an initial comparison of the magnitude of A we shall calculate the steric factor P. Equation (31) represents the A factor for a reaction of structureless particles and hence A_{31}/A_5 will give P.

$$A_{31}/A_5 \approx (Q_{IR}^3)(Q_V^2)/(Q_R^5) \approx 10^{-3} \tag{E 34}$$

where we have substituted the order of magnitude estimates of Q given in Table 3.3. The value of 10^{-3} is rather smaller than that required (0.16) but our calculation is very approximate. In particular, we might expect that our complex will have lower vibrational frequencies than the reactant molecule, reflecting the smaller force constants at the col on the potential energy surface. This would increase the vibrational partition function for $X^{\ddagger}$ and so increase P.

Table 3.4 Spectroscopic data for NOCl and NOCl)$_2$

Geometry		
Bond lengths/nm	a = 0.195	c = 0.12
	b = 0.114	d = 0.2
Bond angles/degrees	$\theta_1 = 116$	$\theta_2 = 120$
Moments of Inertia $I_1 I_2 I_3$ kg^3 m^6		
Figure axis rotation	2.06×10^{-136}	5.26×10^{-134}
Internal rotation		1.9×10^{-138}
Wavenumbers of vibrational modes	1799, 322, 592	(1700) ×2, (600) ×2, 300, 250, 200, 150

It is worth expressing the above argument in words, since it illustrates how TST accounts for the P factor introduced so arbitrarily in collision theory. In our reference reaction between two structureless spheres, formation of the activated complex involves a transformation of three translational degrees of freedom in the reactants into two rotational degrees of freedom in the complex, plus the dissociation mode. Since the translational energy levels are more closely spaced than rotational levels, this involves a reduction in the magnitude of the partition functions. However this is the minimum decrease that we can hope to achieve. In forming all activated complexes, we have to transform translations into rotations. For the NOCl + NOCl reaction, additional transformations take place and the net result is that five rotational modes are replaced by three internal rotations and two vibrations. The typical values of Q illustrated in Table 3.3 show that this leads to a further decrease in the magnitude of the partition function for X, additional to that found for the structureless particles. This extra reduction corresponds to TST's estimate of P.

Our theory and particularly its relationship to experiment deserves much more careful attention and we should really examine the basic equations in greater detail. From transition state theory,

$$k = A_{th}(T)\exp(-\Delta E_{th}^0/RT) \tag{E 35}$$

where

$$A_{th} = \frac{\kappa k_B T}{h} \frac{(Q_X^{\ddagger})}{Q_A Q_{BC}}$$

and

$$\Delta E^0 = L\Delta\varepsilon^0.$$

The experimental rate coefficient is

$$k = A_{expt}\exp(-E_{expt}/RT) \tag{E 36}$$

where, because of the limitations on the accuracy of our data, we usually assume A_{expt} to be independent of temperature. We cannot directly compare the two A factors, but must introduce a term to correct for the temperature dependence of the theoretical A factor. By deriving $d(\ln k)/dT$ for the theoretical and experimental expressions we find that:

$$\Delta E_{\text{th}}^0/RT = E_{\text{expt}}/RT - \{Td(\ln A_{\text{th}})/dT\}. \tag{E 37}$$

We can now substitute for $\Delta E_{\text{th}}^0/RT$ in eqn (35) to eliminate the theoretical activation energy.

$$k = A_{\text{th}}\exp\{Td(\ln A_{\text{th}})/dT\}\exp(-E_{\text{expt}}/RT) \tag{E 38}$$

Now on comparing eqns (38) and (36) we find that A_{expt} is equivalent to $A_{\text{th}}\exp(Td(\ln A_{\text{th}})/dT)$, a term we shall designate as A_{calc}.

$$A_{\text{calc}} = \frac{\kappa k_B T}{h}\frac{(Q_{\ddagger}^{\ddagger})}{(Q_A)(Q_{\text{BC}})}\exp\left[Td\left\{\ln\left(\frac{\kappa k_B T}{h}\frac{(Q_{\ddagger}^{\ddagger})}{(Q_A)(Q_{\text{BC}})}\right)\right\}\Big/dT\right]. \tag{E 39}$$

Converting to molar units

$$A_{\text{calc}} = 3.41 \times 10^{37}\kappa T\prod_i A_i \ \text{dm}^3\,\text{mol}^{-1}\,\text{s}^{-1} \tag{E 40}$$

where $A_i = Q_i\exp(Td\ln Q_i/dT)$, and the product $\prod_i A_i$ is taken over all reactant and complex degrees of freedom, except the dissociation mode. Our comparison must be made between A_{calc} and A_{expt}.

We now run into a crucial difficulty which was glossed over in our earlier treatment by assuming that all partition functions were of equal magnitude. In order to evaluate A_i, we must know the rotational and vibrational constants for the reactants and the activated complex. For NOCl, these are well known from spectroscopy, and are listed in Table 3.4. The spectroscopy of activated complexes, however, is a near virgin research field, and we are forced to guess the values. In Table 3.4 the vibrational frequencies and moments of inertia of the complex have been estimated by comparison with similar stable molecules. Table 3.5 shows the details of the calculation of A_{calc}.

We find $A_{\text{calc}}/\kappa = 6 \times 10^8\,\text{dm}^3\,\text{mol}^{-1}\,\text{s}^{-1}$ compared with the experimental value of $9.4 \times 10^9\,\text{dm}^3\,\text{mol}^{-1}\,\text{s}^{-1}$. At this point we might say that our complex structure

Table 3.5 Data for the calculation of A_{calc} for 2NOCl → 2NO + Cl$_2$ at 470 K

A. Partition Functions	$Q_T^3\,\text{m}^{-3}$	Q_R^3	Q_{iR}^3	Q_v^n	Q_e
NOCl	1.02×10^3	3.19×10^4	–	1.92	1.0
(NOCl)$_2$	2.89×10^3	2.58×10^5	9.7×10^3	26.0	1.0

B. $Td\ln Q_i/dT$	Translation	Rotation	Internal rotation	Vibration (n degrees)
NOCl	1.5	1.5	–	0.97
(NOCl)$_2$	1.5	1.5	1.5	3.53

is incorrect, and undertake to modify it. Increasing the bond lengths by only 10 per cent increases A_{calc} by nearly 80%, whilst lowering the 150 cm^{-1} vibration to 100 cm^{-1} increases it by almost 50%. There is some justification for this procedure since the frequencies were estimated by comparison with stable molecules, and the bonding in a short-lived complex will probably be weaker and the vibrational frequencies lower. Nevertheless, it is clear that, at present, accurate predictions based on TST and on experimentally determined molecular parameters are impracticable. Neither are we able to obtain detailed information about the structure of the activated complex from a comparison of TST and experimental rate coefficients, since there are so many variable parameters in the partition function of the complex. In the next chapter, we shall return to a more detailed discussion of TST, and of its assumptions and implementation. In the rest of this chapter, we shall consider several aspects of bimolecular reactions which are best considered within the general framework of TST.

Study notes
An interesting and readable account of the historical development of TST may be found in an article by Laidler, K. J. and King, M. C. (1983). *Journal of Physical Chemistry* **87**, 2657–64. They divide their discussion into sections on thermodynamic, kinetic and statistical mechanical treatments of TST and end with an absorbing account of the reception the chemical world gave to the new theory; this section includes a brief critique of the equilibrium assumption made in the formulation of the theory, a topic to which we shall return in Chapter 4.

3.5 Thermodynamic formulation of transition state theory

For many reactions, especially those involving larger molecules, it is often difficult to employ the statistical mechanical formulation in a meaningful manner. The number of variable parameters is too large and a unique set cannot be obtained. On the other hand, a comparison of the magnitude of the calculated and experimental rate coefficients does give information about the mechanism of a reaction. The thermodynamic form of TST is designed for use in the interpretation of kinetic data from more complex systems and we shall come across some applications in Chapter 6 on the kinetics of reactions in solution.

We may write eqn (28) as

$$k = \kappa(k_B T/h)\, K_c^{\ddagger} \qquad \text{(E 41)}$$

for ō dissociation mode

where $K_c^{\ddagger}$ is the equilibrium constant for forming the activated complex from the reactants, modified by the removal of one degree of freedom in the complex. Rate coefficients for bimolecular reactions are usually expressed in the form (concentration time)$^{-1}$ so that $K_c^{\ddagger}$ has units (concentration)$^{-1}$. We may use the results of equilibrium thermodynamics for $K_c^{\ddagger}$:

$$-RT\ln(K_c^{\ddagger}c^{\circ}) = \Delta G^{\ddagger} = \Delta H^{\ddagger} - T\Delta S^{\ddagger} \qquad \text{(E 42)}$$

where $\Delta G^{\ddagger}$, $\Delta H^{\ddagger}$ and $\Delta S^{\ddagger}$ are the standard Gibbs free energy, enthalpy, and entropy of activation (once more remembering the slight modification because of the removal of the one degree of freedom). The quantity c° has been introduced because the equilibrium constant is expressed in specific units, related to those of the rate coefficient, and these define the standard states to which $\Delta G^{\ddagger}$, $\Delta H^{\ddagger}$, and $\Delta S^{\ddagger}$ refer. Put another way, when the equilibrium constant is expressed in logarithmic form and related to $\Delta G^{\ddagger}$, each concentration (or pressure) in the equilibrium constant is expressed as a ratio of the standard concentration (or pressure). Thus c° is the standard concentration and if $K_c^{\ddagger}$ has units $dm^3 \, mol^{-1}$ then $c^{\circ} = 1$ mol dm^{-3} and $\Delta G^{\ddagger}$, $\Delta H^{\ddagger}$ and $\Delta S^{\ddagger}$ are referred to this concentration as the standard state. Substituting eqn (42) into (41) gives

[margin note: since K shld be unitless]

$$k = \kappa (k_B T/hc^{\circ}) \exp(\Delta S^{\ddagger}/R) \exp(-\Delta H^{\ddagger}/RT). \qquad (E\ 43)$$

(Note that unless we have decided to choose a bizarre standard state, c° will be unity, but it still serves to define the concentration units in k. In the statistical mechanical derivation of TST the units of k are defined from the units of the molar partition functions).

Following similar arguments to those described above, we may relate the temperature dependence of the rate coefficient to the experimental activation energy:

[margin note: rmb $\dfrac{d \ln k_{\text{opt}}}{dT} = E_{\text{ept}}/RT^2$]

$$E_{\text{expt}} = RT^2 \, d/dT[\ln\{\kappa K_c^{\ddagger} (k_B T/hc^{\circ})\}]$$
$$= RT^2 d/dT\{\ln(\kappa k_B/hc^{\circ}) + \ln T + \ln K_c^{\ddagger}\}$$
$$= RT + \Delta E^{\ddagger} \qquad (E\ 44)$$

since $d \ln K_c^{\ddagger} /dT = \Delta E^{\ddagger}/RT^2$ and $\Delta E^{\ddagger}$ is the energy of activation. From the First Law of Thermodynamics

[margin note: $H = U - PV$]

$$\Delta E^{\ddagger} = \Delta H^{\ddagger} - \Delta(PV) \qquad \text{[margin note: } U = H + PV\text{]} \qquad (E\ 45)$$

If we assume ideal gas behaviour, $\Delta PV = \Delta n^{\ddagger} RT$, where $\Delta n^{\ddagger}$ is the change in the number of moles on forming the complex. For a bimolecular reaction $\Delta n^{\ddagger} = -1$ and thus

$$E_{\text{expt}} = \Delta H^{\ddagger} + 2RT \quad \text{[margin note: } = RT + \Delta H^{\ddagger} + RT\text{]} \qquad (E\ 46)$$

and

$$k = \kappa (k_B T/hc^{\circ}) e^2 \exp(\Delta S^{\ddagger}/R) \exp(-E_{\text{expt}}^{\ddagger}/RT). \qquad (E\ 47)$$

[margin note: arises from +2RT term.]

3.5.1 Structure of the transition state and the A factor

The thermodynamical formulation of the rate coefficient leads to the following expression for the A factor:

$$A_{\text{th}} = \kappa (k_B T/hc^{\circ}) e^2 \exp(\Delta S^{\ddagger}/R) \qquad (E\ 48)$$

Variations in A_{th} for different reactions will arise from variations in the entropy change involved in forming the complex. Well defined relationships relate entropy and molecular structure and so the thermodynamic formulation of TST also allows for the rational and realistic incorporation of molecular complexity into the magnitude of the rate coefficient. An obvious benchmark for comparison is the case when $\Delta S^{\ddagger} = 0$. In this case A_{th} reduces to $A' = \kappa (k_B T/hc^{\circ}) e^2$, which at 500 K is equal to $1.04 \times 10^{13} \, dm^3 \, mol^{-1} \, s^{-1}$. If $A_{\text{th}} > A'$, $\Delta S^{\ddagger}$ must be positive, i.e. there has been an increase in entropy in forming the transition state. This is termed a '*loose transition state*'. The transition state involves the weakening and stretching of a bond, thus making the transition state more 'floppy' than the reactant and hence giving an increase in entropy. Conversely if $A_{th} \nleqslant A'$ $\Delta S^{\ddagger}$ is negative. Consider the following example:

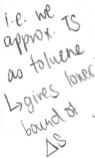

i.e. we approx. TS as toluene

more rigid than TS.

$$CH_3 + C_2H_6 \rightarrow [CH_3 \cdots H \cdots C_2H_5] \rightarrow CH_4 + C_2H_5. \qquad (R\ 6)$$

By approximating the structure of the transition state to that of the closest hydrocarbon $CH_3C_2H_5$ we can obtain a lower bound to $\Delta S^{\ddagger}$. The estimate will be a lower bound because the transition state will contain partially formed bonds and hence be 'looser'. This gives $\Delta S^{\ddagger} = -124\ \mathrm{J\ mol^{-1}\ K^{-1}}$, a decrease in entropy from reactants to transition state. The loss in entropy arises from the reduction in the number of degrees of freedom in forming the transition state. Refined calculations give $A_{th} = 1 \times 10^8\ \mathrm{dm^3\ mol^{-1}\ s^{-1}}$ in good agreement with the experimental value of $1 \times 10^{8.3}\ \mathrm{dm^3\ mol^{-1}\ s^{-1}}$.

L gives lower bound of ΔS

Study notes

1 A more detailed discussion of standard states and their relationship to rate coefficients may be found in Robinson, P. J. (1978). Dimensions and standard states in the activated complex theory of reaction rates. *Journal of Chemical Education*, **55**, 509–10.

2 $\Delta E^{\ddagger}$ is independent of V and therefore of $c°$ for ideal gases and so the units and standard states affect only $\Delta S^{\ddagger}$. This is most readily appreciated by reference to the statistical mechanical formulation, since V enters explicitly into the translational partition functions.

3 By taking the statistical mechanical expression for k, and differentiating $\ln k$ with respect to T (remembering that $\mathrm{d}\ln Q/\mathrm{d}T = \varepsilon/k_B T^2$, where ε is the average thermal energy per molecule for a Boltzmann distribution) show that

$$E_{\mathrm{expt}} = L(k_B T + \varepsilon_{X^+} - \varepsilon_A - \varepsilon_{BC} + \Delta\varepsilon^0). \qquad (E\ 49)$$

$\varepsilon_X{}^{\ddagger}$ corresponds to the average energy in the complex, with one mode removed. The average energy in this mode, assuming that it is a translational mode, is $k_B T/2$. The additional contribution of $k_B T$ in the expression for E_{expt} derives from the temperature dependence of the velocity. To a good approximation, therefore, (i.e. to within $k_B T/2$) we can combine the term $k_B T$ into ε_{X^+}, to obtain the average energy of the complex in all modes. Thus,

E_{expt} = L (average thermal energy of complex molecules − average thermal energy of reactants + difference in zero point energies).

4 Bimolecular rate coefficients may be expressed in molar or molecular units. Table 3.6 gives an explicit summary of the expressions used and their relationships to one another.

3.6 Experimental evidence for transition state theory

The lifetime of the postulated transition state is extremely short ($k_B T/h = \approx 6 \times 10^{12}\ \mathrm{s^{-1}}$ at 300 K and a corresponding lifetime $1/k = 1.6 \times 10^{-13}\ \mathrm{s}$) and hence the steady-state concentration of transition states for any reaction is vanishingly small. However, recent experiments using pulses of laser light of comparable duration to the transition state lifetimes have yielded positive evidence for the existence of the transition state. One example has been a study of the reaction

$$K + NaCl \rightarrow [K\text{–}Cl\text{–}Na] \rightarrow KCl + Na. \qquad (R\ 7)$$

Table 3.6 Summary of transition state expressions

	Statistical mechanics	Thermodynamics	Units	Factor
Molecular Units	$(\kappa k_B T/h)(Q_{X^+}/Q_A Q_B)$ $\exp(-\Delta\varepsilon^0/k_B T)$	$(\kappa k_B T/h10^{-3}Lc^\circ)\exp(\Delta S^{\pm})$ $\exp(-\Delta H^{\pm}/RT)$	$cm^3\,molecule^{-1}\,s^{-1}$	$10^6\,cm^3$
Molar Units	$(\kappa k_B T/h)L(Q_{X^+}/Q_A Q_B)$ $\exp(-\Delta\varepsilon^0/k_B T)$	$(\kappa k_B T/hc^\circ)\exp(\Delta S^{\pm})$ $\exp(-\Delta H^{\pm}/RT)$	$dm^3\,mol^{-1}\,s^{-1}$	$10^3\,dm^3$

(a) $\kappa k_B T/h$ has units s^{-1}. (b) $\Delta\varepsilon^0$ is the difference in zero point energies in molecular units. (c) $\Delta S^{\pm}$ and $\Delta H^{\pm}$ are the entropy and enthalpy of activation, in molar units, referred to a standard state of 1 mol dm^{-3}. (d) $c^\circ = 1$ mol dm^{-3}. (e) The partition functions are the molecular partition functions listed in Table 3.3. The present table assumes that the partition functions have been evaluated in SI units, with $V = 1\,m^3$. To evaluate the statistical mechanical rate coefficient, it must be converted into the relevant units by multiplying by the factor shown in the last column.

The reaction is not exothermic enough to produce electronically excited sodium atoms, but if the transition state could be excited by a pulse of light of the correct frequency, matching an electronic transition in the transition state, electronically excited Na atoms would be produced, whose presence could be observed by their characteristic fluorescence. Sodium fluorescence is indeed observed at a laser frequency which does not match any of the transitions for the reactants or products and hence must arise from excitation of the transition state. As the frequency of the excitation laser is scanned, enhancement of sodium fluorescence is observed whenever the laser matches a resonance in the transition state and hence this technique offers the possibility of probing the structure of the transition state. More recently femtosecond laser pulses have been used to probe the activated complexes in the Na–I and I–Hg–I dissociations (Zewail 1991, 1988).

3.7 Applications of transition state theory

3.7.1 Determination of rate coefficients

The fact that we still need to measure most rate coefficients shows that TST theory has not been a widespread quantitative success! However, accurate calculations are becoming available for several potential energy surfaces, allowing determination of both barrier heights and transition state properties. The results of TST calculations on these surfaces are in reasonable agreement with experiments and the ever increasing power of computers will enable more sophisticated calculations to be performed on increasingly complex systems.

3.7.2 Linear free energy relationships

Transition state theory states that the rate coefficient for a reaction is related to the free energy of activation. Equilibrium constants are related to the overall free energy of the reaction. For a series of similar reactions there may be an observable trend in the equilibrium constants and we might predict that the rate coefficients would be related to this change. Experiment shows that in many cases this relationship is borne out and that there is indeed a relationship between the rate coefficient and the equilibrium constant. We can express this quantitatively in eqn (50)

$$\Delta G_1^{\ddagger} - \Delta G_2^{\ddagger} = \alpha(\Delta G_1 - \Delta G_2) \tag{E 50}$$

or

$$\ln k_2 - \ln k_1 = \alpha(\ln K_2 - \ln K_1); \; \ln(k_2/k_1) = \alpha \, \ln(K_2/K_1) \tag{E 51}$$

where the subscript 1 refers to a reference reaction and 2 to the reaction under study. If we measure k for a variety of similar reactions then a plot of $\ln(k_2/k_1)$ vs. $\ln(K_2/K_1)$ will be a straight line (hence the name linear free energy) of gradient α. An appreciation of the physical basis of this method can be obtained from Fig. 3.12 which shows the potential energies of the reactants and products for a set of similar reactions as a function of reaction coordinate. Because the reactions are similar then we would expect the shape of these curves to be identical. As we are only interested in relative values of the free energy change we have set all the reactants to the same energy. As the release of free energy in the reaction increases, the potential curves of the reactants and products intersect at progressively lower energies, lowering the values of $\Delta G^{\ddagger}$. Whilst most of the work on linear free energy relationships has been in physical organic chemistry, similar effects exist for a number of inorganic reactions such as the replacement of ligands. Figure 3.13 shows the linear relationship obtained in a study of the ligand substitution reaction

$$\text{Co(NH}_3)_5\text{X}^{2+} + \text{H}_2\text{O} \rightleftarrows \text{Co(NH}_3)_5\text{H}_3\text{O}^{3+} + \text{X}^-. \tag{R 8}$$

We can use these relationships in a number of ways. For example the relationship can be used to predict the rate of a reaction, from the equilibrium constant. The fact that a linear relationship is obtained for a number of reactions is good evidence that these reactions proceed via a similar mechanism. Alternatively if a point lies far from the line then an alternative mechanism may exist for that particular reaction.

3.7.3 Isotope effects

According to collision theory, reactions of different isotopomers should occur with almost the same rate, modified only by a small change in the collision frequency due to the differing reduced masses. Experimentally, however, dramatic differences, often greater than an order of magnitude, have been observed in rate

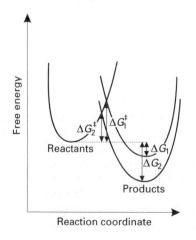

Fig. 3.12 Schematic diagram of the potential surfaces for reactants and products. With increasing reaction exoergicity the free energy of activation decreases.

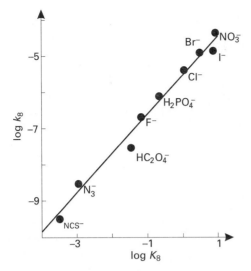

Fig. 3.13 A linear free energy plot for reaction (8) at 298 K.

coefficients for different isotopomers. One of the notable successes of TST has been the ability to explain these effects quantitatively.

The first point we should notice is that the reactions will occur on the same potential energy surface; however, we defined $\Delta\varepsilon^0$ as the energy difference between the lowest occupied levels of the reactant and the transition state. The Heisenberg uncertainty principle requires that molecules have a minimum amount of vibrational energy, equal to $h\nu/2$ per vibrational mode, where ν is the vibrational frequency. This energy is known as the zero point energy. As the frequency of the vibration is dependent on the reduced mass of the vibrating system (eqn 52) we can see that different isotopomers will have varying amounts of zero point energy.

$$\nu = (1/2\pi)(f/\mu)^{\frac{1}{2}} \tag{E 52}$$

where f is the force constant of the bond (independent of isotope) and μ is the reduced mass. The most dramatic effects arise when hydrogen atoms are replaced by deuterium (or more expensively, tritium). For a molecule RH, where R may be a single atom or a group of atoms, the reduced mass for the hydrogen stretching vibration μ_{RH} is $m_R m_H/(m_R + m_H)$ and

$$\frac{\mu_{RH}}{\mu_{RD}} = \frac{m_R m_H (m_R + m_D)}{m_R m_D (m_R + m_H)}. \tag{E 53}$$

Since $m_R \gg m_H, m_D$,

$$(m_R + m_D)/(m_R + m_H) \approx 1 \text{ and } \mu_{RH}/\mu_{RD} = m_H/m_D = \tfrac{1}{2}. \tag{E 54}$$

As the force constants are identical the ratio of frequencies is

$$\nu_{RH}/\nu_{RD} = (\mu_{RD}/\mu_{RH})^{\frac{1}{2}} = \sqrt{2}. \tag{E 55}$$

At normal temperatures, when only the lowest vibrational levels are occupied, it follows that RH has more vibrational energy than RD. Similar calculations can be performed for the transition state and the effect that isotopic substitution will have on the energy barrier can be seen from Fig. 3.14. If the isotopically substituted bond is weakened in the transition state then the energy barrier for the deuterated isotopomer is higher than for the hydrogenated isotopomer. In the extreme, when this bond is completely broken in the transition state, then the energy difference will correspond to the difference in reagent zero point energies. For a C–H bond this is approximately 5 kJ mol^{-1}. The kinetic isotope effect is usually too small to observe for elements other than hydrogen, because the change in reduced mass on substitution produces an activation energy difference which is undetectable within the accuracy of most kinetic determinations.

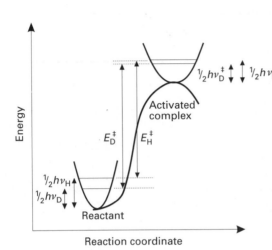

Fig. 3.14 Schematic diagram of the potential wells for the stretching vibration of R^1–H in its unperturbed state, and on formation of the activated complex in the reaction R^1–H + R^2 → R^1 + R^2–H. The diagram shows the zero point energies of the hydrogen (H) and deuterium substituted molecules, and the activation energies for each reaction.

Example 3.1

Hydrogen atoms may be efficiently generated in acid solution by γ radiolysis. They react with alcohols by hydrogen abstraction, and the relative rates of the two reactions

$$H + (CH_3)_2CHOH \rightarrow H_2 + (CH_3)_2COH \qquad (R\ 9)$$

$$H + (CD_3)_2CDOH \rightarrow HD + (CD_3)_2COH \qquad (R\ 10)$$

may be determined by measuring the final $[H_2]/[HD]$ ratios mass spectrometrically as a function of temperature.

$$\ln(k_H) = \ln(A_H) - (E_H/RT) \qquad (E\ 56)$$

$$\ln(k_D) = \ln(A_D) - (E_D/RT). \qquad (E\ 57)$$

Subtracting the two equations gives

$$\ln(k_H/k_D) = \ln(A_H/A_D) - (\Delta E/RT). \qquad (E\ 58)$$

The ratio of the rate coefficients is plotted in Arrhenius form in Fig. 3.15 and from the slope we find that $\Delta E = 4.3$ kJ mol^{-1} i.e that the C–H bond is almost broken in the transition state. The intercept yields the ratio of A factors. For gas phase reactions we should be able accurately to calculate this ratio.

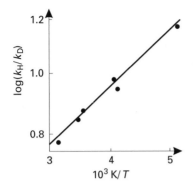

Fig. 3.15 Plot of $\log(k_H/k_D)$ vs. $1/T$ for the reaction of hydrogen atoms with $(CH_3)_2CHOH$ and $(CD_3)_2CDOH$.

$$\frac{A_H}{A_D} = \frac{\kappa_H Q_H^{\ddagger} Q_{A(D)} Q_{B(D)}}{\kappa_D Q_D^{\ddagger} Q_{A(H)} Q_{B(H)}}. \tag{E 59}$$

κ and many of the molecular parameters will be the same for both transition states and generally the ratio of A factors can be calculated more accurately than the absolute value as often the uncertain parameters tend to cancel each other out. Note that the magnitudes of some of the partition functions will be dependent on the isotopomer.

3.7.4 Tunnelling

Quantum mechanics tells us that there is a finite possibility of a wavefunction, ψ, (or because of wave-particle duality, a particle) appearing on the other side of a potential barrier. For a 1-D wavefunction the magnitude of ψ at any value of x within a rectangular barrier is given by:

$$\psi(x) = \alpha \exp[-[2m(V-E)/(h/2\pi)^2]^{\frac{1}{2}}x]. \tag{E 60}$$

Since $\psi(x)$ is not zero there is some probability that the particle may be found in the barrier even though the energy E of the particle is less than the height of the potential barrier V. This penetration into classically forbidden regions is called tunnelling.

An obvious example of such a barrier is the potential energy barrier between reactants and products. This barrier is of finite width and therefore there will be a non-zero probability of the particle appearing on the other side of the barrier even if it classically does not have the energy to react (Fig. 3.16).

At normal temperatures the tunnelling rate is much slower than what we might call the classical rate, but it is much less sensitive to decreasing temperature. As the temperature is lowered, the fraction of energized molecules falls and eventually the classical rate becomes comparable to the tunnelling rate causing an upward curvature of the Arrhenius plot at low temperatures, i.e. the reaction proceeds faster at low temperatures than we would classically expect. Equation (60) shows that the probability of tunnelling is exponentially dependent on the mass of the particle; particles of lower mass (electrons and H atoms) tunnel far more efficiently than heavier particles such as deuterium atoms. This can lead to a significant extra isotope effect at low temperatures. An example is shown in Fig. 3.17, an Arrhenius plot for the reactions

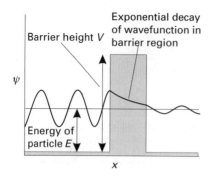

Fig. 3.16 The oscillating wavefunction of the particle incident (reactant) on the barrier decays exponentially inside the barrier, but providing the barrier is not too wide, the wavefunction on exiting the barrier (products) is non-zero.

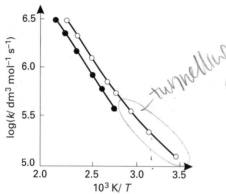

Fig. 3.17 Arrhenius plots for the reactions H + $H_2 \rightarrow H_2$ + H (○) and D + $D_2 \rightarrow D_2$ + D (●).

$$H + H_2 \rightarrow H_2 + H \qquad \text{(R 11)}$$

$$D + D_2 \rightarrow D_2 + D. \qquad \text{(R 12)}$$

Whilst the deuterated plot is linear, that for the hydrogenated reaction shows significant curvature, far greater than would be expected from a 'normal' isotope effect. Tunnelling contributions can be incorporated and accounted for in advanced TST calculations.

Although this present chapter is primarily concerned with gas phase reactions, it is worth noting that tunnelling reactions produce much larger effects in condensed materials at low temperatures. Methyl radicals can be generated in acetonitrile, at low temperatures, by γ-radiolysis. They may then react with a neighbouring CH_3CN molecule by H abstraction:

$$CH_3 + CH_3CN \rightarrow CH_4 + CH_2CN. \qquad \text{(R 13)}$$

The reaction is very slow for $T < 120$ K and may be followed by observing the disappearance of the electron spin resonance (esr) signal for CH_3 and the appearance of that for CH_2CN. Comparative measurements on CH_3CN and CD_3CN show that $k_H/k_D > 2.8 \times 10^4$ at 77 K, whilst a maximum ratio of 1.5×10^3 would be expected on the basis of zero point energy differences alone.

More detailed information on tunnelling can be found in the book *The tunnel effect in chemistry* by Bell (1980). Chapters 1–3 provide a detailed development of the subject which the reader need only skim through. Chapters 4 (The theory of kinetic hydrogen isotope effects) and 5 (Experimental evidence for tunnelling) can be read in detail with great profit. It should be remembered that the overall isotope effect is a combination of tunnelling, variations in the partition functions of the reactants and transition states and zero point energy effects. The results can be quite complex and difficult to disentangle. The reader may find it easier to study Chapter 4 before embarking on any extra study and reading.

3.7.5 Curved Arrhenius plots

We saw in Example 2.3 that the temperature dependence of some reactions is non-Arrhenius in nature. We have already discussed one possible reason for this in the previous subsection on tunnelling but we can see from the transition state derivation that our pre-exponential factor will depend on the variation of the partition functions of the reactants and activated complex with temperature. At high temperatures the vibrational partition functions of both the reactants and transition state will start to become appreciable and may significantly alter the magnitude of the pre-exponential factor. In contrast collision theory predicts a constant $T^{\frac{1}{2}}$ temperature dependence for A. For reaction (14)

$$OH + H_2 \rightarrow H_2O + H \qquad\qquad (R\ 14)$$

the apparent activation energy is predicted to increase by a factor of two from 300 to 2000 K, due to the temperature dependencies of the partition functions and indeed a positive curvature is experimentally observed.

3.8 Conclusions

In this chapter we have examined two theories of bimolecular reactions. Simple collision theory was introduced and modified by the introduction of an arbitrary geometrical factor, P. We formulated transition state theory by making some rather sweeping equilibrium assumptions that enabled us to apply the results of statistical mechanics to the calculation of rate coefficients. We thus demonstrated that the magnitude of the calculated A factor depends on the structure of the activated complex and that the P factor arises because of the replacement of rotational and translational motion in the separated reactants by vibrational or internal rotational motion in the complex. An equivalent thermodynamic formulation relates the A factor to the entropy change on forming the complex and small P factors correspond to unusually large decreases in entropy on complex formation. These broad generalizations and the applications discussed in the last two sections demonstrate the strengths of transition state theory as a general framework for the discussion of bimolecular reactions. When we try to make quantitative predictions or comparisons we run into difficulties because of our ignorance of the detailed structure of the transition state. Furthermore we have not examined the validity of the equilibrium assumption as fully as we might. We shall discuss these problems in the next chapter where we shall also look at the experimental investigation of the dynamics of molecular reactions, a field which enables us to learn more about potential energy surfaces and of the behaviour of reactants on them.

3.9 Questions

3.1 What is the relationship between collision frequency and the mean time between collisions? What is the mean time between collisions experienced by a single argon atom, diameter 0.29 nm, at (a) 100 kPa and (b) 100 Pa of argon gas, and 300 K?

3.2 Estimate the probability factor P for reaction between an atom and a diatomic molecule with (a) a linear and (b) a bent activated complex. Comment on the differences.

3.3 Calculate the number of collisions per cubic centimetre per second between molecules A and B for a mixture containing 100 Torr of each at 300 K. The molecular diameters of A and B are 0.3 and 0.4 nm respectively and the average velocity is 5×10^2 m s^{-1} at 300 K. Given that the experimental rate coefficient for the reaction is 1.18×10^5 mol^{-1} cm^3 s^{-1} at 300 K, and that the activation energy is 40 kJ mol^{-1}, calculate the fraction of collisions at 300 K that occur with sufficient energy for reaction and determine the steric factor of the reaction.

3.4 Give a brief account of the transition state theory of chemical kinetics, highlighting its strengths and weaknesses.

3.5 Explain why a small A factor may be indicative of a tightly-bound activated complex, whilst a large A factor indicates a loosely-bound complex.

3.6 The following parameters have been calculated for the reaction

$$D + H_2 \rightarrow DH + H \qquad\qquad (R\ 15)$$

for a linear transition state $(D \cdots H \cdots H)$.

Separation / Å	Reactants (D, H$_2$)	Transition state
$r(D–H_1)$		0.929
$r(H_1–H_2)$	0.741	0.929
$r(D—H_2)$		1.858
Electronic energy/kJ mol^{-1}	0.00	40.3

Frequencies /cm^{-1}	H$_2$	Transition state
symmetric stretch	4401	1780
bend (doubly degenerate)		861

Calculate the rate coefficient at 1000 K. How would you qualitatively expect the A factor to vary with temperature?

3.7 The following data were recorded for the isotopic variants of the reaction

$$Cl + HI \rightarrow HCl + I \qquad\qquad (R\ 16)$$

Temperature/K	345	295	275	240	223
k_H/k_D	1.55	1.8	1.92	2.24	2.66

Calculate the difference in activation energies for the two reactions. How does this correspond to the maximum possible difference in activation energies? $(\nu(HI) = 2308$ cm^{-1}, $\nu(DI) = 1632$ cm$^{-1})$. The data need not be interpreted in this simple fashion and a fuller discussion of the possible implications of the temperature dependence of the isotope ratios can be found in the original paper by Mei and Moore (1979).

3.8 Estimate the barrier length, l, for H atom transfer 1 kJ mol^{-1} below the maximum in the reaction coordinate for the reaction.

$$Cl + HCl \rightarrow HCl + H \qquad \text{(R 17)}$$

which has an activation energy of approximately 20 kJ mol^{-1}. (Experimentally we might study this reaction by using two different isotopes of chlorine). Assume now that the shape of this barrier is rectangular (height 1 kJ mol^{-1}, length l) and hence estimate the probability of the reaction emerging on the other side of the barrier. You need to think carefully about the reduced mass of the system. Remember the H atom is essentially moving between two fixed chlorine atoms.

The realistic shape of the barrier is parabolic, the potential being given by $V(x) = V_{max} - kx^2/2$ where k is a constant which determines the shape of the parabola ($x = -l/2$ at beginning of barrier, 0 at TS, and $+l/2$ at end of barrier). Calculate the value of k for the parabola above the tunnelling pathway. The solution for the fractional transmission through a parabolic barrier is:

$$T = (1 + e^{-2\pi\varepsilon})^{-1} \qquad \text{(E 61)}$$

where $\varepsilon = (E - V_{max})/\hbar\omega^*$ and $\omega^* = (k/\mu)^{\frac{1}{2}}$. Re-evaluate the tunnelling probability and compare it with your previous estimate.

3.9 Vibrational enhancement of a simple chemical reaction can introduce curvature into an Arrhenius plot. Reaction (14) has vibrational state specific rate coefficients for H_2 ($v = 0$) and H_2 ($v = 1$) given by:

$$k_{v=0} = 9.3 \times 10^{-12} \exp(-18000/RT) \text{ cm}^3 \text{ molecule}^{-1} \text{ s}^{-1} \qquad \text{(E 62a)}$$

$$k_{v=1} = 6.0 \times 10^{-11} \exp(-11000/RT) \text{ cm}^3 \text{ molecule}^{-1} \text{ s}^{-1}. \qquad \text{(E 62b)}$$

(a) Calculate $k_{v=0}$ and $k_{v=1}$ at 300, 500, 1000, 1600 and 2500 K and comment on the relative values.
(b) At each of the above temperatures the Boltzmann populations for H_2 in $v = 0$ and $v = 1$ are:

$$f_{v=0} = 1/Q_{vib} \qquad \text{(E 63a)}$$

$$f_{v=1} = \exp(-\Delta E/RT)/Q_{vib} \qquad \text{(E 63b)}$$

where ΔE is the vibrational energy of H_2 relative to H_2 ($v = 0$) and equals 49.6 kJ mol^{-1} for $v = 1$. For each temperature calculate values for the fraction of molecules in each vibrational level.
(c) Finally calculate the thermal rate coefficient at each temperature using the expression:

$$k(T) = k_{v=0}f_{v=0} + k_{v=1}f_{v=1} \qquad \text{(E 64)}$$

and plot the results on an Arrhenius graph commenting on the shape of the graph. We shall return to the subject of vibrational enhancement of the rate coefficient in Chapter 4.

References

Atkins, P. W. (1994). *Physical Chemistry*, (5th edn), Chapters 1 and 24. Oxford University Press.

Atkins, P. W. (1994). *Physical Chemistry*, (5th edn), Chapters 19 and 20. Oxford University Press.

Bell, R. P. (1980). *The tunnel effect in chemistry*. Chapman and Hall, London.

Gasser, R., and Richards, G. (1979). *Entropy and energy levels*. Oxford University Press.

Hill, T. L. (1962). *An introduction to statistical thermodynamics*. Addison–Wesley, Reading, MS.

Mei, C.-C. and Moore, C. B. (1979). Thermal rate constants, energy dependence, and isotope effect for halogen–hydrogen halide reactions. *Journal of Chemical Physics*, **70**, 1759-64.

Rosker, M. J., Daritus, M. , and Zewail, A. H. (1988). Femtosecond real-time probing of reactions. *Journal of Chemical Physics*, **89**, 6113–40.

Zewail, A. H. (1991). Femtosecond transition state dynamics. *Faraday Discussions of the Chemical Society*, **91**, 207–37.

4 Reaction dynamics

4.1 Introduction

In the preceding chapter we developed two theories which provide a means of estimating, or at least rationalising, the rate coefficient for an elementary bimolecular reaction. In collision theory, we limited ourselves to an idealized model in which the reactants are structureless, incompressible spheres. Our approach in transition state theory was to *assume* equilibrium between the reactants and the transition state, but we made no attempt to justify this assumption.

In the present chapter, we shall take a more detailed look at intermolecular collisions and reactions by discussing the following topics:

1. *The effect of intermolecular potential energy on the collision frequency.*

2. The experimental investigation of the rates of reaction from specific *quantum states of the reactants into specific quantum states of the products*. We shall divide our discussion broadly into two parts firstly focusing on the effects of reagent excitation, answering the question as to how various forms of reagent excitation alter the course of the reaction and, secondly, on the measurement of product distributions and what this can tell us about the shape of the potential energy surface upon which reaction has occurred. Research in these areas has made enormous strides in the last twenty years and has provided us with a very detailed picture of the ways in which reactions occur.

3. *The calculation of reaction potential energy surfaces.* We introduced the idea of a potential energy surface in Chapter 3 p. 66, although our discussion there was very schematic. The detailed dynamics of the reaction and indeed, the success of transition state theory, depend intimately on the shape of the potential energy surface.

4. *Numerical studies of the rates of reaction on theoretical potential energy surfaces.* Experimental reaction dynamics have been supplemented by computer calculations of the behaviour of reactive systems on potential energy surfaces. These studies not only help us to interpret the experimental results, but they also enable us to formulate more satisfactory theories.

5. *The assumptions of transition state theory.* How valid is the equilibrium assumption? As we shall see, the basic assumptions in transition state theory can be reduced to simple requirements related to the transmission coefficient, κ.

A discussion of reaction dynamics is worth a book in itself and we can only begin to scratch at the surface of this rapidly developing subject. In order to focus the discussion we have concentrated mainly on aspects closely related to the calculation of rate coefficients. However, the boundary between reaction dynamics and kinetics is often rather artificial. In both elementary reaction kinetics and reaction dynamics the ultimate goal is the same—a thorough understanding of the potential energy surface upon which reaction occurs.

4.2 Collisions of real molecules

We know from the pressure–volume behaviour of real gases, that atoms and molecules are very different from the incompressible spheres we used as our

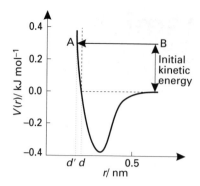

Fig. 4.1 Intermolecular potential energy functions: (- - - - -) hard sphere potential; (——) Lennard-Jones potential for neon.

initial model in Chapter 3. As the molecules approach, they at first attract one another (the van der Waals attraction); at closer proximity repulsive forces take over. This repulsive part of the potential is not infinitely steep, i.e. the molecules are compressible.

Figure 4.1 shows a plot of potential energy versus internuclear separation for hard spheres:

$$V(r) = \infty, r < d; \ V(r) = 0, r > d, \tag{E 1}$$

and for two neon atoms. The most widely used simple intermolecular potential function is the Lennard-Jones 6-12 potential:

$$V(r) = 4 \, \varepsilon_{LJ}\{(d/r)^{12} - (d/r)^{6}\}. \tag{E 2}$$

This has an attractive limb varying as r^{-6} (for neutral molecules) and a steep repulsive limb varying as r^{-12}. The well depth is ε_{LJ} and the curve crosses the r-axis ($V = 0$) at $r = d$.

Attractive forces arise from electrostatic attractions of one form or another. The most obvious case is the Coulomb attraction of oppositely charged ions (in which case the attractive limb varies as r^{-1}). The presence of the strong electric field from an ion will induce a dipole in an otherwise uniform neutral molecule, with the electrons of the neutral molecule either being attracted to or repelled from the approaching ion. The oppositely charged regions of the ion and neutral molecule will then attract, the force varying as r^{-4}. Finally, neutral–neutral interactions depend on the strengths of the dipoles in the molecules. In some cases, e.g. HF and H_2O, strong permanent dipoles exist and the electrostatic interaction is strong; at the other extreme interactions between noble gas atoms depend on the presence of minute, fluctuating, temporary dipoles.

Table 4.1 lists typical values for ε_{LJ} and d, obtained from gas viscosity measurements for a number of neutral species. At 300 K, $RT = 2.5$ kJ mol^{-1}, so the potential wells of these colliding species are much too shallow to overcome the relative kinetic energy and trap them in close proximity. Consequently, liquefaction does not occur until much lower temperatures. Even so, we can appreciate why CO_2 shows such large deviations from Boyles law.

Let us consider the effect of this realistic potential function on our treatment of intermolecular collisions. First of all we find that, for a head-on collision (where the impact parameter, b, $= 0$), the distance of closest approach depends on the relative velocity. The total energy is conserved in a collision, and we can repre-

Table 4.1 Lennard-Jones parameters for some neutral species

Gas	$L\varepsilon_{LJ}/$ kJ mol^{-1}	d/nm
He	0.085	0.26
Ne	0.30	0.28
H$_2$	0.32	0.29
CO$_2$	1.58	0.40

sent this by the horizontal line AB in Fig. 4.1. The relative kinetic energy is the difference between the potential energy curve and AB. As the atoms approach they attract one another and the kinetic energy increases, until the repulsive limb is reached, when the kinetic energy decreases and reaches zero at A. At this point the motion is reversed, and so the distance of closest approach is d'. This is less than d, the distance of closest approach for a collision with zero relative kinetic energy.

Figure 4.2 shows a collision trajectory for a large impact parameter ($b > d$). The long range attractive force pulls the incident molecule in, deflecting it from its original path. By our original definition (Chapter 3) a collision has taken place, despite the fact that the distance of closest approach r is greater than d. The total collision cross-section has increased since

$$\sigma = \pi b_{max}^2 > \pi d^2 \qquad \text{(E 3)}$$

where b_{max} is the largest impact parameter for which a deflection occurs. This effect is more important in elastic than reactive collisions. Viscosity, for example, depends on momentum transfer, and this can occur at large separations. In most reactions the molecules must approach to small separations where they encounter the repulsive limb of the potential; chemical forces are inoperative until these separations are reached. In this region, the hard sphere potential is a reasonable approximation (Fig. 4.1), but because the activation energy is usually quite high (> 10 kJ mol^{-1}) the distance of closest approach is less than d (cf. d' in Fig. 4.1). Thus, the compressibility of the molecules leads to a reaction cross-section which is generally smaller than that based on viscosity measurements (πd^2), where the mean effective collision energy is much smaller.

For reactions with small activation energies, attractive collisions can lead to larger rate coefficients. In Chapter 3 we found that the reaction

$$K + Br_2 \rightarrow KBr + Br \qquad \text{(R 1)}$$

has a probability factor P of 4.8. This reaction is thought to occur by a *harpoon mechanism*. At an internuclear distance of 0.8 nm K transfers an electron (the harpoon) to Br$_2$; K$^+$ and Br$_2^-$ then attract one another strongly and approach to smaller distances (the harpoon is drawn in) where reaction takes place, K$^+$

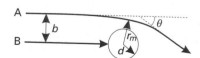

Fig. 4.2 High impact parameter trajectory for real molecules.

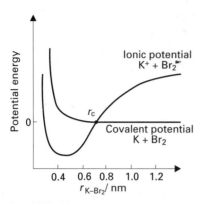

Fig. 4.3 Potential energy diagram to illustrate the harpoon model for the $K + Br_2$ reaction. Two schematic one-dimensional potential energy curves are shown, for the ionic $(K^+ + Br_2^-)$, and covalent $(K + Br_2)$ interactions. At large separation, the ionic curve is of higher energy: the diabatic curves cross at r_c where the difference between the ionization energy of K and the electron affinity of Br_2 is matched by the Coulombic attraction between the ions. Switching of potential curves at r_c leads to reaction with cross-section $\approx \pi r_c^2$.

picking up Br^- and leaving Br behind. Once the large separation electron transfer has taken place, reaction is ensured (Fig. 4.3)

Study notes
Chapters 1 and 2 of *The Forces between Molecules* by Rigby, M., Smith, E. B., Wakeham, W. A., and Maitland, G. C. (1986) (Oxford Science Publications) gives a clear and detailed account of the field of molecular scattering and how such measurements yield information about the interactions between real molecules. The field of *elastic scattering*, in which the probability of deflecting A through an angle θ (cf. Fig. 4.2) is determined is discussed in Chapter 2, and the term elastic scattering is used because A and B interchange only translational energy. The scattering pattern (the dependence on θ), depends on, and can be used as a probe of the intermolecular potential. Alternatively, consult Levine and Bernstein (1987) Sections 3.1 and 3.2 (Levine, R. D. and Bernstein, R. B., *Molecular Reaction Dynamics,* Oxford University Press).

4.3 Experimental reaction dynamics

If we take our schematic bimolecular reaction:

$$A + BC \rightarrow AB + C \qquad (R\ 2)$$

as a model, the major questions, to which experimental studies of reaction dynamics seek to provide an answer, are:
1. How does the rate coefficient depend on the vibrational and rotational energy of BC and on the relative translational energy of A and BC? Our treatment of collision theory assumed that only translational energy could be used to overcome the energy barrier. In transition state theory we assumed a Boltzmann distribution of energy in the reactants and the complex. Here we are moving away from such a distribution and examining rates of reaction from specific quantum states.
2. When the reaction system passes through the transition state, which we have looked on as the highest energy point on the reaction coordinate, the potential energy falls and is converted into kinetic energy. How is this energy distributed amongst the products? Is the energy channelled into vibrational and rotational energy of AB, and the energy of relative translational motion of AB and C,

according to some simple statistical law (like equipartition) or does it pre-ferentially enter one form of motion? This was one of the first questions to be probed by reaction dynamicists and, as we shall see, the answers depend on the reaction involved and, in particular, on the potential energy surface.

3. Is it possible to obtain information on the activated complex? How long does it last? Is there any hope of probing the activated complex spectroscopically?

4. When a reaction takes place, there is, of necessity, some rearrangement of the electrons in the reaction system. When and how does this take place and how does it affect the reaction dynamics?

5. The possible importance of specific reactive orientations was discussed in our treatment of collision theory. Can we obtain experimental evidence of a steric requirement?

Experimental techniques broadly fall into two categories. So called *bulb experiments* take place in an initially thermal distribution of reactant gases. State selection is provided by laser excitation. In order to observe the effects of reaction from, or to, a particular quantum state we need to probe the reaction before either the reactant or product has been collisionally deactivated.

Molecular beams are an alternative approach. Here beams of reactants are made with uniform velocities so that the molecules in the same beam do not collide. The two beams of reactants cross at a certain angle and the products are detected as a function of their scattering angle. One of the major advantages of the molecular beam technique is that greater spatial information, as well as product quantum state distributions, are obtained. However, beam experiments are generally very expensive to set up and bulb experiments provide a greater degree of flexibility. Limited spatial information can be obtained through polarization and Doppler measurements. Very often the techniques provide complementary information.

4.3.1 Bulb methods

The reactants are contained in a bulb or flow tube, and so the experimental method involved, and questions that may be addressed, differ from those using molecular beam techniques, which we shall describe later. We shall firstly con-sider reagent specification and then look at the measurement of nascent product distributions.

Reagent specification

(a) Translational excitation

Whenever we change the temperature of our reagents we are altering the mean translational energy of collisions, however, because the distribution of velocities is so large it is difficult to be precise about the role played by any particular col-lision energy. As we shall see below, molecular beam techniques are an ideal way of generating uniform reagent velocities which can be continuously varied to probe the effect of translational excitation. In bulb experiments we are limited to laser photolysis as a method of generating translationally excited species. For example HI and HBr may be photolysed by excimer lasers at 193 nm giving atomic products with translational energies of 327, 283 (HI), 260 and 169 (HBr) kJ mol^{-1}, the two differing values depending upon which electronic state of the halogen is produced. Conservation of momentum ensures that over 99 per cent of this translational energy is imparted to the departing H atoms (see p. 284).

Of course for any experiment on reagent excitation it is essential that the reaction occurs before any relaxation of the reagents can occur. For translational and rotational excitation, this means working under *single collision* conditions. By this we mean that we probe for products during the approximate time of the initial collision of the reacting species, ensuring that the excited reactant and their corresponding products are unaffected by subsequent collisions with other species. Using kinetic theory it is possible to calculate the average time for the first collision (either with reagent or bath gas). Only products appearing before this time can be considered to have arisen from the nascent distribution of excited reagents.

(b) Rotational excitation

The effects of rotational excitation are exceedingly difficult to study. Firstly there are few convenient excitation sources to promote significant quantities of reactants to higher rotational states and secondly, rotational relaxation is extremely rapid. Because of the low energies generally associated with rotational excitation any effects are expected to be small and are difficult to pick out.

One area where rotational effects are important is in rotational energy transfer. As the rotational energy of a molecule scales with $BJ(J+1)$ the energy spacing between rotational levels increases dramatically with J. For molecules such as HF which have large B values, the energy separation between highly rotationally excited states (e.g. $J = 16$, 17) can become appreciable (≈ 6 kJ mol^{-1}) in comparison to that at lower J values. The efficiency of rotational relaxation is related to the energy gap between rotational states and will therefore decrease with increasing J. This can lead to bimodal distributions of product rotational states. Those molecules formed with low rotational energy are rapidly relaxed into a Boltzmann distribution by collisional energy transfer, whereas the molecules formed with high J remain comparatively unaffected. An example of this is shown in Fig. 4.4 for HF$(v = 1)$ produced from reaction (3)

$$CF_3 + H \rightarrow HF + CF_2 \tag{R 3}$$

where the second maximum is due to unrelaxed rotationally excited HF molecules.

(c) Vibrational excitation

This has been a fruitful area of reaction dynamics research. For most chemical reactions bonds have to be broken. Putting energy into vibration, which stretches

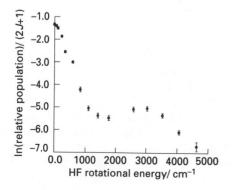

Fig. 4.4 Boltzmann plot (ln(relative population)$/(2J + 1)$ vs. E_{ROT}) of HF $(v = 1)$ produced from the reaction of CF$_3$ with H atoms. The arrested relaxation of higher rotational levels of HF can clearly be seen.

the bond increasing the mean atomic separation, will surely aid this process. As examples of such systems we shall look at two reactions:

$$O + HCl \rightarrow OH + Cl \qquad\qquad (R\ 4)$$

$$OH + C_2H_6 \rightarrow H_2O + C_2H_5 \qquad\qquad (R\ 5)$$

and consider the effects of exciting either of the diatomic reactants.

HCl excitation A thermal distribution of O atoms is generated by a microwave discharge of molecular oxygen. HCl can be excited by a pulse from an HCl chemical laser (see p. 99 and Lin *et al.* 1983). This experiment completes an interesting circle of interactions; in the chemical laser the $H + Cl_2$ reaction generates highly vibrationally excited HCl. The topography of the exit channel of the potential energy surface allows for a population inversion and laser action between vibrational levels (see p. 97). The resulting chemical laser can now be used to study the effects of changing the energy of the HCl reactant, probing the entrance valley of a different potential energy surface!

As might be expected there is a considerable enhancement of the rate of reaction (4) with vibrational excitation of HCl, $k_4(\text{HCl } v = 0) = 1.4 \times 10^{-16}$ molecule $cm^{-3}\ s^{-1}$, $k_4(\text{HCl } v = 1) = 1.0 \times 10^{-13}$ molecule $cm^{-3}\ s^{-1}$, almost a 1000-fold increase.

OH excitation OH radicals (both ground and vibrationally excited) can be produced by the photolysis of nitric acid vapour. We can look at the effect of OH reagent excitation by probing the rate of disappearance of vibrationally excited OH produced in the dissociation. Our problem is slightly more complex as $OH(v = 1)$ can be removed by vibrational relaxation as well as reaction. However, it is possible to disentangle the two effects, the final result being that there is very little increase in the rate of reaction (5) with OH excitation.

How can we explain these results? Both OH and HCl vibrational excitation generate a set of reactants energetically closer to the activation barrier. Obviously the answer is that it is not only important how much energy the system has, but more importantly, where that energy is located. In the first reaction the HCl bond has to be broken. As we postulated above, vibrational excitation will aid this process. In the second reaction the excited OH bond remains intact throughout the reaction. The energy stored in this vibration cannot get across to 'help' with the reaction and the OH bond is effectively a 'spectator' of the C–H bond breaking and new O–H bond forming process.

Box 4.1 Bond selective chemistry

One of the questions posed by the preparative chemist is 'how can we control the outcome of a multichannel reaction?' We have already discussed how vibrational excitation of a breaking bond promotes reaction. A logical extension of this process is to polyatomic species where more than one reaction channel is accessible. The relatively simple example which we shall consider is:

$$H + HOD \rightarrow H_2 + OD \qquad\qquad (R\ 6a)$$

$$\rightarrow HD + OH \qquad\qquad (R\ 6b)$$

but the principle can be applied to more complex systems and a possible *eventual* goal would be to dramatically increase the efficiency of certain syntheses. The realization of this aim is still a long way off, but in the mean time such studies yield more information into the details of elementary reactions.

In these experiments by Crim and co-workers (1990, 1991), H atoms are generated in a microwave discharge and HOD is excited using a pulsed near IR laser, initially into the fourth overtone (|0,4> state) of the O–H stretch. To a good approximation this is a well-defined local mode, i.e. the energy remains in the O–H stretch for a considerable period. Laser induced fluorescence (see Section 2.4 and p. 99) is used to probe for either OH or OD products after the vibrational excitation. The observed selectivity of the reaction is enormous with at least a 100 fold excess of OD over OH. As would be expected the vibrationally excited bond is the one most readily broken.

In another experiment the |0,4> and |1,3> (i.e. 3 quanta of stretch in one bond, one in the other) modes of H_2O are excited and the experiment is repeated. Figure 4.5 shows the results. For the |0,4> excitation no OH ($v = 1$) is observed, however, for |1,3> reagent, a considerable excess of OH ($v = 1$) product is observed. In this case the H atom still preferentially attacks the bond with the highest vibrational energy, the one most likely to break. The remaining quantum of vibration is retained in the reagent appearing as vibrational excitation in the product.

Zare has extended the experiments to look at the effects of a single quantum of excitation on reaction (6) (In this case H atoms are generated by laser photolysis). O–H or O–D excitation is again produced by a tunable IR laser which is triggered just before the H atom photolysis laser. For OH excitation the OH:OD product ratio is 1:25 showing a considerable enhancement of the reaction cross-

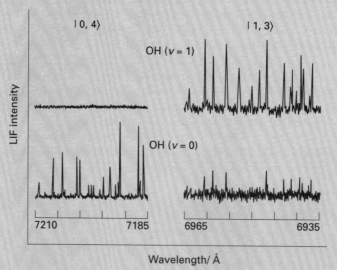

Fig. 4.5 Vibrational overtone excitation spectra for transitions to the |0,4> and |1,3> states of water obtained by probing OH($v = 0$) and OH($v = 1$) products resulting from the H_2O |a,b> + H reaction. Reaction from the |0,4> produces essentially no detectable OH($v=1$), in contrast to the results from the |1,3> state. Reproduced from Sinha *et al.* (1991) *Journal of Chemical Physics*, **94**, 4928.

section for breaking the vibrationally excited bond, even though in this case the vibrational energy is considerably below the activation barrier for the reaction.

Further details on all these reactions may be found in the following references, all of which are very easily digestible. Baggott, J. (1992). Shaking out the right reaction. *New Scientist*, **22 August,** 27–30; Sinha, A., Haiao, M. C., and Crim, F. F. (1991). Bond selective chemistry. *Journal of Chemical Physics,* **92,** 6333–4; Sinha, A., Haiao, M. C., and Crim, F. F., (1991). Controlling bimolecular reations. *Journal of Chemical Physics*, **94,** 4928–35; Bronikowski, M. J., Simpson, W. R., Girard, B., and Zare, R. N. (1991). Bond specific chemistry. *Journal of Chemical Physics*, **95,** 8647–8.

Product distributions

(a) Infrared chemiluminescence

Hydrogen atoms may be produced by passing hydrogen gas through an electric discharge. If we add chlorine to the gas stream, the exothermic reaction

$$H + Cl_2 \rightarrow HCl(v, \mathcal{J}) + Cl \tag{R 7}$$

occurs. Any vibrationally excited HCl molecules produced may be detected by observing the infrared radiation which they emit:

$$HCl(v, \mathcal{J}) \rightarrow HCl(v-1, \mathcal{J} \pm 1) + h\nu. \tag{R 8}$$

The intensity of emission from a given $(v, \mathcal{J})$ level is proportional to the number of molecules in that level, which, in turn, is proportional to the rate coefficient , $k_{v\mathcal{J}}$, for formation of HCl in that level, provided the HCl distribution can be 'frozen' in its initial form. In other words, the experimentalist must ensure that the HCl molecules are removed from the region, from which the infrared chemilumines-cence is observed, before a significant number of them undergo transitions to other states. These transitions can occur on collision with other molecules, in which energy is interchanged between the colliders, or by radiation (the infrared chemiluminescence itself). This relaxation, as it is known, is avoided by keeping the pressure low by rapid pumping, so that the time between collisions is long, and so that diffusion to the walls is rapid. The walls themselves may be cooled by liquid nitrogen so that the molecules stick on them, thus increasing the effective pumping rate by 'cryogenic pumping', also ensuring that no HCl molecules return to the gas phase after suffering deactivating collisions at the walls. Some infrared chemiluminescence must occur otherwise we should have no means of monitoring the products, but it is sufficiently slow, compared with the loss by pumping, that the initial distribution is effectively undisturbed. Figure 4.6 shows a schematic diagram of an infrared chemiluminescence apparatus used by Polanyi and co-workers (1972) at the University of Toronto. The mirrors serve to increase the collection of infrared radiation by the detector. Figure 4.7 shows some of the results of $k_{v,\mathcal{J}}$ for the reaction

$$Cl + HI \rightarrow I + HCl(v, \mathcal{J}). \tag{R 9}$$

The exothermicities of these and related reactions are well known and, from the relative $k_{v,\mathcal{J}}$, the fraction of this total energy channelled into vibration, rotation and, by difference translation, may be determined. The results of four reactions are shown in Table 4.2. There is a wide variation in the fraction of energy

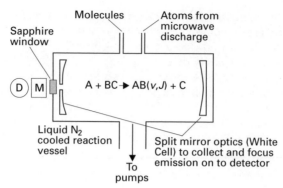

Fig. 4.6 Infrared chemiluminescence system for studying the kinetics of vibrationally excited molecules. D, IR detector; M, monochromator.

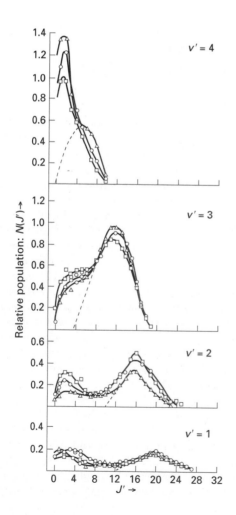

Fig. 4.7 Rotational distributions for the Cl + HI reaction for vibrational levels 1–4. The data are shown for a number of experimental conditions to investigate the possibility of relaxation. Even with the rapid pumping available rotational relaxation has occurred and the dotted lines in the figures indicate the extrapolated shapes of the initial rotational distributions. The relative vibrational distributions for this reaction are given in Table 4.2.

Table 4.2 Product energy distributions by infrared chemiluminescence

Reaction	k_v (relative)	Fraction of energy released as:		
		Vibration	Rotation	Translation
$H + Cl_2 \rightarrow HCl + Cl$	$k_1 = 0.28, k_2 = 1.00;$ $k_3 = 0.92; k_4 = 0.08$	0.39	0.07	0.54
$H + Br_2 \rightarrow HBr + Br$	$k_1 = 0.10, k_2 = 0.19; k_3 = 1.00;$ $k_4 = 0.88; k_5 = 0.27; k_6 = 0.05$	0.56	0.05	0.39
$F + H_2 \rightarrow HF + H$	$k_1 = 0.31, k_2 = 1.00; k_3 = 0.48;$	0.67	0.07	0.26
$Cl + HI \rightarrow HCl + I$	$k_1 = 0.18, k_2 = 0.32; k_3 = 1.00;$ $k_4 = 0.74$	0.71	0.13	0.16

channelled into internal motion of HX (i.e. into vibration and rotation). For H + Cl_2, more than half the available energy appears as translation; we shall return to this point later.

It is particularly noticeable that in all these cases $k_2 > k_1$: e.g. for H + Cl_2 $k_{v=1}$ = 0.28 and $k_{v=2}$ = 1.00. This means that immediately following reaction there are more molecules in the second vibrational level than in the first. This is one of the requirements needed for laser action and an HCl laser, based on these principles, has been constructed.

These experiments employed a Boltzmann distribution of reactant molecules at ambient temperatures. Because the reactant diatomic BC has a high vibrational frequency, the vast majority of the molecules are in the ground vibrational state. The effect of vibrational excitation of the reactants on the vibrational distribution of the products can be investigated in a number of ways. Firstly we might use the laser techniques described above to excite the diatomic reactant into a specific vibrational level. Alternatively we can generate BC *in situ* in a pre-reaction chamber, via another atom–molecule reaction. Figure 4.8 shows the arrangement for such an investigation. Analysis of the results is more complex in this case as the diatomic reactant will contain a range of vibrational energies.

(b) Laser induced fluorescence (LIF)

An alternative technique, developed more recently following the production of tunable dye lasers, uses laser induced fluorescence as a means of detecting product molecules in specific quantum states. It is used not only in bulk experiments, employing both flow and pulsed methods, but also in the molecular beam experiments which we shall discuss below. We have already encountered the technique on pp. 34 and 47; the dye laser is tuned over the product absorption spectrum and the total unresolved fluorescence monitored. The different (v, J) levels absorb at different laser wavelengths and so, provided the excited AB molecules produced are not deactivated before they emit, and also that the dye laser intensity is constant over the absorption spectrum, the fluorescence intensity is proportional to the population of the ground (v, J) state from which it was excited. This population is, in turn, proportional to $k_{v, J}$ provided the conditions enumerated above for infrared chemiluminescence are observed. The lifetimes of excited electronic states are generally much shorter than those of excited

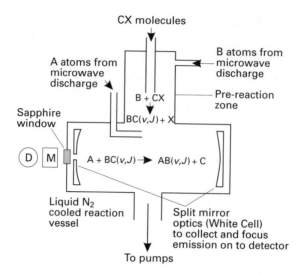

Fig. 4.8 Infrared chemiluminescence system for studying the kinetics of vibrationally excited molecules by the 'pre-reaction' technique. D, IR detector; M, monochromator.

vibrational states, so that the reaction may be probed at much shorter times. In fact the limiting factor on the timescales of processes which can be studied is the delay between the photolysis and probe pulses. The sensitivity of the technique allows lower collision densities to be used leading to much less exacting pumping requirements than was the case with infrared chemiluminescence.

4.3.2 Molecular beams

With so-called bulb methods, great care has to be taken to avoid collisional relaxation of the product molecules. In addition, the reactants generally conform to a Boltzmann distribution. Although the reactant vibrational distribution can be changed in a specific manner, for example by laser excitation, the translational and rotational distributions may only be varied by changing the temperature, and the range of energies in the resulting distribution is thus quite broad. In an ideal experiment, we should like to prepare A and BC with very specific energies, allow them to undergo a *single* collision, and then analyse the product states. This ideal is most closely approached in crossed molecular beam experiments, in which two reactant beams are directed into a scattering region, S (Fig. 4.9), where single reactive (or unreactive) collisions occur between the component molecules of the beams. The scattered products are then detected, usually by mass spectrometry, as a function of the angle Θ.

Reagent specification

Figure 4.9 shows an apparatus where one of the beams is produced by effusive flow from an oven. Such a beam has a range of energies, since it derives from a Boltzmann distribution of molecules within the oven. It also covers a wide range of angles and must be collimated by the use of slits. By means of notched wheels, reminiscent of the Fizeau velocity of light experiment, a given velocity may be selected from the Boltzmann distribution albeit with a large loss of beam

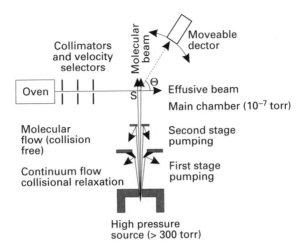

Fig. 4.9 Schematic diagram of a crossed beam apparatus with both effusive and supersonic molecular beams. Reaction detection is generally by a mass spectrometer which is moved about by a small crane. The apparatus is very bulky, occupying a large room.

intensity. The other beam shown is known as a seeded nozzle beam and has many advantages. The nozzle beam is created by expanding a gas through a pinhole, from a high pressure oven, into a series of rapidly pumped chambers. It is collimated as it passes from one chamber to the next and eventually forms a well collimated beam. Because the flow is far from effusive (high oven pressure, combined with a comparatively large pinhole) there are many collisions in the early section of the beam and random thermal motion is converted into translational motion, with a relatively narrow velocity distribution, in the beam direction. Thus the intensity of the beam is high and, at the same time, its temperature (random motion) is very low. If a beam of a light gas, such as H_2 or He, is seeded with a heavy reactant, very high reactant translational energies, along the beam direction, can be achieved, since the reactant is dragged along at the same velocity as the light gas yet its energy is given by $\frac{1}{2}mv^2$ where m is its mass. By varying the fraction of reactant in the beam and the temperature of the nozzle, the velocity of the reactant may be varied. At the same time its rotational and vibrational energies are considerably reduced in the general conversion of thermal into directed motion. Thus beams with near ideal conditions (high intensity, variable velocity, low internal excitation) may be produced. The reactant may then be vibrationally excited by laser irradiation (Brooks, 1976). For example, laser excitation can be used to study the effects of vibrational excitation on reaction (10).

$$K + HCl \ (v = 1) \rightarrow KCl + H. \qquad (R \ 10)$$

It is even possible to select rotational states using electric or magnetic fields, and to select molecules with a specific orientation. Finally, unstable reactants, atoms or radicals, can be incorporated into beams after formation in discharges or simply by pyrolysis.

Box 4.2 Supersonic molecular beams

As we noted above a supersonic molecular beam consists of a high pressure region with a nozzle through which the gas expands into a low pressure region. There are three important points to note.

1. At the nozzle the gas is compressed and behaves as a liquid being squeezed out of a hole. The random motion of the gas molecules in the oven is converted into translational motion in the beam direction by the many collisions that occur in the nozzle region.

2. As the gas moves out from the nozzle it expands and as it does so it cools. This process forms the basis of refrigeration and we may have experienced the effect from spray can toiletries or medication.

3. The expansion is adiabatic—no energy is lost from the beam.

As the expansion increases the collision rate falls and we are left with a beam of molecules, all moving in approximately the same direction. Skimmers and slits further collimate the beam. As no energy has been lost from the beam, the average velocity of the beam remains the same—however the distribution of velocities is drastically narrowed because the temperature has fallen—remember that temperature corresponds to random motion. The velocity distribution of the beam is given by

$$n(v) \propto v^2 \exp\{-(v-v_{\mathrm{s}})/\alpha_{\mathrm{s}}{}^2\} \tag{E 4}$$

where vs is the most probable velocity, given approximately by the expression $\{2\gamma k_{\mathrm{B}} T_0/(\gamma-1)m\}^{\frac{1}{2}}$ (where T_0 is the temperature in the high pressure source), $\alpha_{\mathrm{s}} = \{2k_{\mathrm{B}} T_{\mathrm{s}}/m\}^{\frac{1}{2}}$ (where T_{s} is the translational temperature of the beam) and γ is the ratio of heat capacities, Cp/Cv. The local speed of sound is given by $(\gamma k_{\mathrm{B}} T/m)^{\frac{1}{2}}$ and therefore as T rapidly decreases in the beam the flow will become supersonic with a Mach number $Ma = v_{\mathrm{s}}/ (\gamma k_{\mathrm{B}} T/m)^{\frac{1}{2}}$. A limited analogy of a molecular beam occurs when the occupants of a crowded passenger train try to pass through the ticket barrier. At the 'high pressure' region the passengers jockey for position with lots of random motion, but in the barrier region, the nozzle, they move with a constant speed in one direction. Unfortunately, the analogy begins to breakdown in the low pressure region, the station concourse, as the passengers can change their speed and direction and do not continue at a constant velocity from the barrier, however, the number of collisions is reduced.

So far we have only considered translational motion, however, energy transfer between rotation and translation is efficient and the rotational temperature will tend to follow that of translation. The transfer of vibrational to translational energy is much less efficient and vibrational cooling is much reduced.

So-called seeded molecular beams can also be used. Here a small quantity of the reagent is mixed with an excess of a low molecular weight carrier gas, typically H_2 or He. The properties of the molecular beam derive from the average ratio of specific heats and average molecular mass, i.e. essentially that of the carrier gas. The stream velocity is inversely proportional to the square root of the molecular mass and hence the heavy reagent molecules will be carried along at the faster carrier gas velocity.

Major technical problems associated with molecular beam sources are the high pumping speeds required, the volume of gas used and the possible formation of clusters. Because of the large pumps required to maintain such high vacuums, molecular beam experiments occupy large areas of several tens of square metres. The use of pulsed valves on the nozzle limits the amount of gas used and can provide a degree of time resolution. Cluster formation, where two or more atoms or molecules form weakly bound complexes, can generally be avoided with the correct conditions. Clusters are essentially an intermediate state between the gas and liquid phases and as such have interesting properties. An experiment on cluster reactions is briefly described on p. 168.

Product detection

The products are usually detected as a function of scattering angle Θ by means of a mass spectrometer which incorporates a quadrupole mass filter. Their translational energy distribution may be determined by passing the reactants through a chopper wheel, and the distribution of times from the passage of the reactants through the chopper to the arrival of products at the detector determined. Knowing the reactant velocity, the velocity distribution of the products may be found, from this distribution of arrival times. The rotational and vibrational energy distribution may be determined in some cases by LIF. The experimental data may be plotted in the form of product density maps (see Fig. 4.10b for H + Cl_2). The scattering centre lies at the origin of angular coordinates, which describe the direction in which HCl is scattered. The angle plotted is not the one measured in the laboratory.

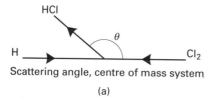

Scattering angle, centre of mass system

(a)

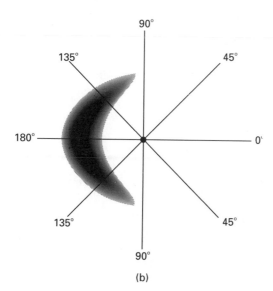

(b)

Fig. 4.10 (a) Scattering angle for the centre of mass system. (b) Product density map for reactive scattering of HCl from H + Cl_2. The scattering centre lies at the origin of the angular coordinates, which describe the scattering in the centre of mass system. The distance from the origin denotes the velocity of the scattered product, and the density of shading is a measure of the probability that a reactive collision will lead to scattering at a given angle and velocity. Note the pronounced backward scattering.

It turns out to be much easier to interpret the data if we imagine that we are sitting on the centre of mass of the colliding system, and watching the reaction take place (see Fig. 4.10b). Instead of approaching at right angles, H and Cl_2 now appear to be approaching at an angle of 180°. The incident direction of H is taken as 0° and that of Cl_2 is, consequently, 180°. We now allow reaction to occur at S, and we measure both the direction θ, and the velocity of the product HCl in our new centre-of-mass coordinate system (or rather we measure them in laboratory coordinates and convert them to centre-of-mass coordinates). Next we draw a line from S whose direction is the same as that of the scattered HCl molecule, and whose length is proportional to the product velocity. At the end of this line we make a mark, and we repeat this process for a great many reactive collisions, eventually building up a density map as shown in Fig. 4.10b. The dense area represents the most probable direction and velocity of HCl, and we see that it is scattered preferentially around 180°—the hydrogen atom collides with Cl_2 and the product molecule moves back in the direction from which the hydrogen atom came. The distribution of distances of our marks from the origin corresponds to the product velocity distribution and hence the amount of energy channelled into translation. Knowing the exothermicity of the reaction, we can find, by difference, the amount which goes into internal excitation. Alternatively, the internal energy distribution may be measured with higher resolution using direct techniques such as LIF.

The angular pattern of the reactive scattering is sensitive to the reaction mechanism. The results of five reactions are shown in Table 4.3. The term *forward scattering* means that, following reaction between atom M and molecule RX, the product MX emerges in roughly the direction in which M was travelling before the collision (i.e. $\theta < \pi/2$, see Fig. 4.10b), whilst for *backward scattering* $\theta > \pi/2$. We can rationalise the scattering patterns for the first three reactions in terms of their cross-sections. We met the K + Br_2 reaction earlier (p. 91). It takes place on a strongly attractive surface by electron transfer at a large distance, and reaction is possible even with a large impact parameter (hence the large cross-section). Provided there is no energy barrier for reaction (which is the case here), the fraction of all reactive collisions which take place with an impact parameter between b and $b + db$ is $2\pi b db / \pi b_{max}^2$; the large b values make the greatest contribution to the reaction cross-section. Thus, most reactive collisions do not sample the repulsive wall of the potential, and scattering is mainly in a forward direction (cf. trajectory A Fig. 4.2). K + CH_3I and H + Cl_2 are more conventional reactions. The cross-sections are smaller and the effective collisions are more nearly head on, leading to backward scattering (cf. trajectory B Fig. 4.2).

Table 4.3 Molecular beam results

Reaction	σ/nm^2	Scattering direction
K + Br_2 → KBr + Br	2	Forward
K + CH_3I → KI + CH_3	0.4	Backward
H + Cl_2 → HCl + Cl	0.1	Backward
Cs + RbCl → CsCl + Rb	1.5	Forward and backward
O + Br_2 → BrO + O	0.1	Forward and backward

A collision complex will undergo significant rotational motion if its lifetime τ is comparable to or greater than the rotational period. As it dissociates the rotating complex will throw out product molecules in all directions; both forward and backward scattering will be observed. Marked forward or backward scattering patterns demonstrate that the complex 'remembers' (or does not have time to forget) the direction of approach of the reactants, i.e. its lifetime is much shorter than its rotational period ($\tau < 10^{-12}$ s). Many reactions conform to this pattern, though two which do not are the last two reactions (4) and (5) in Table 4.3. Here both forward and backward scattering take place and the complex is long lived ($\tau > 5 \times 10^{-12}$ s).

Study notes
This section has of necessity been brief and qualitative. These notes provide a means of gaining reasonable access to a more detailed discussion.
1   Infrared chemiluminescence is discussed in greater detail by Smith, I. W. M. (1980), *Kinetics and dynamics of elementary gas reactions,* pp. 243–6. Butterworths, London. Alternatively consult Bernstein (Bernstein, R. B. (1982) *Chemical dynamics by molecular beam and laser techniques,* Oxford University Press.) pp. 55–57.
2   In Chapter B 2.4 of *Modern Gas Kinetics* (Simons, J. P. (1987) *Modern Gas Kinetics* (ed. M. J. Pilling and I. W. M. Smith). Blackwell, Oxford). Simons discusses in greater detail a number of topics briefly outlined above, regarding both state selection and molecular beam techniques. The final section 'some case histories' illustrates relevant examples of the techniques in action. References at the end of the chapter provide further access to the extensive literature on reaction dynamics.
3   Bernstein discusses nozzle beams on pp. 27–33, giving schematic diagrams and more quantitative explanation of the high reactant energies produced in seeded nozzle beams. Alternatively consult Smith pp. 313–316 or Levine and Bernstein (Levine, R. D. and Bernstein, R. B. (1987). *Molecular reaction dynamics and chemical reactivity.* Oxford University Press.) pp. 233–247.
4   A discussion of the use of centre of mass angular distributions to obtain dynamical information about reactions is discussed in Levine and Bernstein pp. 247–60. The reaction $K + I_2$ is discussed in detail and a more qualitative analysis given for $K + CH_3I$. This section also examines the $D + I_2$ reaction, where the DI is scattered *sidewards,* a consequence, it is argued, of a bent transition state with D approaching I_2 from a sideways direction. The $O + Br_2$ reaction is provided as an example of a reaction with forward–backward symmetry and a scattering diagram is given.
5   Levine and Berstein present a detailed discussion of the $F + H_2$ (D_2) reaction (pp. 396–411). The section concludes with a brief discussion of a theoretical analysis of the reaction using an *ab initio* potential energy surface (see below).
6   A quantitative analysis of angular scattering diagrams for reactions proceeding through longer-lived complexes is presented by Levine and Bernstein on pp. 411–29. The discussion is illustrated by reference to the $Cs + RbCl$ and $F + C_2H_4$ reactions. The latter shows symmetric sideways scattering, a consequence of its proceeding through a complex which is an oblate top while the $Cs + RbCl$ complex is a prolate top. (See Atkins, P. W. (1994). *Physical Chemistry* (5th edn), p. 558. Oxford University Press, for a discussion of oblate and prolate tops.) This section

also presents a discussion of the energetics and kinetics of complex formation based on unimolecular rate theory (Chapter 5).

7   Grice, R. (1982) Molecular Beam Scattering, *Chemical Society Review*, **11**, 1– 15, gives a general review of advances in the study of O atom reactions using seeded nozzle beams crossed with a molecular nozzle beam. Note particularly the discussion of the O + Cl_2, Br_2, I_2 reactions on p. 4 and the dependence of the mechanism of the reactant translational energies. Note also the discussion of the detailed mechanism of the reactions of oxygen atoms in terms of electronic states on pp. 9–13.

8   More detailed reviews of individual techniques may be found in a number of chapters of Liu, K. and Wagner, A. (ed.) (1996). *Kinetics and Dynamics of Small Radical Species*. (World Scientific, New York).

4.4 Reaction dynamics and potential energy surfaces

Later in this chapter we shall discuss the calculation of potential energy surfaces and their relationship to the dynamics of a reaction. Some useful qualitative generalizations can be made before then, on the basis of the experimental results we have encountered. As we shall see, a reaction may be simulated by the motion of a point mass over the potential surface. The K + Br_2 reaction takes place on an attractive potential surface, i.e. much of the exothermicity is released as the reactants approach following the electron transfer. Let us see what type of behaviour we might expect from simple kinematic arguments. The potential surface is shown schematically in Fig. 4.11a together with a possible trajectory of the representative point. The point is accelerated as it moves through the head of the entrance valley, it overshoots and exits with an oscillatory trajectory, corresponding to vibration of KBr. Energy is preferentially located in the product internal modes, as was found experimentally. Alternatively, we might use a less dynamic explanation of the production of excess vibrational of energy. Since the KBr bond is 'made' on electron transfer, its initial bond length is much longer than the equilibrium value and the molecule is formed high up on its Morse-type potential well , i.e. with an excess of vibrational energy.

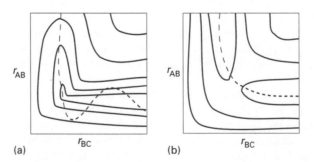

(a)   (b)

Fig. 4.11 Trajectories for (a) attractive and (b) repulsive potential energy surfaces. The reaction dynamics can qualitatively be simulated by a ball, mass $m_A m_{BC}/(m_A + m_{BC})$, rolling over the surface although more accurate simulations require skewed coordinates see Box 4.3.

Fig. 4.12 Molecular orbital for the H + Cl$_2$ reaction.

At the other extreme, the exothermicity might appear solely in the exit valley (Fig. 4.11b). Now the representative point stays close to the reaction coordinate, and gets a push as it leaves, leading to an excess of relative translational motion of the products. H + Cl$_2$ approximates to this type of behaviour, with over half the available energy appearing as translation. A simple molecular orbital description of the HCl$_2$ transition state suggests that the odd electron occupies an orbital which is bonding between H and the adjacent Cl but antibonding between Cl and Cl see Fig. 4.12. This leads to strong repulsion between Cl and the newly formed HCl, and accounts for the form of the surface and the high proportion of translational energy evident in the products.

In between the two theoretical extremes of attractive and repulsive surfaces, we can envisage a whole range of 'mixed energy release' surfaces.

We can use similar kinematic arguments to understand the energy requirements of the reverse reactions in Fig. 4.11(a) and (b). In Fig. 4.11(a) the energy barrier is in the exit valley leading to A + BC, and motion perpendicular to the reaction coordinate is required. If we roll a ball bearing with high velocity along the entrance valley, it will tend to run up the wall at the end of the valley and show little inclination to turn the corner. We need to give it a velocity component perpendicular to the entrance valley in order to overcome the energy barrier efficiently, i.e. the reactants require an excess of energy in vibration, and translational energy is of little use in surmounting the barrier. In Fig. 4.11(b), the energy barrier occurs in the entrance valley, and translational energy is now more useful. These arguments and the results of molecular beam experiments, emphasize the oversimplification of collision theory, where only relative translational motion is permitted to contribute towards the reaction.

We have seen in this section how a qualitative consideration of the topography of the potential energy surface (PES) predicts experimental observables such as vibrational or rotational excitation. The interaction of theory and experiment is generally of this symbiotic form with theoretical calculations provoking new experiments, but calculations are always being updated and refined in the light of new experimental results.

Box 4.3 Skewing angles

We have seen that a great deal of mechanistic insight into the reaction can be obtained by representing the reaction as a ball rolling across the potential energy surface. Whilst this approach is qualitatively correct, we need to carry out a couple of mathematical transformations to the two axes r_{AB} and r_{BC} in order to make this simulation quantitatively valid. The result of these operations is that the two axes are skewed at an angle β to each other in the new scaled cartesian axes given by eqns (5) and (6):

$$X = ar_{AB} + br_{BC} \cos \beta \qquad \text{(E 5)}$$

$$Y = br_{BC} \sin \beta \tag{E 6}$$

where

$$a = [m_A(m_B+m_C)/M]^{\frac{1}{2}} \tag{E 7}$$

$$b = [m_C(m_B+m_A)/M]^{\frac{1}{2}} \tag{E 8}$$

$$\cos^2 \beta = m_A m_C / \{(m_B + m_C)(m_A + m_B)\} \tag{E 9}$$

$$M = m_A + m_B + m_C. \tag{E 10}$$

The degree of skewing and scaling depends on the ratios of the reagent masses. Many systems reduce to 'heavy–light–heavy' (HLH) or 'light–heavy–heavy' (LHH) combinations. An example of the former case would be the reaction

$$Cl + HCl \rightarrow ClH + Cl \tag{R 11}$$

Here the skewing is considerable. Alternatively for the $H + Cl_2$ reaction, a LHH combination, β is much closer to 90°.

The skewing angle and scaling can have considerable effects on the quantitative dynamics of the reaction. For example, reaction (9)

$$Cl + HI \rightarrow HCl + I \tag{R 9}$$

has a late barrier, located in the product valley, such a potential energy surface would be expected to lead to little vibrational excitation when viewed as an unskewed representation. However, considerable vibrational excitation is observed due to the HCl product having to turn the tight corner to get over the activation barrier (Fig. 4.13). Further information can be found in Appendix 4A of Levine and Bernstein (Levine, R. D. and Bernstein, R. B. (1987). Molecular reaction dynamics and chemical reactivity. Oxford University Press.)

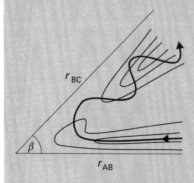

Fig. 4.13 Skewing the atom separation axes can have a dramatic effect on the reaction dynamics. Here a late transition state, which would normally channel energy into translation on an unskewed surface, produces considerable vibrational excitation.

4.5 Potential energy surfaces

We learned in the last section that the dynamics of a reaction are intimately related to the potential energy surface on which the reaction occurs. In the last chapter, we introduced transition state theory with a brief discussion of the potential energy surface for our model reaction A + BC. Thus a knowledge of potential energy surfaces and, if possible, a means of calculating them, would aid our understanding of, and our ability to predict, both the kinetics and dynamics of chemical reactions.

There are two widely differing approaches to the calculation of potential energy surfaces. In the first, a purely empirical potential function, which contains some adjustable parameters, is 'tuned' until it fits known experimental results (such as the threshold energy). As we shall describe below, this surface may then be used in computer calculations of the dynamics of reactions. The alternative approach, which has very different aims, is to perform *ab initio* calculations in which the Schrödinger equation is solved numerically by methods that have been well established for stable molecular systems and are finding increasing application in reaction kinetics. Since these surfaces are not empirically adjusted, they provide independent information of the shape of the surface and on the mechanism of the reaction. Most applications refer to triatomic systems, and the most successful calculations involve $H + H_2$ and $F + H_2$. However, an increasing number of calculations on larger systems are being published, although for such systems, some empirical tuning may be necessary. We shall briefly discuss the method employed in constructing potential surfaces for one particular, and popular, empirical method. A discussion of *ab initio* calculations is beyond the scope of this book, however an excellent introduction to the field may be found in Section 2.3 of '*A Computational Approach to Chemistry*' by Hirst (1990) which describes the development and application of *ab initio* calculations. Whilst the emphasis is very much on the PES of stable molecules (and in fact the latter sections on TST are rather short) the same principles apply to the calculation of the complete PES for a reaction.

Study notes

An example of the level of recent *ab initio* calculations for reactive systems can be found in Walch, S. P. (1993). Characterization of the PES for $CH_3 + O_2 \rightarrow$ products. *Chemical Physics Letters*, **215**, 81–6. He examines the reaction $CH_3 + O_2$ which has two product channels giving $CH_3O + O$ and $H_2CO + OH$. He shows that the reaction proceeds on two distinct potential energy surfaces and calculates vibrational frequencies for activated complexes and bound intermediates.

Box 4.4 An example of a PES

The LEPS (London–Eyring–Polanyi–Sato) surface is widely used in reaction dynamics. Although it is essentially an empirical surface, it does have some basis in theory. The valence bond treatment of H_2 (the Heitler–London theory) gives the energies of the lowest singlet and triplet states as:

$$E_{singlet} = (Q + J)/(1 + S^2) \quad E_{triplet} = (Q - J)/(1 - S^2) \quad \text{(E 11 a, b)}$$

where Q, J and S are known as the Coulomb, exchange and overlap integrals.[1] The LEPS surface expresses the potential energy for the reaction

---

[1] For H_2 containing nuclei A and B and electrons 1 and 2, a valence bond treatment of Schrödinger equation in terms of atomic orbitals (e.g. 1s orbital on nuclei $_A$ containing electron 1 is written as $1s_A(1)$) gives:

$$Q = \int 1s_A(2) 1s_B(1) \mathcal{H} 1s_A(2) 1s_B(1) d\tau \quad \text{or} \quad \int 1s_A(1) 1s_B(2) \mathcal{H} 1s_A(1) 1s_B(2) \, d\tau$$
$$J = \int 1s_A(2) 1s_B(1) \mathcal{H} 1s_A(1) 1s_B(2) d\tau \quad \text{or} \quad \int 1s_A(1) 1s_B(2) \mathcal{H} 1s_A(2) 1s_B(1) \, d\tau$$
$$S = \int 1s_A(2) 1s_B(1) d\tau \quad \text{or} \quad \int 1s_A(1) 1s_B(2) \, d\tau$$

where $\mathcal{H}$ is the Hamiltonian for the molecule.

$$H_A + H_B H_C \rightarrow H_A H_B + H_C \tag{R 12}$$

as

$$E(r_{AB}, r_{BC}, r_{CA}) = (Q_{AB} + Q_{BC} + Q_{CA}$$
$$- [\tfrac{1}{2}\{(\mathcal{J}_{AB} - \mathcal{J}_{BC})^2 + (\mathcal{J}_{BC} - \mathcal{J}_{CA})2 + (\mathcal{J}_{CA} - \mathcal{J}_{AB})^2\}]^{\tfrac{1}{2}})/(1 + \varDelta) \tag{E12}$$

where the Qs and $\mathcal{J}$s are the Coulomb and exchange integrals for the three pairs of atoms. (In many texts the potential energy is often denoted by the symbol V.) Note that, although this has some formal similarity to eqn (11), it is an empirical expression, as is emphasized by the presence of $\varDelta$, which is an adjustable parameter. In fact an equivalent expression for H_3 was proposed by London, however, this expression, which is similar to eqn (12) can only be evaluated if certain terms in the valence bond expression are neglected! Further development of this subject may be found in a worked example by Child, (Child, M. S. (1987). *Modern Gas Kinetics* (ed. M. J. Pilling and I. W. M. Smith), p. 41, A. 1.1. Blackwell, Oxford.)

For the $H + H_2$ reaction, the Q and $\mathcal{J}$ values show an identical dependence on r, the pair separation. They are evaluated by recourse to two empirical functions, (1) the Morse function, which describes the energy of the diatomic molecule, $H_B H_C$ (or equivalently, $H_A H_B$, or $H_C H_A$) in its singlet ground state.

$$E_M(r) = D_e/4 \{\exp(-2\beta\Delta r) - 2 \exp(-\beta\Delta r)\} \tag{E 13}$$

where $\Delta r = r - r_e$ (r_e is the equilibrium bond length for the diatomic), D_e is the bond dissociation energy and β is a constant which depends on D_e and the vibrational frequency. (2) The anti-Morse function, which applies to the repulsive curve for the diatomic molecule in its triplet state:

$$E_{AM}(r) = D_e/2 \{\exp(-2\beta\Delta r) + 2 \exp(-\beta\Delta r)\} \tag{E 14}$$

where Δr, D_e, and β are defined as before.

Thus by analogy with eqns (11a) and (11b) we have:

$$Q + \mathcal{J} = E_M(r) (1 + \varDelta) \tag{E 15}$$

$$Q - \mathcal{J} = E_{AM}(r) (1 - \varDelta) \tag{E 16}$$

where $\varDelta$ has been substituted for S^2. We now have two equations and two unknowns, Q and $\mathcal{J}$, which depend on the separation, r, of the appropriate pairs of atoms. $\varDelta$ is assumed to be independent of r and is used as a variable parameter to adjust the energy barrier to that observed experimentally. Although our discussion has been limited to $H + H_2$, the LEPS approach has been applied to a great many reactions.

Study notes

A general discussion of empirical potential energy surfaces (notably the LEPS and BEBO (bond-energy–bond-order) surfaces) may be found in Smith pp. 22–49 (Smith, I. W. M. (1980). *Kinetics and dynamics of elementary gas reactions.* Butterworths, London). See also Johnston, Chapter 4 and pp. 171–183, 209–214 (Johnston, H. S. (1966). *Gas phase reaction rate theory.* Ronald, New York) and Laidler pp. 67–71 (Laidler, K. (1987). *Chemical Kinetics* (3rd edn). Harper and Row, New York).

4.6 Calculations of kinetic properties from potential energy surfaces

The calculation of the potential energy surface is not necessarily an end in itself. From the kineticist's point of view, the ultimate aim is the use of the surface for the calculation of certain aspects of the kinetics of the reacting system (although other important molecular parameters such as dipole moments can also be obtained). There are three principal applications:

(1) determination of the position and properties of the activated complex (or transition state) on the surface and the calculation of the transition state theory rate coefficient;

(2) calculation of the dynamics of the reacting system on the surface and their comparison with experimental results;

(3) comparison of transition state theory and dynamical results *on the same surface.*
   We shall deal with each of these applications in turn.

4.6.1 Transition state theory

As we noted in the previous chapter, a major difficulty in making *a priori* calculations of rate coefficients using transition state theory derives from our ignorance of the molecular parameters of the activated complex. Whilst direct experimental data are still hard to come by (see, however, Box 5.2 on p. 133) we can obtain these parameters from *ab initio* calculations. In Chapter 3 we located the transition state at the maximum of the minimum energy route between reactants and products, the saddle point, so that its geometry can be determined by finding the position of the saddle point on the calculated surface. The vibrational frequencies may be determined from the *shape* of the surface at the saddle point. For a diatomic molecule, the force constant is given by

$$f = (d^2 V / dr^2) \tag{E 17}$$

where V is the potential energy function and r is the bond distance. The force constant is equal to the curvature of the potential energy at the equilibrium bond distance. The vibrational frequency; v is then given by:

$$v = (1/2\pi)(f/\mu)^{\frac{1}{2}}. \tag{E 18}$$

Similar relationships can be determined, albeit with greater effort, for polyatomic systems. Thus, by investigating the shape of the potential surface around the saddle point, the vibrational frequencies may also be determined and these are shown in Table 4.4 for the reaction

$$H_2 + OH \rightarrow H_2O + H. \tag{R 13}$$

Figure 4.14 shows a comparison between the calculated and experimental rate coefficients. The agreement is improved considerably by incorporating suitable corrections for tunnelling (see Section 3.7) and is then exceedingly good over a very wide range of temperatures.

4.6.2 Dynamics of reacting systems

The aim of the investigations we shall describe in this section is to calculate the outcome of a single collision on a potential energy surface. Calculations of this

Table 4.4 Molecular parameters for the transition state in the $H_2 + OH \rightarrow H_2O + H$ reaction (transition state: $HO - H' - H''$)

Bond distances/nm and Angles/degrees	Barrier heights/ kJ mol^{-1}	Vibrational frequencies/ cm^{-1}
$r_{OH} = 0.1335$	$\Delta\varepsilon_e = 31^a$	3370
$r_{H'-H''} = 0.0857$	$\Delta\varepsilon_o = 30^b$	*1940*
$r_{OH'} = 0.0986$		1650
$\angle (H'OH) = 97.6$		1250
$\angle (OH'H'') = 165$		690
		440^c

^{a}$\Delta\varepsilon_e$ is the height of the barrier calculated from the potential energy surface—it is a purely electronic energy.
^{b} $\Delta\varepsilon_o$ is the difference in zero point energies between reactants and activated complex. $\Delta\varepsilon_o = \Delta\varepsilon_e +$ (zero point energy of activated complex) − (zero point energy of reactants).
^{c} This is the imaginary frequency corresponding to the reaction co-ordinate, which is used to calculate the tunnelling correction.

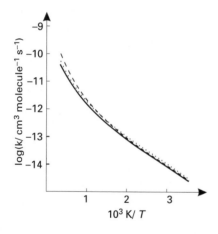

Fig.4.14 Arrhenius plot for the OH + H_2 reaction. The solid curve shows experimental results; the dashed curve (indistinguishable from the experimental results on this scale below ≈ 600 K) is the semi-empirical transition state calculation using BEBO and LEPS surfaces; the dotted line is the *ab initio* calculation with a tunnelling correction.

sort helped to establish the general principles we discussed on p. 106. The equations of motion derived from classical mechanics are generally employed and these dynamical equations (essentially Newton's laws of motion) are solved for particular initial conditions. Thus, if we confine ourselves to the *collinear* reaction of A + BC, our initial conditions are: (1) the velocity of A relative to BC, i.e. the relative translational energy of A and BC, (E_t); (2) the vibrational energy of BC (E_v); (3) the phase of the vibration (i.e. the specific bond extension at the start of the calculations); (4) a starting separation between A and BC generally chosen to be sufficiently large that there is little interaction between the reactants.

Since we are confining our attentions to motion in one dimension, the rotational motion of the colliding molecules and the bending vibration of the complex are not considered.

The positions and momenta of the atoms are then found as a function of time. Fig. 4.15 shows an example of the outcome of one such calculation for H + H_2. The calculation shows that, for this surface, there is a simple exchange of atom A

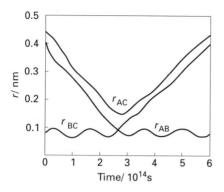

Fig. 4.15 Typical direct reactive classical trajectory for an H + H_2 collision. (Modification of Karplus, Porter and Sharma 1965).

for atom C and the mechanism is an impulsive one (the atoms do not stick together for many vibrational periods of the collision complex).

Of course not all collisions will lead to reaction and this will be evident from the final relative positions we compute for A, B and C. We can therefore study the effect on the reaction probability (the number of collisions leading to reaction divided by the total number considered) of: (1) the relative translational energy E_t, so obtaining the threshold for reaction and the dependence of the probability on relative translational energy above this threshold; (2) the vibrational energy by varying E_v. Thus we can determine the relative abilities of vibrational and translational energies to overcome the energy barrier for a particular surface.

In addition, by examining the motion of the products as they go down the exit valley, we can evaluate the partitioning of the available energy between the product modes.

Although we have restricted our discussion to a collinear reaction, this is not necessary but we then have to consider additional variables (e.g. the impact parameter, b, and the rotational energy of BC). Dynamical calculations have been an important component in the interpretation of the experimental results obtained by molecular beams and infrared chemiluminescence. By using simple empirical potentials, the effect on the dynamics of modifying the shape of the potential can be readily studied. The increasing availability of *ab initio* potentials has led to their use in dynamical calculations so that the dynamics of a specific reaction may be examined on a realistic surface. The approach here is generally to fit the surface to an approximate analytic form before performing the dynamics calculations.

Study notes
1 The calculations need not be confined to purely classical treatments and full quantum mechanical calculations have been performed for H + H_2. In addition, many approximate quantum calculations have been made for more complex systems.
2 Molecular dynamics calculations may also be used to calculate a thermal rate coefficient by studying the probability of reaction of a great number of collisions in which the initial conditions are determined by a so-called Monte Carlo sampling procedure, which ensures that they are drawn at random, but in a correctly weighted manner, from the relevant distribution e.g. a Boltzmann distribution. Examples may be found in Sections A 2.4 and A 2.6 of *Modern Gas Kinetics* (Pilling, M. J. and Smith, I. W. M. (ed.) (1987). *Modern Gas Kinetics.* Blackwell, Oxford).

4.6.3 Dynamics calculations vs. transition state theory

It is very difficult in most cases, if not impossible, to test transition state theory experimentally, because we are ignorant of the structure of the activated complex. We can, however, compare the results of dynamics calculations with the predictions of transition state theory *for the same surface.* We could calculate $k(T)$ as described in study note 2. above, but a much easier and more instructive way to carry out the calculations is to take a specific total energy, E, and to carry out the dynamics calculations at this total energy. It is also possible to formulate transition state theory so that it estimates $Pr^T(E)$, the probability of reaction at total energy E. This formulation is known as microcanonical transition state theory, as distinct from the canonical treatment we derived in Chapter 3 which gives us $k(T)$ (i.e. k as a function of temperature). For a collinear classical collision between A and BC it turns out that $Pr^T(E)$, is very simple:

$$Pr^T(E) = (\nu_{BC}/\nu_{ABC})\{(E - E_c)/E\}, \qquad (E\ 19)$$

where ν_{BC} and ν_{ABC} are the frequencies of the molecule BC and the complex and E_c is the barrier height. ν_{ABC} and E_c are determined directly from the surface, the superscript T signifies that we have calculated the reaction probability using TST. Figure 4.16 shows the surfaces for two reactions, H + H$_2$ and H + HBr, and plots of $Pr^T(E)$ vs. $(E - E_c)$ determined from TST via eqn (19), are shown as the bold solid line in Fig. 4.17. Also included are the results of dynamics calculations . These are shown in two ways, $P_{TRANS}(E)$ (light solid line) which is the probability of reaching the transition state at energy E; according to TST, all collisions which satisfy this requirement go on to form products and, $Pr(E)$ (broken line) which is the probability of forming the products. The overall reaction probability $Pr(E)$ is less than $P_{TRANS}(E)$ or $Pr^T(E)$ if some of the trajectories go through the transition state towards the products and then turn round and return to the reactants.

Figure 4.17(a) shows very good agreement between all three estimates up to quite high energies for H + H$_2$ demonstrating that the assumptions of TST are valid for thermal reactions, at least for our very simple model. Fig. 4.17(b) shows much larger deviations, even at low energies and TST, as we have formulated it,

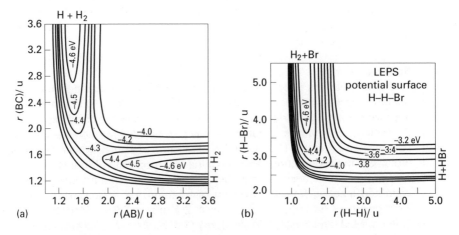

Fig. 4.16 Potential energy surfaces for the (a) H + H$_2$ and (b) H + HBr reactions.

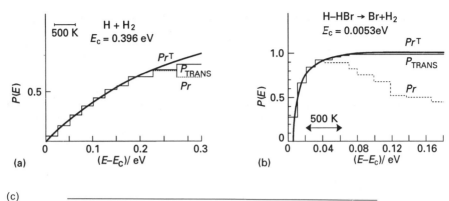

Fig. 4.17 Comparison of TST and dynamics calculations for the (a) H + H$_2$ and (b) H + HBr reactions. $Pr(E)$ (- - -) reaction probability (from dynamics calculations), $Pr^T(E)$ (—) probability of reaction from TST, $P_{TRANS}(E)$ (—) probability of forming the transition state. The table in Fig. 4.17(c) shows the effective transmission coefficient, κ, for the H + HBr reaction.

(c)

Excess total energy (eV)[a]	Effective transmission Coefficient for H + HBr reaction $\kappa(E)$
0.035	1.00
0.070	0.84
0.105	0.68
0.180	0.46

[a] Total energy—barrier height.

overestimates $Pr(E)$. It is very important for our subsequent discussion to examine the system more closely to see just why this failure has occurred. Note that $P_{TRANS}(E)$ and $Pr^T(E)$ agree very well, but that $Pr(E)$ falls as E increases. The reason for this is very easy to appreciate from the potential energy surface: the saddle point is located in the entrance valley (Fig. 4.16(b)) and, as we increase the energy, it becomes increasingly likely that the reactants will fail to turn the corner into the product valley and instead will run up the wall at the end of the entrance valley and return through the transition state to the reactants. Thus TST is failing and is overestimating the reaction probability because not all the collisions that go through the transition state go on to form products. This is the dynamical basis for the transmission coefficient, κ, that was rather arbitrarily introduced in Chapter 3 and transmission coefficients have been calculated as a function of excess of reagent energy for the H + HBr reaction (Fig. 4.17c).

From the significant discrepancy for the H + HBr reaction we might think the introduction of the transmission coefficient (which we can only calculate if we have a complete PES) is a yet further nail in the coffin of quantitative TST predictions. However, in general activation barriers are high in comparison to the thermal energy of the reagents; most reactions only just creep over the barrier and are therefore able to adjust to the topographical features of the PES and turn the corner to products. Only in molecular beam, or similar experiments, where we are able to give the reagents significant energy over and above the energy barrier, are these effects likely to be observed.

4.7 Variational transition state theory

The results of the last section contain a very important conclusion. The estimate of $Pr^T(E)$ made by TST provides an *upper limit* for the true classical probability. The quality of this limit (i.e. how close it comes to the true probability) depends on the collision dynamics of the reacting system i.e. on the shape of the potential energy surface (and also, it turns out, on the masses of the reactants see Box 4.3). Provided no trajectory goes through the transition state and then returns to the reactants, microcanonical transition state theory is exact. If we turn to the canonical formulation which seeks to calculate k at a given temperature, then there is the additional constraint that the reactants must be maintained in a Boltzmann distribution. This is generally the case, because in bulk experiments the reactants undergo many collisions with the bath gas which maintains this Boltzmann distribution. It is possible that, in very fast reactions, molecules in high energy states react more rapidly than these states can be replenished by collision, but this is comparatively rare. Thus provided (1) no trajectory goes through the transition state and then returns through it to the reactants and (2) the *reactants* are maintained in a Boltzmann distribution, then TST is exact, at least for a reaction following classical dynamics. Note that there is no need to assume that the transition state is in an equilibrium distribution.

Since these two conditions are not always fulfilled and since the TST estimate is an upper limit, so called variational transition state theory has been developed in which the position of the transition state is varied until the minimum value of $Pr^T(E)$ is obtained. In canonical variational transition state theory, where $k(T)$ is minimized, this amounts to varying the position of the transition state until the maximum value of $\Delta G^{\ddagger}$ is found.

Table 4.5 compares the rate coefficients estimated by TST, variational TST (VTST) and exact classical dynamics for the reactions $H + H_2$ and $F + H_2$, as a function of temperature. Note that VTST always provides an equal or a better estimate of the rate constant than TST, but it too may overestimate k, especially at high temperatures, where high energies are involved. In addition, the quality of the estimates varies from reaction to reaction.

4.7.1 Adiabatic theory of reactions

An alternative approach to VTST is to require that the reaction system behaves adiabatically. We can illustrate this theory best by reference, once again, to a collinear reaction between A and BC and to Fig. 4.18. Because of the collinear constraint, we do not have to worry about the bending vibration or of rotation of

Table 4.5 A comparison of rate coefficient estimates by TST (k_{TST}), VTST (k_{VTST}) and exact, classical dynamics (k_c)

	$H + H_2 \to H_2 + H$		$F + H_2 \to HF + H$	
T/K	k_{TST}/k_c	k_{VTST}/k_c	k_{TST}/k_c	k_{VTST}/k_c
300	1.00	1.00	1.02	1.00
1000	1.11	1.03	1.22	1.16
4000	1.89	1.32	3.14	2.68

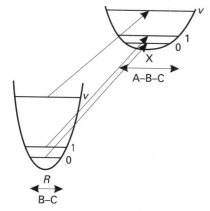

Fig. 4.18 Correlation diagram for a vibrationally adiabatic reaction. The figure shows the potential energy wells for vibration of BC and for the symmetric stretching vibration of ABC.

BC. The adiabatic theory assumes that motion along the reaction coordinate can be separated from motion perpendicular to it. The latter motion corresponds to the vibration of BC for large r_{AB} and to the symmetric stretching vibration of the complex in the transition state. Stated another way, the theory requires that if the molecule BC starts in quantum state v, then it stays in this quantum state as it approaches A. The transition state is formed in state v of the symmetric stretching vibration. As a result, we can construct a new potential energy surface for each quantum state simply by adding the vibrational energy to the electronic potential energy. If the reaction behaves adiabatically, then the representative point, describing the motion of the reacting system in that quantum state, will move along the reaction coordinate to the saddle point for that specific surface. If the shape of the trajectory is such that it does not stick to the reaction coordinate, but develops a component of motion perpendicular to it, then the reaction is not behaving adiabatically and the theory becomes inapplicable. Direct evidence that this condition is not being met comes from experimental studies of state-to-state dynamics. Adiabatic reaction theory can, of course, be extended to non-collinear systems and to more than three atoms. Interestingly, it is identical to microcanonical VTST, despite its very different approach. Like VTST it provides an improved estimate of the rate coefficient over conventional TST, but it too can fail for exactly the same dynamical reasons.

Study notes
This topic is of active research interest and many of the advances have been made over the past few years. It is difficult, therefore, to find readable review articles. Some feel for the subject may be found by a selective reading of one of the seminal papers in the field; Garrett, B. C. and Truhlar, D. G. (1979), Generalized transition state theory. *Journal of Physical Chemistry*, **83**, 1052–79. Section I, the Introduction, may be read with profit, but sections II–V contain theoretical development and require a more detailed knowledge of rate theory than the reader is likely to have. Section VI contains the results of tests of different theories and the diagrams and tables in this section are of interest; the commentary on them may also be read. The diagrams show plots of $N_c(E,s)$ vs s and of $\Delta\Gamma$ vs. s, where s is the reaction coordinate, $N_c(E,s)$ is effectively the number of ways of arranging the energy E, amongst the various modes of motion of the reaction system. The microcanonical variational transition state is located at the minimum value of $N_c(E,s)$. Note that at

low energies the minimum is at the saddle point ($s = 0$) but, as the energy increases, multiple minima arise. ΔG is the free energy of the system and in canonical VTST, the transition state is located at the maximum. This is generally close to the saddle point except at very high temperatures. In the tables, $k_c^*(T,D)$ is the TST estimate of the rate constant, $k_c^{CVT}(T,D)$ the canonical VTST estimate and $k_c(T,D)$ the exact classical value. $k_c^*(T,D)$ is the microcanonical VTST estimate obtained by evaluating $Pr(E)$ using variational theory for each energy and then averaging over a Boltzmann distribution. In this theory, the position of the transition state is allowed to vary with energy. $k_c^{CVT}(T,D)$ provides estimates of the rate coefficients closest to the exact values.

Child (Child, M. S. (1987). *Modern Gas Kinetics* (ed. M. J. Pilling and I. W. M. Smith), Chapter A1. Blackwell, Oxford) provides an interesting and readable article which gives a modern view of transition state theory. He discusses some general background additional to that presented above, but also discusses the idea of periodic orbit dividing surfaces (PODS), which are a means of defining and locating the transition state. Some background in simple classical mechanics is of advantage when reading this article.

4.8 Questions

4.1   Molecular beam experiments show that the reaction

$$Cs + Br_2 \rightarrow CsBr + Br \qquad (R\ 14)$$

proceeds via a stripping or harpoon mechanism with a reaction cross-section of 150 Å. Explain this statement, especially the meaning of the terms: stripping mechanism and reaction cross-section. Given that the ionization energy of Cs is 3.9 eV and the electron affinity of Br_2 is 1.8 eV, confirm that an electron-jump mechanism is consistent with the given value of the reaction cross-section ($\varepsilon_0 = 8.854 \times 10^{-12}\ J^{-1}\ C^2\ m^{-1}$, $e = 1.602 \times 10^{-19}\ C$).

4.2   Calculate the velocity of a hydrogen and deuterium atom following the photolysis of HBr or DBr at 193 nm. The bond energy $D(H–Br)$ is 366 kJ mol^{-1}; assume that all the bromine atoms are produced in their lowest electronic state and that $D(D–Br) = D(H–Br)$. Which of the noble gases would be most effective in relaxing the translationally excited H atoms?

4.3   The terminal velocity of a supersonic beam, v_t, is given by, $v_t = \alpha\{\gamma/(\gamma - 1)\}^{\frac{1}{2}}$, where $\alpha = (2k_B T_0/m)^{\frac{1}{2}}$ and γ is the constant pressure, constant volume specific heat ratio. Calculate the average translational energy for a beam of H_2 molecules expanded from a nozzle at 300 K. [$m_H = 1.67 \times 10^{-27}$ kg, $k_B = 1.38 \times 10^{-23}$ J K^{-1}]. Calculate the average energy of a small fraction of ethene molecules seeded into the beam.

4.4   The reaction surfaces in Fig. 4.19 show typical potential energy surfaces for two types of 'A + BC' reactions. How might one measure the final product state distributions? What factors determine the relative applicability of each of the methods you have chosen? Describe the qualitative differences you would expect in the vibrational state distributions from reactions over surfaces (a) and (b).

4.5   Outline two possible mechanisms for reaction (15), one of these will share a common intermediate with reaction (16)

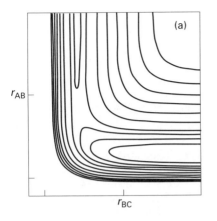 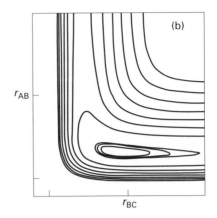

Fig. 4.19 Potential energy surfaces for question 4.4.

$$H + CH_2CF_3 \rightarrow HF(v) + CF_2CH_2$$
$$\Delta H = -331 \text{ kJ mol}^{-1} \tag{R 15}$$

$$CF_3 + CH_3 \rightarrow HF(v) + CF_2CH_2$$
$$\Delta H = -271 \text{ kJ mol}^{-1}. \tag{R 16}$$

What qualitative vibrational distributions might you expect from either of your proposed mechanisms? From the following vibrational distributions calculate the fraction of energy appearing as HF vibrational energy for the reactions (take the average energy of each vibrational level of HF to be 3900 cm^{-1}):

	$P(v=0)$	$P(v=1)$	$P(v=2)$	$P(v=3)$	$P(v=4)$
H + CF$_3$CH$_2$	0.430	0.302	0.182	0.068	0.017
CF$_3$ + CH$_3$	0.471	0.331	0.150	0.041	0.011

From the shape of the vibrational distributions and the fraction of energy channelled into vibration, which mechanism do you think is applicable.

4.6 Account for the following product distribution map of DF produced from the reaction of F + D$_2$ studied in a crossed molecular beam apparatus (Neumark *et al.* 1985), at a collision energy of 1.82 kcal mol^{-1} (Fig. 4.20).

(a) Explain qualitatively the observed distribution and relate this to the mechanism of the reaction. Why does the contour map contain several peaks? Qualitatively sketch out the distribution of the vibrational energy within the DF product.

(b) What is the exothermicity of the reaction? What is the maximum vibrational level of DF which can be accessed and HF for the isotopic variant F + H$_2$?

(c) The dashed circles represent the maximum translational energy release available from each vibrational level. From the observed distribution in each level what can you say about the rotational excitation of the molecule?

(d) The relative populations of DF from the reaction are:

Vibrational level	0	1	2	3	4
Relative population	0	0.19	0.67	1.00	0.41

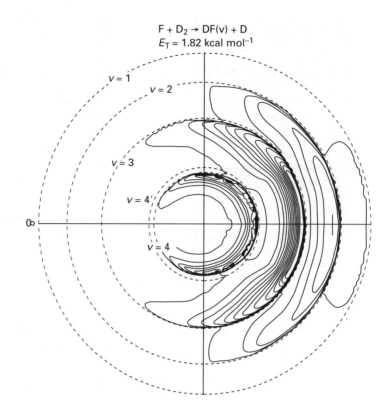

Fig. 4.20 Centre-of-mass flux-velocity contour map for the F + D$_2$ reaction at E$_T$ = 1.82 kcal mol^{-1}. (From Levine and Bernstein, 1987)

Calculate the average fraction of energy channelled into vibration. Compare your answer to that from question 4.5. (ΔH_f (F) = 18.36 kcal, ΔH_f (HF) = −65.13 kcal, ΔH_f (H) = 51.6 kcal, v(DF) = 2800 cm^{-1}, v(HF) = 3900 cm^{-1}).

4.7   Calculate the skewing angles for the reactions F + H$_2$; F + D$_2$; H + H$_2$; H + Cl$_2$; Cl + HI; (RMM F = 19, H = 1, D = 2, Cl = 35, I = 127).

References

Brooks, P. R. (1976). Reactions of oriented molecules. *Science*, **193**, 11–17.
Hirst, D. M. (1990). *A computational approach to chemistry*. Blackwells, Oxford.
Karplus, M., Porter, R. N., and Sharma, R. D. (1965). Exchange reactions with activation energy. I Simple barrier potential for (H, H$_2$). *Journal of Chemical Physics*, **43**, 3259.
Mayotte, D. H., Polanyi, J., and Woodall, K. B. (1972). Energy distribution among reaction products. *Journal of Chemical Physics*, **57**, 1547–60.
Neumark, D. M., Wodtke, A. M., Robinson, G. N., Hayden, C. C., Shobatake, K., Sparks, R. K., *et al.* (1985). Molecular beam studies of F + D$_2$ and F + HD reactions. *Journal of Chemical Physics*, **82**, 3067–77.

Unimolecular and association reactions

5.1 Introduction—The Lindemann theory

At high pressures, some reactions, such as the decomposition of azomethane:

$$CH_3N_2CH_3 \rightarrow C_2H_6 + N_2 \qquad\qquad (R\ 1)$$

show first order kinetics, i.e the rate of reaction is proportional to the concentration of azomethane to the first power:

$$d[CH_3N_2CH_3]/dt = -k_1[CH_3N_2CH_3]. \qquad\qquad (E\ 1)$$

For a long time the mechanism of these reactions was a mystery. The rate increases with temperature, showing that the reactant must surmount an energy barrier, but the first-order kinetics apparently preclude collisional activation. In the 1920s Lindemann proposed the following kinetic scheme, which explained this apparent dichotomy and predicted that the rate coefficient should decrease with pressure, with the reaction eventually becoming second order overall (Fig. 5.1), a fact which was experimentally verified by Ramsperger in 1927.

$$A + M \underset{k_{-2}}{\overset{k_2}{\rightleftarrows}} A^* + M \qquad\qquad (R\ 2, -2)$$

$$A^* \overset{k_3}{\rightarrow} P. \qquad\qquad (R\ 3)$$

A is a reactant molecule, which can be excited on collision with M, which is either another A molecule or an added diluent (or bath) gas molecule, to form an *energized* molecule A^*. A^* may either be collisionally de-activated, or react to form products. Note that the first steps in the mechanism (R 2 and R −2) are second order, the second step (R 3) is first order. The overall rate of formation of the products $(d[P]/dt)$ is equal to $k_3[A^*]$ and to evaluate this quantity we must find some way of calculating $[A^*]$. The differential equation for the rate of formation and removal of A^*:

$$d[A^*]/dt = \underbrace{k_2[A][M]}_{\text{formation of } A^*} - \underbrace{k_{-2}[A^*][M] - k_3[A^*]}_{\text{removal of } A^*} \qquad\qquad (E\ 2)$$

is difficult to solve exactly, and we utilize the *steady-state approximation* (see Section 1.8.2 and Chapter 8) to solve eqn (2). A^* has a short lifetime, and unlike the products does not accumulate during the reaction. After a short build-up period A^* remains at a steady concentration, i.e. it is consumed at the same rate at which it is produced or $d[A^*]/dt = 0$. Thus:

$$k_2[A][M] = k_{-2}[A^*][M] + k_3[A^*] \qquad\qquad (E\ 3)$$

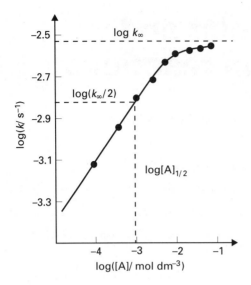

Fig. 5.1 The fall off in the unimolecular rate coefficient, k, with decreasing pressure, for the decomposition of azomethane at 603 K.

and

$$[A^*] = k_2[A][M]/(k_{-2}[M] + k_3) \tag{E 4}$$

therefore

$$d[P]/dt = k_3[A^*] = k_3k_2[A][M]/(k_{-2}[M] + k_3). \tag{E 5}$$

If we write $d[P]/dt = k_{uni}[A]$, where k_{uni} is the formal unimolecular rate 'constant' (the inverted commas indicating that the word constant is singularly inappropriate!), we see that

$$k_{uni} = k_3k_2[M]/(k_{-2}[M] + k_3) \tag{E 6}$$

and that k_{uni} is a function of pressure. Figure 5.1 indicates that there are two extreme pressure dependencies of k_{uni}: at high pressure k_{uni} is independent of pressure and at low pressures k_{uni} is directly proportional to the total pressure. By making the appropriate substitutions into eqn (6) we can appreciate the physical basis behind these extremes of behaviour.

At high pressures $k_{-2}[M] \gg k_3$ i.e. collisional deactivation is much faster than the unimolecular reaction of A*. Collisions maintain an equilibrium, (Boltzmann) distribution of A throughout its energy levels, including those represented by A*. k_{uni} reduces to k_3k_2/k_{-2} i.e. it becomes independent of pressure. The rate-determining step is the first order reaction of A* and the overall kinetics are also first order (but note that the rate coefficient is not simply k_3).

Conversely, at low pressures, the rate determining step is bimolecular excitation. Once the activated molecule has formed it is more likely to react rather than be deactivated, therefore $k_{-2}[M] \ll k_3$ and eqn (6) now reduces to

$$k_{uni} = k_2[M]$$

and the overall rate coefficient is pressure dependent.

5.2 Comparison of the Lindemann theory with experimental data

The Lindemann model explains qualitatively the behaviour shown in Fig. 5.1. We must now examine its ability to model unimolecular reactions quantitatively.

5.2.1 Comparison of experimental and calculated values of $[M]_{\frac{1}{2}}$

We may rewrite eqn (6) as

$$k_{uni} = k_\infty / \{1 + (k_3/k_{-2}[M])\} \qquad \text{(E 7)}$$

where the rate coefficient at high (or infinite) pressure, $k_\infty = k_3 k_2 / k_{-2}$. If we define $[M]_{\frac{1}{2}}$ as the third body concentration at which the experimental rate coefficient falls to half of its high pressure value then

$$[M]_{\frac{1}{2}} = k_\infty / k_2. \qquad \text{(E 8)}$$

If we further assume the collision theory rate coefficient for activation (see Chapter 3 p. 61),

$$k_2 = \mathcal{Z} \exp(-E/RT) \qquad \text{(E 9)}$$

where E is the activation energy and $\mathcal{Z}$ the collision frequency factor, then we can also estimate $[M]_{\frac{1}{2}}$ from the experimental values of k_∞. Table 5.1 shows rate data for several unimolecular reactions and compares the calculated values of $[M]_{\frac{1}{2}}$ with those obtained directly for 'fall-off' plots such as Fig. 5.1. For all the cases shown, $[M]_{\frac{1}{2}}^{expt}$ is less than $[M]_{\frac{1}{2}}^{calc}$ and the discrepancy gets worse as the reactant molecule gets larger. Accepting, as we must, that $[M]_{\frac{1}{2}}^{expt}$ is correct, we see that we must have underestimated k_2, in some cases by up to ten orders of magnitude! Our previous discussions on theories of bimolecular reactions should have warned us about the simplistic application of collision theory and as we shall see k_2 turns out to be significantly larger than $\mathcal{Z} \exp(-E/RT)$.

5.2.2 Extrapolation to infinite pressure

Further rearrangement of eqn (7) (substituting k_∞/k_2 for k_3/k_{-2} and inverting) gives eqn (10)

Table 5.1 Rate parameters for unimolecular reactions, showing the discrepancy between experimental and calculated values of $[M]_{\frac{1}{2}}$

Reaction	A /s^{-1}	E/kJ mol^{-1}	$[M]_{\frac{1}{2}}^{expt}/$ $mol\ dm^{-3}$	$[M]_{\frac{1}{2}}^{calc}/$ $mol\ dm^{-3}$	T/K
Cyclopropane → propene	3×10^{15}	275	3×10^{-4}	1.5×10^4	760
Cyclobutane → 2 ethenes	4×10^{15}	267	10^{-5}	2×10^4	720
Methylcyclobutane → ethene + propene	2.5×10^{15}	256	10^{-6}	10^4	670
$CH_3NC \rightarrow CH_3CN$	4×10^{13}	161	4×10^{-3}	2×10^2	500
$C_2H_5NC \rightarrow C_2H_5CN$	6×10^{13}	160	3.5×10^{-5}	3.5×10^2	500
$N_2O \rightarrow N_2 + O$	8×10^{11}	256	0.8	4	890

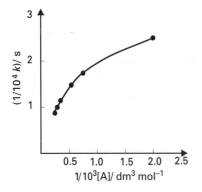

Fig. 5.2 Plot of $1/k$ versus $1/[A]$ for the reaction *trans*-$C_2H_2D_2 \rightarrow$ *cis*-$C_2H_2D_2$, at 794 K, showing the departure from linearity.

$$1/k_{uni} = 1/k_2[M] + 1/k_\infty \qquad (E\ 10)$$

Thus a plot of $1/k_{uni}$ vs. $1/[M]$ should be a straight line and we may obtain k_∞ as the reciprocal of the intercept on extrapolating to $1/[M] = 0$. Figure 5.2 shows that rate coefficient for the *cis–trans* isomerization of $C_2H_2D_2$ treated in this way: it demonstrates that the plot is far from linear and that there is therefore a further problem with the simple Lindemann mechanism.

5.3 Further developments

In the next few sections we shall examine the failures of the Lindemann mechanism, as such a critique gives an insight into the physical processes involved in unimolecular reactions, and develop a somewhat simplified method of dealing with them. The study of unimolecular reactions is not just of mechanistic interest. Pressure dependent unimolecular and termolecular (recombination) reactions play important roles in phenomena as diverse as ion–molecule chemistry of the upper atmosphere, combustion and high pressure detonation. The ability to predict the rate coefficient as a function of pressure and temperature is of obvious practical importance.

5.3.1 Rate of activation

The first failure of the simple Lindemann mechanism arose because of an underestimation of the rate of activation. We can trace this back to a complete neglect of the internal degrees of freedom of the molecule, and in particular of its vibrational modes. To a first approximation, a diatomic molecule behaves as a harmonic oscillator with energy levels:

$$E_v = (v + \tfrac{1}{2})h\nu \qquad (E\ 11)$$

where ν is the vibrational frequency and v the quantum number. The levels are evenly spaced and non-degenerate. The fraction of molecules in the vth level is

$$n_v/n = \exp(-v h\nu/k_B T)/Q \qquad (E\ 12)$$

where Q is the vibrational partition function, $(1 - \exp(-h\nu/k_B T))^{-1}$, n is the total number of molecules, and n_v the number of molecules, in the vth level.

The situation is more complex for a polyatomic molecule. In order to simplify the problem, let us first consider a hypothetical molecule with s equivalent harmonic oscillators, all of frequency v. We now find that there are several ways of distributing a given number of quanta, v, amongst the various oscillators. For example with two oscillators, we can distribute two quanta in three ways: $(2,0)$, $(1,1)$, $(0,2)$. The general expression for the number of ways of distributing v quanta amongst two oscillators is $v + 1$ and the degeneracy, g_v of the vth level (i.e. the number of distinguishable states at that vibrational energy) will thus be $(v + 1)$. For a molecule with s oscillators, then

$$g_v = (v + s - 1)!/(v!(s - 1)!).$$ (E 13)

As in the two oscillator case, g_v corresponds to the number of distinct states at total energy $(v + \frac{1}{2})hv$. Thus n_v/n now becomes

$$n_v/n = g_v \exp(-vhv/k_B T)/Q$$ (E 14)

where

$$Q = (1 - \exp(-hv/k_B T))^{-s}.$$ (E 15)

We now make the so-called strong collision assumption in which it is assumed that the state of the molecule after a collision is totally uncorrelated with its state before. Since collisions promote equilibrium, the relative probability of forming state v in a collision is simply given by $n_v/n = g_v \exp(-vhv/k_B T)/Q$, the ratio of the contribution of state v to the Boltzmann distribution, to the contribution of all states. Furthermore, since chemical energies are generally much higher than $k_B T$, the major contribution to Q comes from states below the critical energy, and the probability of deactivation on collision is effectively one. Thus $k_{-2} = Z$ and

$$k_2^v = Z g_v \exp(-vhv/k_B T)/Q$$ (E 16)

where k_2^v is the rate of activation to state v. The overall rate of activation (k_2) is obtained by summing k_2^v over all levels which can dissociate:

$$k_2 = \sum_{v = m}^{\infty} Z g_v \exp(-vhv/k_B T)/Q$$ (E 17)

where $mhv = E_0$, the critical energy the molecule needs to react. k_2 is the rate coefficient for collisional excitation to all levels above E_0. The energies involved are normally large, and $E_0 \gg hv$. Hinshelwood developed equations for the case where the energy levels may be assumed to be continuous, i.e. $k_B T \gg hv$. Equation (17) now becomes:

$$dk_2 = Z N(E) \exp(-E/k_B T) \, dE/Q$$ (E 18)

where $N(E) \, dE$ is the number of energy levels in the range $E - (E + dE)$ and $N(E)$ is called the *density of states* and dk_2 is the rate of activation into this energy range. The total rate of activation may be found by integrating eqn (18) over all energies greater than E_0, whence

$$k_2 = Z/(s - 1)!(E_0/k_B T)^{s-1} \exp(-E_0/k_B T)$$ (E 19)

Equation (19) differs from eqn (9) by the introduction of the factor $1/(s - 1)!(E_0/k_B T)^{s-1}$. Since $E_0 \gg kT$ this term is very much greater than unity and leads to an increased theoretical value of k_2. Furthermore, this increase is more pronounced for larger molecules, which have more oscillators. This is exactly what we need to overcome the first failure of the Lindemann theory, since a larger value of k_2

reduces the calculated value of $[M]_{\frac{1}{2}}$, bringing it more into line with experiment. This modification stems entirely from our recognition of the large density of vibrational states which exist in energized polyatomic molecules, and which arise from the degeneracy terms of eqn (13).

5.3.2 Rate of dissociation

We now turn to the second failure of Lindemann theory, and to an examination of k_3, the rate of dissociation. A unimolecular reaction specifically involves one particular form of motion, e.g. rotation about the double bond for a *cis–trans* isomerization. A minimum amount of energy must be located in this form of motion before reaction can take place. So far, we have considered only general excitation of the molecule. We must now examine the conversion of our generally excited molecule, A* (the *energized molecule*) into the specifically excited molecule A (the *activated complex*), which has sufficient energy localized in the required degree of freedom, and is about to undergo conversion into products:

$$A* \xrightarrow{k_{3a}} A^{\ddagger} \xrightarrow{k^{\ddagger}} P \qquad\qquad (R\ 3a,\ 3b)$$

$k^{\ddagger}$ is of the order of a vibrational frequency, and as we shall see k_{3a} is generally much smaller than $k^{\ddagger}$. The conversion of A* to $A^{\ddagger}$ is thus rate determining, and the overall rate coefficient for conversion of A* to products (the k_3 of the Lindemann scheme) is equal to k_{3a}. Since $k_{3a} \ll k^{\ddagger}$, $[A^{\ddagger}]$ is very small, and we may apply the steady state approximation, $d[A^{\ddagger}]/dt = 0$ to obtain:

$$k_{3a} = k^{\ddagger}[A^{\ddagger}]/[A*]. \qquad\qquad (E\ 20)$$

The most successful early treatment of this aspect of the problem was initiated by Rice and Ramsperger, and later developed by Kassel (RRK theory), who assumed that energy can flow freely from one vibrational mode to another. Figure 5.3 shows the potential energy surface for a stable linear molecule ABA. It is similar to the surfaces of Chapter 3, except that there is a deep well in place of the saddle since now the triatomic molecule is stable. Motion along QR and ST corresponds to the symmetric and asymmetric vibrations. If the vibrations are harmonic, then once energy has been placed in a particular mode it stays there. Anharmonicity causes a slight curvature in the potential energy surface and classical motion initially in one mode degenerates into a Lissajous motion, with energy transferring rapidly from one mode to another. Anharmonicity may be said to couple the oscillators. The free flow assumption of RRK theory seems reasonable, since vibrations are highly anharmonic at chemical energies, but we shall look at the validity of this hypothesis later in the chapter.

For a molecule with s equivalent oscillators, the degeneracy of the vth vibrational level is: $(v + s - 1)!/v!(s - 1)!$ (E 13). Let us suppose that for dissociation we must locate at least m quanta in one particular mode ($mhv = E_0$). This simply reduces our choice of distributable quanta to $(v - m)$, and the total number of ways of arranging these is:

$$(v - m + s - 1)!/\{(v - m)!\ (s - 1)!\}. \qquad\qquad (E\ 21)$$

The probability, P_m^v of locating at least m quanta out of v in the dissociation mode is the ratio of these quantities, i.e. (E 21)/(E 13).

(a)

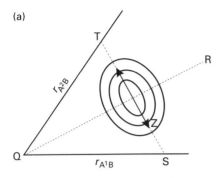

(b)

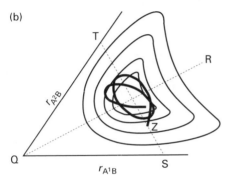

Fig. 5.3 Vibrational motion of a linear molecule for (a) a harmonic oscillator and (b) an anharmonic oscillator. Both diagrams show the motion of the molecule following initial displacement to the point Z, corresponding to initial excitation of the asymmetric stretching mode only (zero point energy is neglected).

$$P_m^v = (v - m + s - 1)!v!/[(v - m)!(v + s - 1)!]. \tag{E 22}$$

The quantum numbers involved are very large $(v, m \gg s)$, and $(v - m + s - 1)!/(v - m)!$ reduces to $(v - m)^{s-1}$ and similarly $v!/(v + s - 1)!$ reduces to $v^{-(s-1)}$, so that

$$P_m^v = (1 - m/v)^{s-1} \tag{E 23}$$

or

$$P_{Eo}^E = (1 - E_0/E)^{s-1} \tag{E 24}$$

since $E = vhv$ and $E_0 = mhv$. P_{Eo}^E is the probability of locating a minimum energy E_0 out of the total energy E, in the dissociation mode. If energy randomization takes place sufficiently rapidly for the vibrational energy to be distributed statistically, $[A^+]^E/[A*]^E = P_{Eo}^E$, and substituting into eqn (20) we find:

$$k_3(E) = k^+(1 - E_0/E)^{s-1}. \tag{E 25}$$

The bracketed E is used to indicate the dependence of k_3 on the energy of the molecule which is reacting, i.e. that we are considering only energized and activated molecules in the range $E - E + dE$. k_3 increases with energy because the probability of localizing a given amount of energy E_0 in one particular mode increases as E gets larger. The localization probability decreases as s increases (there are more modes to hide the energy in), with the result that k_3 is smaller for more complex molecules (in contrast to the rate of activation). Figure 5.4 shows a plot of $k_3(E)/k^+$ vs. E for several values of s. In the case of a single oscillator, $k_3(E) = k^+$ for all energies greater than E_0 (there is nowhere else for the energy to go except into the bond being broken). For polyatomic molecules, $k_3(E)$

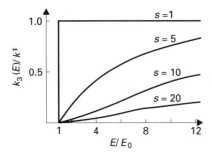

Fig. 5.4 Plot of $k_3(E)/k^{\ddagger}$ for the RRK model, showing the slow increase in $k_3(E)$ with energy for larger molecules.

approaches $k^{\ddagger}$ asymptotically, its rate of increase being slower for the larger values of s. Lindemann theory implicitly assumes $s = 1$.

5.4 Contributions to the rate of reaction

We have seen that the rate of dissociation is energy dependent and hence we need to reformulate the mechanism in the light of that discovery.

$$
\begin{array}{c}
dk_2(E) \\
A + M \rightleftarrows M + A^*(E) \\
k_{-2}
\end{array}
\tag{R 2}
$$

$$
\begin{array}{c}
k_3(E) \\
A^*(E) \rightarrow P
\end{array}
\tag{R 3}
$$

where $dk_2(E)$ is the rate of activation to the energy range $E - (E + dE)$. Solution of steady-state eqns similar to (E 3) but for each energy range yields (Box 5.1):

$$
k_{uni} = \int_{E_0}^{\infty} k_3(E)\,[M]\,dk_2(E)\,/\{\,k_{-2}[M] + k_3(E)\}
\tag{E 26}
$$

Box 5.1 Derivation of eqn (26)

Assuming a steady state for $[A^*(E)]$ then

$$
dk_2(E)[A][M] = k_{-2}[A^*(E)][M] + k_3(E)[A^*(E)].
\tag{E 27}
$$

$$
[A^*(E)] = dk_2(E)[A][M]/\{k_{-2}[M] + k_3(E)\}
\tag{E 28}
$$

As before $d[P(E)]/dt = k_3(E)[A^*(E)]$ so

$$
d[P(E)]/dt = k_3(E)dk_2(E)[A][M]/\{k_{-2}[M] + k_3(E)\}.
\tag{E 29}
$$

The unimolecular rate coefficient, k_{uni} for the energy range $E - E + dE$, will be

$$
k_{uni}(E) = k_3(E)\,dk_2(E)[M]/\{k_{-2}[M] + k_3(E)\}.
\tag{E 30}
$$

To obtain the thermal rate coefficient we need to integrate this expression over the energy range E_0 to ∞.

$$k_{uni} = \int_{E_0}^{\infty} k_3(E) \, dk_2(E)[M] / \{k_{-2}[M] + k_3(E)\}. \tag{E 31}$$

This equation can be rearranged by dividing top and bottom by $k_{-2}[M]$ to give

$$k_{uni} = \int_{E_0}^{\infty} k_3(E) \, dk_2(E)/k_{-2}\{1 + k_3(E)/k_{-2}[M]\}. \tag{E 32}$$

At high pressures $k_3(E)/k_{-2}[M]$ tends to zero and the high pressure formula for k_{uni} becomes $k_{uni}(E) = \int_{E_0}^{\infty} k_3(E) \, dk_2(E)/k_{-2}$. Note that term $d \, k_2(E)/k_{-2}$ is essentially an equilibrium constant, often referred to in the literature as $P(E)$ (where P in this case stands for a probability and not for products) and can be calculated from statistical mechanics.

The overall rate of reaction will be a product of the energy dependent rate of dissociation and the number of molecules with energy $E - (E + dE)k_3(E)n(E))$. Figure 5.5(a) shows the distribution of $n(E)$ as a function of energy E without any dissociation reaction to perturb the equilibrium (Boltzmann) distribution. Two extreme scenarios can exist for this distribution once we allow for unimolecular reactions. As long as the equilibrium distribution of reagents is maintained by collisions (high pressure), the shape of the distribution remains essentially identical (Fig. 5.5 (b)). (The shape of this distribution will remain identical throughout the reaction, although the absolute number of reagent molecules will obviously decrease.) Figure 5.6 shows how under these conditions the rate of reaction/energy distribution is a bell shaped curve; low at energies close to the critical energy (because of the small value of $k_3(E)$), rising with

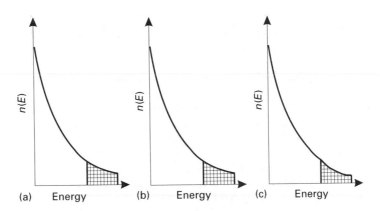

(a)   Energy   (b)   Energy   (c)   Energy

Fig. 5.5 (a) Boltzmann distribution of reactants with no reaction occurring. (b) Reaction is allowed to occur but under high pressure conditions such that the unperturbed Boltzmann distribution of reactants (although obviously not the concentration of reactants) remains the same. (c) Reaction occurs at low pressures such that reaction occurs faster than collisional excitation. The higher (reactive) energy levels of the reactant distribution are significantly depleted.

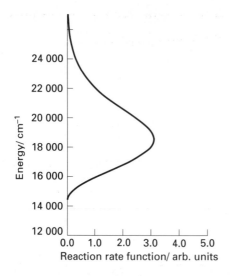

Fig. 5.6 Plot of reaction rate vs. reactant energy for the decomposition of the isopropyl radical. The reaction rate is the product of the energy dependent rate coefficient and the concentration of reactant. Although the rate coefficient increases with energy the relative concentration of reactant decreases, the net result being a peak in the rate approximately 3500 cm^{-1} ($\approx$ 42 kJ mol^{-1}) above the threshold energy for reaction.

energy as $k_3(E)$ increases, before decreasing due to the falling population of reactants at very high energy.

As the pressure falls the shape of the reactant distribution alters as collisions can no longer maintain the equilibrium (Fig 5.5(c)). The overall rate will decrease as only levels just above the dissociation limit (and hence with slow decomposition rates) are populated. As the pressure increases, higher, faster reacting levels are populated and the net rate of reaction increases more rapidly with pressure than would be expected on the basis of the low pressure behaviour, accounting for the negative curvature of the plot 5.2 and the second failure of the Lindemann mechanism.

5.5 Summary of the Lindemann mechanism and Hinshelwood–RRK modifications

The Lindemann mechanism provides a simplified representation of the physical processes which are involved in unimolecular reactions. All subsequent models are built on the foundation of second-order activation to an energized molecule, followed by a first-order reaction. The failures of the Lindemann mechanism are similar to those of collision theory; these simple theories take no account of real molecular structure.

In our derivations of the Hinshelwood–RRK formulation we have seen how firstly the rate of activation was modified to take account of the large number of states, $\mathcal{N}(E)\,\mathrm{d}E$, in the range $E - E + \mathrm{d}E$, which arise from the number of ways of distributing the energy throughout the molecule. $\mathcal{N}(E)$ is the *density of states* i.e. the total number of states per unit energy range. The rate coefficient for activation into this range is energy dependent. Secondly, the rate coefficient for dissociation is also energy dependent. In contrast to activation, the rate of dissociation increases with energy. Equally importantly we have seen how this rate coefficient can be treated in a statistical fashion; the rate of dissociation is proportional to the number of arrangements of the vibrational energy in which enough energy (E_0) is located in the critical mode, divided by the total number of arrangements. This ratio is dependent on the total energy in the energized molecule.

5.6 More recent developments

The Hinshelwood–RRK treatment of unimolecular reactions formed a very important developmental stage and it has served well as a means of understanding the basic concepts. Its approach originates from an era when computers were not available and gross approximations had to be made. The main two simplifications are the assumptions that all the oscillators have the same frequency and that this frequency is very small, so that a classical approximation can be made. As a result of these assumptions, the model is not totally successful. In particular it is usually necessary to reduce s by a factor of approximately two in order to get reasonable agreement with experiment. This error stems from the two main approximations both of which tend to overestimate $\mathcal{N}(E)$, the density of states, and we have to reduce the effective size of the molecule in order to compensate.

More recent developments, which date from the work of Marcus, rely on computational methods and can lead to very good agreement with experiment. We shall outline later some sources for additional reading; for the present we shall simply outline the general approach used in the so-called Rice–Ramsperger–Kassel–Marcus (RRKM) theory.

1.   The real frequencies of the molecule are used to evaluate $\mathcal{N}(E)$, using 'direct count' methods.

2.   The RRK approach is adapted and $k_3(E)$, which is equal to $k^{\neq}[A^{\neq}(E)]/[A^*(E)]$, becomes $W(E')/\mathcal{N}(E)h$, where $\mathcal{N}(E)$ is the density of states *of the reactant molecule* at energy E, h is Planck's constant and $W(E')$ is called the sum of states *at the activated complex* for energy E. E' is simply the energy that remains once the energy barrier has been surmounted (Fig. 5.7), thus $E' = E - E_0$, and $W(E')$ is the total number of ways in which this energy can be arranged in the complex. Once again $W(E')$ is determined by direct-count methods. An important point here is that we do not know the frequencies of the complex (but see below), and we are

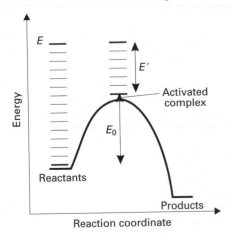

Fig. 5.7 Potential energy diagram for an isomerization reaction at total energy E. For reaction to occur a certain amount of energy E_0 must be located within the bonds to be broken, the remainder, E', is distributed throughout the activated complex.

left with an adjustment problem not dissimilar to that found in transition state theory.

3. Finally, account is taken of the conservation of angular momentum, which requires that as the molecule dissociates, it does not change its total angular momentum quantum number J. The activated complex is usually larger than the stable molecule, since at least one of the bonds extends as the molecule dissociates; thus the activated complex has larger moments of inertia ($I = \mu r^2$ for a diatomic molecule, bond length r and reduced mass μ) and the total rotational energy of the complex ($E_R = (h^2/8\pi^2 I)J(J+1)$ for a diatomic) is less than in the reactant molecule, since the molecule is unable to change J. Energy must be conserved and therefore the difference in rotational energy becomes available to help the molecule dissociate, i.e. it contributes to E'.

Study notes
An excellent and thorough account of RRKM theory may be found in Holbrook, K. A., Pilling, M. J., and Robertson, S. H. (1996) *Unimolecular Reactions*. Wiley, London. Sections 1–5 explain the basic problem (which is essentially that outlined in our discussion of the Hinshelwood–RRK model), set up a scheme which relates E' to E and then evaluates $dk_{-2}(E)/k_{-2}$ and $k_3(E)$ (denoted by $k_a(E^*)$). The method used to evaluate $k_3(E)$ may be understood on the basis of our discussion of transition state theory.

A more recent text is that of R. Gilbert and S. Smith (1990) *Theories of Unimolecular and Recombination Reactions*. Blackwells, Oxford. Chapters 1 and 2 may be read with profit and an excellent feature of this book is the number of exercises and answers provided by the authors. In addition, a set of computer programs are available to illustrate the calculations outlined in the text. Subsequent chapters give detailed accounts of the current status of all aspects of unimolecular reaction theory.

5.7 The high pressure limit

Under high pressure conditions we saw that an equilibrium distribution of reactants is maintained. This was one of our criteria for application of transition state theory and we should therefore expect the classic TST formula

$$k_\infty = \kappa(k_B T/h)(Q^{\ddagger}/Q)\exp(-E_0/k_B T) \tag{E 33}$$

to be applicable. Here $Q^{\ddagger}$ is the molecular partition function for the activated complex and Q that for the reactant molecule. At normal temperatures $k_B T/h \approx 10^{13}\,\text{s}^{-1}$ and since $Q^{\ddagger}$ and Q refer to the same molecule, albeit in different configurations, we might expect that, as a first approximation $Q^{\ddagger} = Q$, giving a limiting high pressure A factor, A^∞, of $\approx 10^{13}\,\text{s}^{-1}$. Some unimolecular reactions do indeed approximate to such behaviour. Some reactions do not and A factors can range from $10^{11} - 10^{16}\,\text{s}^{-1}$.

Abnormally large A factors may arise if there is a gross change of geometry on activation e.g cyclopropane $\rightarrow$ propene, which has an A^∞ of $10^{15.5}\,\text{s}^{-1}$, whilst simple bond fission reactions, in which the bond may be extended by a factor of two to three in the complex, also have large A factors (e.g. $C_2H_6 \rightarrow 2CH_3$, $A^\infty = 2 \times 10^{16}\,\text{s}^{-1}$) These large A factors arise because the activated complex has a much higher density of states than the reactant i.e. $Q^{\ddagger} > Q$, since it has a looser structure. The converse occurs for the reaction

Table 5.2 *A*-factors for *cis–trans* isomerization reactions

Reactant	Group	A_∞ / s^{-1}	$E/\text{kJ mol}^{-1}$	T/K
(H, COOH / H, COOH)	1	1.7×10^4	66	410–420
(H, CO₂CH₃ / H, CO₂CH₃)	1	1.3×10^5	111	540–650
(H, H / CN, phenyl)	2	6×10^{12}	179	550–610
(H, H / phenyl, phenyl)	2	4×10^{11}	192	580–650

$$CH_2{=}CHOC_2H_5 \rightarrow C_2H_4 + CH_3CHO \qquad (R\,4)$$

where $A_\infty = 10^{11.4}\,\text{s}^{-1}$. The reaction is thought to proceed via a cyclic activated complex.

An interesting series of *cis–trans* isomerization reactions is shown in Table 5.2. The reactions divide into two groups; (1) those with abnormally low A_∞ factors and low activation energies and (2) those with normal A_∞ factors and high activation energies. In *cis–trans* isomerizations the molecules generally require considerable activation, until they have sufficient energy to overcome the barrier to internal rotation; group (2) reactions conform to this model. The molecules in group (1) have low lying triplet states, in which the C–C bond has more single than double bond character, and in which free rotation can take place. The *cis* and *trans* forms consequently share the same triplet state. Reaction involves activation followed by (i) crossing from the *cis* singlet state to the triplet, (ii) rotation, and (iii) crossing from the triplet to the singlet state of the *trans* form. Singlet-triplet transitions have a low probability for molecules composed of light atoms (cf the absence of singlet-triplet transitions in the atomic spectrum of helium), and the group (1) reactions have a low transmission coefficient. We should therefore insert $\kappa \ll 1$ into eqn (33) to account for the abnormality of the group (1) isomerizations.

Since k_∞ can be calculated using transition state theory, it is also appropriate to use a thermodynamic formulation. Using the same sort of approach adopted in Chapter 3, by setting $k_\infty = (k_B T/h) K^{\ddagger}$, it may be shown

$$A_\infty = \kappa\ (k_B T/h)\ \text{e} \exp(\Delta S/R). \qquad (E\,34)$$

Box 5.2 Further evidence for the transition state.

An elegant experiment by the group of Moore and co-workers (Lovejoy, E. R., Kim, S. K., and Moore, C. B. (1992). Observation of transition state vibrational thresholds in the rate of dissociation of ketene. *Science*, **256**, 1541–2) has shown up confirmatory evidence for the validity of both RRKM theory and the transition state. The essential equation of RRKM is (p. 131)

$$k(E) = \text{W}(E')/\mathcal{N}(E)h. \qquad (E\,35)$$

At energies very close to the critical energy, $k(E)$ should behave in a stepwise manner as a function of (E). For example in a transition state with only one stable frequency of 2000 cm^{-1} (energy $\approx$20 kJ mol^{-1}) then W(E'), the sum of states in the activated complex, will increase by unity for every 2000 cm^{-1} increase in energy. If the activation barrier is large in comparison to this step size then $\mathcal{N}(E)$, the density of reactant states will remain approximately constant. Measurement of $k(E)$ will map out vibrational spectrum of the transition state.

Figure 5.8 shows a schematic diagram of the processes involved in the dissociation of ketene:

$$CH_2CO \rightarrow CH_2 + CO. \hspace{2cm} (R\ 5)$$

A very high resolution pulsed dye laser is used to excite the ketene molecule t the first excited singlet state, S$_1$. The molecule undergoes rapid intersystem crossing (see Chapter 12) to form the triplet state, T$_1$, just above the dissociation barrier. The frequency of the laser can be varied to change the energy of the triplet state.

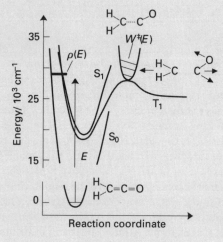

Fig. 5.8 A Schematic representation of the dissociation of triplet ketene. The reaction coordinate corresponds to the distance between the carbon atoms. The solid lines represent the potential energy for the ground singlet (S$_0$), the first excited singlet (S$_1$) and the first triplet (T$_1$) electronic states (see Chapter 12 p. 273). The transition state, which separates the highly vibrationally excited reactant from the products is depicted as a single potential well perpendicular to the reaction coordinate. This well is meant to schematically represent the eight bound vibrations of the transition state.

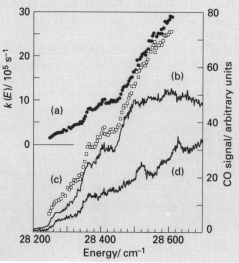

Fig. 5.9 Ketene dissociation rate cofficients and CO production rates. Curve (a) shows the rate coefficient for disappearance of ketene as a function of the energy of the exciting laser pulse. Note that this is not a smooth progression but shows a number of sharp steps. Curves (b) and (d) show data for the rate of production of CO in differential rotational levels, whilst curve (c) is a calculation of the predicted CO results. There is a very good correlation between the disappearance of ketene and the appearance of CO and between the experimental results and theoretical calculations.

The time taken for CO to appear (detected by a second pulsed laser) will give a value for the dissociation rate coefficient. Figure 5.9 shows that indeed $k(E)$ does increase in a stepwise fashion with increasing energy. Analysis of the energies at which the steps occur gives information on the vibrational frequencies of the transition state. The experimental results are in very good agreement with theoretical calculations.

5.8 The low pressure rate coefficient

The limiting low pressure rate coefficient, k_0, can be obtained from eqn (26) in the limit $[M] \rightarrow 0$:

$$k_0 = \int_{E_0}^{\infty} \mathrm{d}k_2(E)[M]. \tag{E 36a}$$

As was pointed out above, $\mathrm{d}k_2(E)/k_{-2} = P(E)$ is effectively an equilibrium constant and corresponds to the probability of locating a molecule in the energy range $E - E + \mathrm{d}E$ in a Boltzmann distribution. Thus

$$P(E) = \mathcal{N}(E)\exp(-E/k_B T)\,\mathrm{d}E/Q_v$$

where Q_v is the vibrational partition function for the reactant molecule. Setting $k_{-2} = Z$ (the strong collision assumption) we obtain

$$k_0 = \int_{E_0}^{\infty} [M]Z\mathcal{N}(E)\exp(-E/k_B T)\,\mathrm{d}E/Q_v. \tag{E 36b}$$

The rate coefficient is now independent of the structure of the activated complex and simply depends on the rate of activation to levels above the threshold energy. Thus for similar threshold energies, large molecules have larger values of k_0, because of their larger densities of states. The independence of k_0 from the structure of the transition state has made the comparison of calculated and experimental values of k_0 an excellent test bed for theory (see for example Pilling and Smith (1987) Chapter A4, Section A4.4).

5.9 The strong collision assumption

In some instances thermal unimolecular reactions confirm the validity of the strong collision assumption (i.e. that the energized molecule is completely deactivated by a single collision). This is especially so when the reaction is taking place without any diluent gas, i.e. the reactant is its own bath gas. However, comparisons of experimental data with RRKM calculations for monatomic or small polyatomic bath gases show quantitative discrepancies in the shape of the fall-off curve. These features can only be explained if the deactivation of the energized molecule occurs in a stepwise fashion. Fortunately we do not need to know the amount of energy removed in every collision (which will differ), rather, an average value of the energy transferred on each collision provides enough information to fit the experimental data.

Table 5.3 shows the experimental values for the root-mean-square amount of energy transferred from toluene per collision for a number of bath gases. In general this value depends more strongly on the structure of the bath gas than on the structure of the reactant. A number of qualitative trends can readily be observed.

Firstly, for monatomic bath gases $\langle\Delta E^2\rangle^{\frac{1}{2}}$ increases with atomic mass reflecting the degree of momentum transfer that can occur at each collision. Transfer of vibrational energy is more complicated than for translational energy which may be modelled on the basis of linear momentum exchange. Larger atoms such as xenon are thought to undergo 'sticky' collisions, the two collision partners remain in contact for several vibrational periods allowing for efficient energy transfer. Secondly $\langle\Delta E^2\rangle^{\frac{1}{2}}$ increases dramatically with the structure of the bath gas. Larger molecules tend to make less elastic collisions, the molecules are in contact for a longer period of time and the vibrational and rotational degrees of freedom in the bath gas are able to soak up the energy of the energized molecule. Finally $\langle\Delta E^2\rangle^{\frac{1}{2}}$ appears to show little temperature dependence, however, recent precise experiments on energy transfer have shown a small positive correlation of $\langle\Delta E^2\rangle^{\frac{1}{2}}$ with temperature.

Table 5.3 Variation of $\langle\Delta E^2\rangle^{\frac{1}{2}}$ with bath gas structure (data for toluene, from Hippler *et al.* (1983))

Bath gas	$\langle\Delta E^2\rangle^{\frac{1}{2}}/cm^{-1}$	Bath gas	$\langle\Delta E^2\rangle^{\frac{1}{2}}/cm^{-1}$
He	215	CO_2	535
Ar	305	H_2O	820
Xe	320	SF_6	705
H_2	245	CH_4	505
CO	350	C_3H_8	880
N_2	305	$C_{11}H_{24}$	2040

5.10 Vibrational energy redistribution

A basic assumption of both RRK and RRKM theories is that the vibrational energy flows freely from one mode to another so that statistical arguments may be employed. Just how valid is this assumption? A very revealing experiment, involving chemical activation, was devised to test this assumption. Addition of methylene to $2[^2H_2]$ hexafluorovinylcyclopropane produces an energized molecule which can dissociate in two ways as shown in Fig 5.10.

Route 1 involves dissociation close to the site of addition. In route 2, dissociation is well removed from this site. The energy of chemical activation must be extensively redistributed if it is to be effective in promoting reaction by route 2. The ratio 1:2 was measured mass spectrometrically as a function of pressure. The product molecules dissociate in the ion source of a mass spectrometer to give the substituted cyclopropyl ions, $C_3F_3H_2^+$ (mass 95) and $C_3F_3D_2^+$ (mass 97). At low pressures, the ratio of 97:95 was unity, indicating efficient energy randomization. At higher pressures the ratio began to increase. The collision frequency and hence the rate of stabilization increases with pressure, and at higher pressures any

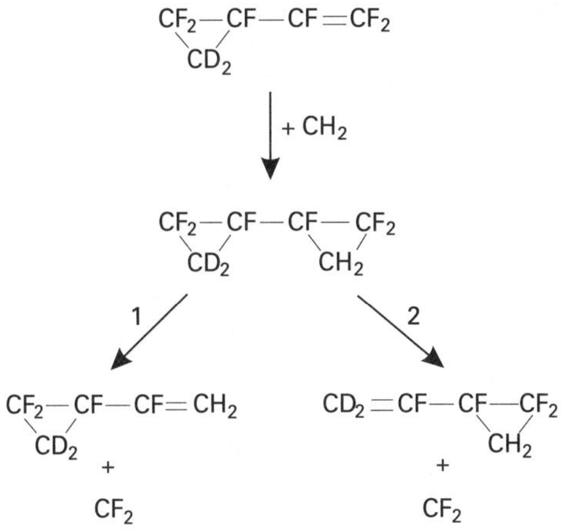

CF$_2$—CF—CF=CF$_2$
  CD$_2$

↓ + CH$_2$

CF$_2$—CF—CF—CF$_2$
  CD$_2$ CH$_2$

1 ↙   ↘ 2

CF$_2$—CF—CF=CH$_2$ CD$_2$=CF—CF—CF$_2$
  CD$_2$ CH$_2$
    +   +
  CF$_2$ CF$_2$

Fig. 5.10 Possible reaction schemes following the addition of methylene to $2[^2H_2]$hexafluorovinylcyclopropane.

dissociation products must originate from molecules decomposing very soon after the initial addition. As the pressure is raised, the time period during which dissociation can occur becomes smaller and smaller, and eventually energy randomization cannot occur. A full analysis of the results showed that the rate coefficient of energy randomization is $\approx 10^{12}$ s^{-1}, justifying, under normal conditions, the basic assumption of RRKM theory.

Study notes
The validity of the free energy flow hypothesis has been the subject of extensive experimental investigation and several interesting articles have been published:
1   Oref, I. and Rabinovitch, B. S. (1979) (Do highly excited reactive polyatomic molecules behave ergodically? *Accounts of Chemical Research*, **12**, 166–8) examine the evidence for so-called ergodic behaviour, which corresponds to free energy flow. The sections on chemical activation and photochemistry are particularly relevant to our earlier discussions.
2   Parmenter, C. S. (1982) (Vibrational energy flow within excited electronic states of large molecules. *Journal of Physical Chemistry* **86**, 1735–50) reviews elegant experiments in which the behaviour of excited electronic states of aromatic molecules is probed using high resolution (narrow band), tunable dye lasers. These enable the experimental elucidation of the differences in behaviour between molecules with only low levels of vibrational excitation (where $N(E)$ is small and the vibrational levels are essentially isolated so that free energy flow does not occur) and high excitation, where $N(E)$ is high and the behaviour ergodic. The first section of the paper discusses these ideas. The most striking experiments are described on pp. 1749–50. The preceding discussion is rather complex, but the basic ideas can be readily appreciated. When a molecule is excited by a narrow band laser, a small number of vibrational levels in the excited electronic state (S$_1$) are populated. If the level of vibrational excitation in S$_1$ is reasonably high, the resulting emission spectrum is broad and structureless, indicating that emission takes place from many vibrational states; other isoenergetic levels, not directly populated by absorption, are populated by vibrational energy redistribution before emission. Oxygen electronically deactivates S$_1$ states and Parmenter added large pressures of oxygen to p-fluorotoluene, so reducing the S$_1$ lifetime considerably

(Fig. 12, p. 1749). When the lifetime was reduced to ≈ 9 ps, the remaining weak emission was much more structured because vibrational energy redistribution did not have enough time to occur before the molecule emitted a photon of light or, more probably, was deactivated by O_2.

A related problem is discussed on pp. 1746–50, which demonstrates that the emission spectrum of 1-azaindolazine is highly structured for low vibrational levels, but becomes broader as the vibrational energy increases. At the higher energy levels, $N(E)$ is sufficiently large for ergodic motion to occur. Note that the levels of excitation are quite small, 1536 cm^{-1} corresponds to only 18.4 kJ mol^{-1}. This exciting paper demonstrates very clearly the absence of vibrational energy redistribution at low vibrational energies and its rapid occurrence for even quite modest vibrational excitation.

3 The experiments of Crim and Zare discussed in Box 4.1 on bond selective chemistry also give information about the nature of vibrational energy distribution within a reacting molecule.

5.11 Association reactions[1]

Unimolecular and association reactions, such as reaction $(6, -6)$,

$$C_2H_6 + M \leftrightarrows CH_3 + CH_3 + M \qquad \text{(R 6, -6)}$$

are linked via the equilibrium constant for the reaction.

$$K_{c6} = [CH_3]^2/[C_2H_6] = k_6/k_{-6}. \qquad \text{(E 37)}$$

The pressure dependence of the association reaction must therefore follow that of the unimolecular reaction in order to maintain the value of the equilibrium constant K_c. As we shall see the pressure dependence of association reactions (as for unimolecular reactions) is due to the competition between activation/deactivation and dissociation of an energized reaction complex.

A schematic one-dimensional cut through the potential energy surface for reactions such as the recombination of methyl radicals is shown in Fig. 5.11. We can see that there is no potential energy barrier for the recombination process and

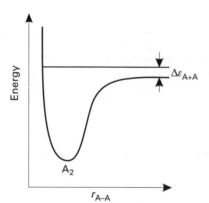

Fig. 5.11 Potential energy diagram for recombination reactions.

---

[1] Also known as termolecular reactions, although as we shall see this is an incorrect designation as no part of the reaction mechanism involves the collision of three bodies.

that therefore the temperature dependence will be very weak. In contrast unimolecular reactions have exceedingly high activation energies and hence a correspondingly large temperature dependence. This feature has important implications for the experimental study of pressure dependent reactions. Most experimental techniques are applicable over limited timescales of reaction and it is therefore hard to study unimolecular reactions over a wide range of temperatures. For example the high pressure rate coefficient for reaction (6) can be characterized by a modified Arrhenius expression of $k_{\infty 6} = 1.8 \times 10^{21} \, T^{-1.24} \exp(-45\,700/T) \, \text{s}^{-1}$ and over the limited temperature range 1000–1200 K will increase by a factor of 1500. Obviously it is also very difficult to study the unimolecular reactions of such stable species as ethane at low temperatures, the half-lives of such reactions would be measured in millennia! On the other hand the association reaction, methyl recombination, can be studied over a wide range of temperatures, including low temperatures, on a single piece of apparatus (see Box 2.2 p. 44).

5.12 The physical basis for association reactions

As the two methyl radicals combine they start to form a new carbon–carbon bond. Potential energy is released on bond formation, and as the energy of the $CH_3 + CH_3$ system must be conserved, the newly formed ethane molecule will be highly excited. Energy will flow rapidly around the internal modes of the ethane complex until at some point enough energy is located in the C–C bond and unimolecular dissociation occurs reforming the methyl reactants. This can only be prevented if some the energy of the nascent ethane molecule can be removed, a process which can be achieved by collisional deactivation either by another reactant molecule or by a bath gas (M). The excess of energy is channelled into the relative motion of ethane and M, or for polyatomic bath gases, into the internal degrees of freedom of the bath gas.

The lifetime of the energized molecule will be short (it will be rapidly removed either by deactivation or dissociation) and therefore we can apply the steady-state approximation to the kinetic scheme:

$$CH_3 + CH_3 \rightleftarrows C_2H_6^* \qquad\qquad (R\ 7, -7)$$

$$C_2H_6^* + M \rightarrow C_2H_6 + M. \qquad\qquad (R\ 8)$$

The rate of formation of $C_2H_6^* = k_7[CH_3]^2$ and the rate of removal of $C_2H_6^* = k_{-7}[\,C_2H_6^*] + k_8[C_2H_6^*][M]$.

Under the steady-state approximation the rate of formation is equal to the rate of decay and hence:

$$k_7[CH_3]^2 = k_{-7}[\,C_2H_6^*] + k_8[\,C_2H_6^*][M]. \qquad\qquad (E\ 38)$$

$$[C_2H_6^*] = k_7[CH_3]^2/\{k_{-7} + (k_8[M])\}. \qquad\qquad (E\ 39)$$

The rate of formation of stabilized product $(d[C_2H_6]/dt)$ is $k_8[\,C_2H_6^*][M]$, therefore

$$d[C_2H_6]/dt = k_8\,k_7[CH_3]^2[M]/\{k_{-7} + (k_8[M])\}. \qquad\qquad (E\ 40)$$

The overall order of the reaction will depend on the relative importance of the two terms in the denominator, k_{-7} and $k_8[M]$. At high pressures $k_{-7} \ll k_8[M]$ and eqn (40) reduces to $d[C_2H_6]/dt = k_7[CH_3]^2$. The rate of stabilization is so great that the rate-determining step in the reaction is the formation of the complex; once formed it is always stabilized before it can redissociate. As the rate-determining step is the bimolecular formation of the complex, the reaction is, as expected, overall second order. Conversely at low pressures $k_{-7} \gg k_8[M]$ stabilization is very slow and eqn (40) now becomes $d[C_2H_6]/dt = k_8k_7[CH_3]^2[M]/k_{-7}$, i.e. overall third order. The rate-determining step is the deactivation of the complex.

The pressure at which second-order kinetics is reached will depend on the lifetime $(1/k_{-7})$ of the complex with respect to dissociation. This in turn will depend on the structure and complexity of the complex and its internal energy which is the sum of the reverse bond dissociation energy and the kinetic energy of the collision. The larger the complex, the more modes there are to redistribute the reaction exothermicity into, and the lower the probability of enough energy being localized in the dissociative mode. Table 5.4 illustrates the variation in lifetime of the adduct as a function of molecular complexity. The pressures at which deactivation will compete with dissociation are also dependent on the efficiency of the third body in removing the excess of energy. This will increase as a function of atomic weight for monatomic bath gases and with molecular complexity (see Table 5.3).

Association reactions are unusual in that they generally have negative temperature dependencies. As we can see from Fig. 5.11 there is no barrier to formation of the energized adduct but as we noted above the lifetime of the adduct is dependent on its internal energy, in turn dependent on the kinetic energy of the collision which will increase with temperature (the molecules will be moving faster). Collision complexes formed with higher internal energy are more likely to redissociate before stabilization giving the observed negative temperature dependence.

As we have seen association reactions and unimolecular decompositions are very closely linked and theories such as RRKM can be equally applied to association reactions in order to calculate the pressure dependence. Association reactions are equally as important as unimolecular reactions in such phenomena as combustion and atmospheric chemistry. Very often they act as chain termination steps (see Chapter 9) i.e. the final step in a chain reaction mechanism. Chain mechanisms are generally carried by radical species, an example being the pyrolysis of ethane to form ethene and hydrogen

$$H + C_2H_6 \rightarrow H_2 + C_2H_5 \qquad \text{(R 9)}$$

Table 5.4 Data for radical recombination reactions at 300 K

Reaction	Lifetime of adduct/s	Pressure/ Pa for $k_{-7} = k_8[M]$
$I + I \rightarrow I_2$	10^{-12}–10^{-13}	Not attained
$CH_3 + H \rightarrow CH_4$	10^{-10}	$\approx 10^5$
$CH_3 + CH_3 \rightarrow C_2H_6$	2×10^{-8}	$\approx 10^3$

$$C_2H_5 \rightarrow H + C_2H_4. \tag{R 10}$$

The radical chain carriers H and C_2H_5 are regenerated and the reaction will continue as long as there is ethane present. However, the chain can be terminated if two ethyl radicals recombine

$$C_2H_5 + C_2H_5 \rightarrow C_4H_{10}. \tag{R 11}$$

An atmospherically important association reaction involves the generation of stratospheric ozone. Oxygen atoms are generated by short wavelength UV photolysis in the stratosphere. They then combine with oxygen molecules to generate ozone (R 13). The interaction of photolysis rates for reactions (12) and (14), and the pressure dependence of reaction (13) (which will be a function of altitude) results in the concentration of ozone at a specific range of altitudes, the ozone layer.

$$O_2 + h\nu \rightarrow 2O \tag{R 12}$$

$$O + O_2 + M \rightarrow O_3 + M \tag{R 13}$$

$$O_3 + h\nu \rightarrow O + O_2 \tag{R 14}$$

$$O + O_3 \rightarrow 2O_2 \tag{R 15}$$

5.13 Questions

5.1 Show that at high pressures the steady-state concentration of A^*, $[A^*]_{eq} = k_2[A]/k_{-2}$ (where the subscript numbers refer to reactions in the main text) and therefore that

$$[A^*]/[A^*]_{eq} = \{1 + (k_3/k_{-2}[M])\}^{-1}$$

Plot $[A^*]/[A^*]_{eq}$ vs. $k_{-2}[M]/k_3$ (vary the latter from 0.1 to 10).

5.2 (i) Derive eqn (5.13) by considering the number of ways of putting v objects in s boxes and then allowing for the fact that the v objects are identical as are the s boxes. (ii) Show that eqn (5.13) reduces to $(v + 1)$ for $s = 2$. (iii) Plot g_v vs. v for $s = 12$ and $v = 1-10$. Note the dramatic increase in g_v with v.

5.3 The Lindemann mechanism for the gas phase isomerization of *cis*- to *trans*-but-2-ene is

$$cis\text{-but-2-ene} + M \rightleftarrows cis\text{-but-2-ene*} + M \tag{R 16, -16}$$

$$cis\text{-but-2-ene*} \rightarrow trans\text{-but-2-ene} \tag{R 17}$$

where *cis*-but-2-ene* is an energized (internally excited) molecule of the olefin. Show that this mechanism leads to the following expression for the first order rate constant in pure but-2-ene

$$k = k_{16}k_{17}\,[cis\text{-but-2-ene}]/(k_{-16}\,[cis\text{-but-2-ene}] + k_{17})$$

The first-order rate constant for the isomerization of *cis*- to *trans*-but-2-ene at 742 K has the limiting high pressure value $k_\infty = 1.9 \times 10^{-5}\ \text{s}^{-1}$. The half-pressure for the reaction is 5 Pa. Taking k_{-16} to be $10^{14}\ \text{dm}^3\ \text{mol}^{-1}\ \text{s}^{-1}$ estimate k_{16} and k_{17}.

5.4 The following data were obtained for the *cis-trans* isomerization of but-2-ene at 740 K:

Concentration $\times 10^5$/mol dm^{-3}	0.25	0.3	0.6	1.2	5.9
$10^5 \, k/s^{-1}$	1.05	1.14	1.43	1.65	1.82

Evaluate k_∞ and k_0.

5.5 Methylene reacts with methane to form an excited ethane molecule, which can decompose to form methyl radicals. Using RRK theory estimate the rate of decomposition of the newly formed ethane. The activation energy for the decomposition of ethane is 200 kJ mol^{-1}, whilst the heats of formation of CH_2, CH_4 and C_2H_6 are 380, -75 and -84 kJ mol^{-1} respectively. (Hint: calculate the energy of the newly formed $C_2H_6^*$ and use eqn (25) to calculate k_3 taking $k^\ddagger$ to be $\approx 10^{13} \, s^{-1}$.)

5.6 From the following data on A-factors in the limit of high pressure, estimate the entropies of activation for these unimolecular reactions. How can the values that you obtain be related to the nature of the transition states?

$$A_\infty/s^{-1}$$

$C_2F_6 \rightarrow 2CF_3$	4×10^{17}
$CH_3CHO \rightarrow CH_3 + CHO$	1×10^{17}
$CH_2CHF \rightarrow C_2H_2 + HF$	2.5×10^{14}

5.7 The following kinetic data have been obtained for the unimolecular isomerization of cyclopropane to propene at 800 K:

p/torr	84.1	11.0	2.89	0.569	0.120	0.067
$10^4 \, k_{obs}/s^{-1}$	2.98	2.23	1.54	0.857	0.392	0.303

Devise a test of the Lindemann theory with these data.

5.8 The following rate constants were determined for the reaction of OH with ethene as a function of pressure at 290 K

helium pressure/torr	3	6	10	20
k/cm^3 molecule^{-1} s^{-1}	2.24×10^{-12}	3.05×10^{-12}	3.63×10^{-12}	4.06×10^{-12}

(i) Explain why k depends on pressure and propose a mechanism for the reaction.
(ii) Estimate the rate constant for the reaction at atmospheric pressure.

5.9 Which of the rate coefficients for the following two reactions would you expect to reach the high pressure limit (k_∞) at the lower pressure?

$$CH_3 + CH_3 \rightarrow C_2H_6$$
$$CH_3 + H \rightarrow CH_4$$

Why?

References

Hippler, H., Troe, J., and Wendelken, H. J. (1983). Collisional deactivation of highly vibrationally excited polyatomic molecules *Journal of Chemical Physics*, **78**, 6709–17.

Smith, I. W. M. (1987). *Modern gas kinetics* (ed. M. J. Pilling and I. W. M. Smith), Chapter A4. Blackwell, Oxford.

6 Reactions in solution

6.1 Introduction

Our discussion so far has been limited to reactions in the gas phase, where our knowledge of molecular motion and of molecular energy levels is well developed. In the present chapter we shall consider how our approach is to be modified when dealing with reactions in solution.

In nitrogen gas, at 298 K and 1 atm, the molecules occupy only 0.2 per cent of the total volume. In a liquid, this figure typically rises to more than 50 per cent. In consequence, the largely unhindered character of motion in the gas phase is lost in liquids where the molecules must squeeze past one another if they are to make substantial displacements. Figure 6.1(a) is a graphical display of a computer calculation of the instantaneous positions of the molecules in a liquid. It is, in effect, a simulated snapshot picture, on a molecular scale, of a liquid in cross-section. It illustrates the small fraction of free volume available. Figure 6.1(b) shows the paths taken by the molecules during the period 2×10^{-12} s following the snapshot. Most of the molecules perform a rather haphazard vibrational motion, with little net movement through the body of the liquid. A few molecules have been substantially displaced, and now find themselves next to a new set of nearest neighbours. Thus the motion of molecules in solution is complex and specific account must be taken of this complexity in any model we develop.

A further difference between gas phase and solution phase reactions is the close proximity between the activated complex, be it formed in a unimolecular or bimolecular reaction, and the surrounding solvent molecules. Indeed we might think of the reactant(s) and the surrounding solvent molecules as a super-molecule in which energy can flow between the different vibrational modes. Thus the solvent molecules can act as a heat bath and pass energy on to the reactants to enable them to overcome the activation barrier or remove energy in an association reaction, thus stabilizing the nascent molecule (cf. Chapter 5 p. 139). The

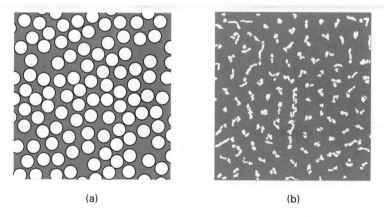

(a)   (b)

Fig. 6.1 (a) 'Snapshot' picture of the instantaneous configuration of a liquid, (b) trajectories of the particles in the 2×10^{-12} s following the snapshot.

surrounding solvent molecules can also affect the energetics of the reaction and this is especially important in polar solvents where the high permittivity lowers the energy required for the formation of ions so that new reaction channels, which require impossibly high activation energies in the gas phase, may become available.

Our experience with bimolecular and unimolecular gas phase reactions has taught us that, in order to develop reasonably tractable models of the reactions, we must make approximations. On the other hand oversimplification and neglect of the effects of molecular structure can make these models physically unrealistic. In transition state theory, we assumed that the concentration of the activated complex could be calculated using statistical mechanical expressions which presuppose equilibrium. Our treatment of unimolecular reactions was based on rapid vibrational energy redistribution in the molecule and on a simplified description of its energy levels. Our approach with liquids will be yet more approximate, simply because the problem is even more complex.

We shall concentrate mainly on bimolecular reactions and shall adopt three different approaches. In the first of these we shall simply recognize that the two reactants, once they have approached one another, are held in close proximity for some time by the surrounding solvent molecules (the solvent cage) before they can separate. We shall call this proximate pair of reactants an *encounter complex* and set up a *steady-state* treatment of this complex in which the component steps are the approach and separation of the reactants and their reaction within the solvent cage. This approach is phenomenological, but it is very helpful in forming an overall picture of bimolecular reactions in solution.

Our second method will pay much more attention to the motion of reactant molecules. This is the approach which any detailed theory of solution phase reactions must adopt at various levels of sophistication. All we shall use is a simplified version, but we shall make reference, through study notes, to more detailed treatments. In our model the molecules are assumed to perform a series of jumps through the solution—the frequency of the jumps is very high $(10^{12}-10^{13} \text{ s}^{-1})$ and the associated distances, d, very short $(d \approx 0.1 \text{ nm})$. After each jump, the molecules are assumed to 'forget' their original direction of motion, so that the new directions in which they jump are uncorrelated with the original directions. The molecules are assumed to perform 'random walks' through the solution. This type of motion, which is sometimes termed Brownian motion, may be treated using Fick's laws of diffusion, which provide a good description provided the jumps are small.

The third approach makes use of the thermodynamic formulation of transition state theory (Chapter 3 p. 76) and enables us, very simply, to understand the effects on solution phase reactions of external variables, such as pressure and temperature.

Finally, we shall see how these models apply to some important reactions characteristic of the liquid phase.

6.2 The encounter pair

We consider a reaction between two molecules A and B, which approach by diffusion to form an encounter pair {AB}. The central role of the encounter pair is recognized in the following kinetic scheme:

$$\overset{k_d}{} \qquad \overset{k_r}{}$$
$$A + B \rightleftarrows \{AB\} \rightarrow \text{products} \qquad\qquad (R\ 1, -1, \text{ and } 2)$$
$$\underset{k_{-d}}{}$$

where $\{AB\}$ is the encounter pair, and k_d, k_{-d} and k_r are the rate coefficients for diffusive approach, for separation, and for reaction respectively. The processes occurring during the reaction are shown schematically in Fig. 6.2. Note that k_d is a second-order rate coefficient, while k_{-d} and k_r are first order. Applying the steady state approximation (Chapter 1 pp. 24–25) to the encounter pair (i.e rate of formation of $\{AB\}$ is equal to its rate of destruction):

$$d[\{AB\}]/dt = k_d[A][B] - (k_{-d} + k_r)[\{AB\}] = 0. \qquad (E\ 1)$$

The overall rate of reaction ρ is given by: $\qquad\qquad$)sub. in $\left[\text{[}AB\text{]} \right]$

$$\rho = k_r[\{AB\}] = k_d k_r[A][B]/(k_{-d} + k_r) \qquad (E\ 2)$$

and the overall second-order rate coefficient by:

$$\boxed{k = \rho/[A][B] = k_d k_r/(k_{-d} + k_r)} \qquad (E\ 3)$$

Our approach to this problem has been somewhat similar to that used in pressure dependent reactions. As collisions with the solvent (bath gas) are not explicitly involved in the kinetic scheme the overall order of the reaction remains constant, however, we can again recognize two limiting conditions, depending on the relative contributions of the terms in the denominator.

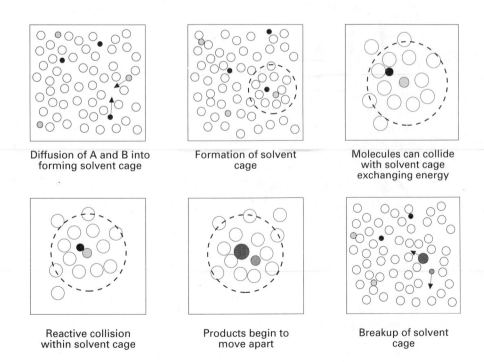

Diffusion of A and B into
forming solvent cage

Formation of solvent
cage

Molecules can collide
with solvent cage
exchanging energy

Reactive collision
within solvent cage

Products begin to
move apart

Breakup of solvent
cage

Fig. 6.2 Schematic representation of some of the processes involved in solution reactions.

1. $k_r \gg k_{-d}$. If separation of A and B is relatively difficult, e.g. in a viscous solvent, or if the reaction has a small activation energy, the kinetics become diffusion-controlled. The rate–determining step is the approach of the reactants; once they are in the solvent cage, reaction is assured. Under these conditions $k = k_d$.
2. $k_r \ll k_{-d}$. For reactions with large activation energies ($E \geq 20$ kJ mol^{-1} for reactions in water) the kinetics are activation-controlled, and $k = k_r(k_d/k_{-d})$. Since k_d/k_{-d} is the equilibrium constant K_{AB} for formation of an encounter pair, we may write

$$k = k_r K_{AB}. \tag{E 4}$$

The reaction rate is now determined by the equilibrium concentration of encounter pairs, and by the rate of passage over the reaction energy barrier.

It is useful to have some idea of the order of magnitude of the component rate coefficients. For a solvent, such as water at room temperature, with a viscosity of 10^{-3} kg m^{-1} s^{-1}, $k_d \approx 10^9$–10^{10} dm^3 mol^{-1} s^{-1} (solution phase kineticists generally employ molar units and we shall follow their practice). As we shall show on p. 156, $K_{AB} \approx 0.1$ dm^3 mol^{-1}, so that $k_{-d} = 10^{10}$–10^{11} s^{-1}. Thus if a reaction is to be diffusion-controlled in water, we need $k_r > 10^{12}$ s^{-1}, whilst for $k_r < 10^9$ s^{-1} the reaction is activation controlled.

6.3 Diffusion-controlled reactions

We start our discussion of the more detailed theories of reactions in solution by considering diffusion-controlled reactions, where the reactive step (described by the rate coefficient k_r) is very rapid and the rate of reaction is controlled by the diffusive approach of the reactants. We might expect considerable quantitative success for our theory, for we need only to model a physical process, diffusion, and hence require no knowledge of the transition state for the reaction. However, we shall see that the complexity of the liquid state leads to new and different problems. To make our treatment simpler, we consider first the case where k_r tends to infinity, so that the reactants A and B are removed instantaneously on encounter.

Initially we begin by splitting the whole system into a set of cells, each centred on an A molecule, and examine the distribution of B molecules in each spherically symmetric cell. If we then average these distributions over all the cells, we can determine $[B]_r$, the averaged concentration of B, as a function of r, the distance from the centre of the average cell. For $r = r_{AB}$, the encounter distance, $[B]_r = 0$, since any A and B molecules which encounter one another react instantaneously and so are removed from our consideration. At r_∞, the A and B molecules have no influence on one another, so that $[B]_r$ tends to its bulk value, $[B]$. At intermediate values of r, the concentration varies smoothly between these two extremes, in a manner we shall presently determine. Thus there is a gradient in the averaged concentration of B around any A molecule, which leads to a flux of B molecules down the concentration gradient. In other words, if we look at all the cells, each centred on an A molecule, there is a net movement of B molecules from large distances towards the central A molecules. It is this net diffusive motion which determines the rate of the overall reaction.

The flux, $\bar{J}$, towards the central A molecules may be evaluated from Fick's first law of diffusion:

$$\bar{J} = 4\pi r^2 D_{AB}\, d[B]_r/dr \tag{E 5}$$

where D_{AB} is the coefficient of relative diffusion of A and B. ($D_{AB} = D_A + D_B$). The flux is the amount of material passing through a spherical shell, centred on A, in unit time. It has units of mols per second. D is the diffusion coefficient (units $m^2 s^{-1}$) which determines the magnitude of the flux for a given concentration gradient; species with large diffusion coefficients, like H^+, respond rapidly to concentration gradients. A discussion of Fick's laws may be found in Atkins. Equation (5) is more frequently encountered without the arrow and with a minus sign; such an equation refers to the 'natural' direction for the flux, i.e. from small to large r. We are specifically interested in the flux towards A, i.e. from large to small r and it will make life a lot easier if we define the flux $\overleftarrow{J}$ in this way.

We may use eqn (5) to determine the averaged concentration profile $[B]_r$, which we discussed above. Rearranging eqn (5) and integrating between r and ∞

$$\left[\frac{-J}{4\pi r D}\right]_r^{\infty} = +\frac{J}{4\pi r D} = \int_r^{\infty}(\overleftarrow{J}/4\pi r^2 D_{AB})\mathrm{d}r = \int_{[B]_r}^{[B]_{\infty}}\mathrm{d}[B]_r \quad [B] - [B]_r$$

$$[B]r = [B] - \overleftarrow{J}/4\pi r D_{AB} \qquad (E\,6) \quad \text{2}$$

$\overleftarrow{J}$ may be found from the above eqn since when $r = r_{AB}$, $[B]_r = 0$, so that

$$\overleftarrow{J} = 4\pi r_{AB} D_{AB}[B]. \qquad (E\,7) \quad \text{3}$$

Substituting $\overleftarrow{J}$ back into eqn (6) we find

$$[B]_r = [B]\{1 - (r_{AB}/r)\} \qquad (E\,8) \quad \text{k}$$

and this distance dependent concentration distribution is shown in Fig. 6.3.

Our ultimate objective is the calculation of the rate of reaction. Once B molecules cross the spherical shell, radius r_{AB}, centred on A, they react instantaneously. Thus, in the averaged cell, we may determine the rate of removal of B from the flux of B across this shell. This flux may be found from eqn (5) after $(\mathrm{d}[B]_r/\mathrm{d}r)_{r = r_{AB}}$ has been found from eqn (6); more simply, it is given by eqn (7). $\overleftarrow{J}$ is constant and independent of r (provided $r > r_{AB}$); it must be otherwise B would build up in some regions at the expense of others. The overall rate of reaction (in all cells) may be found by multiplying the averaged flux across the reaction boundary by the total number of cells per unit volume, $[A]$, thus

if rate rxn↑, flux of B will ↑ too.

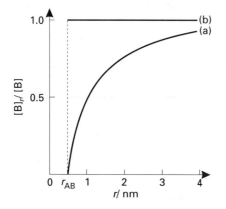

Fig. 6.3 Averaged distribution of B molecules around A: (a) diffusion-controlled reaction, where $[B]_r/[B] = 1 - \{\overleftarrow{J}/(4\pi D_{AB}[B])\} = 1 - r_{AB}/r$, $r_{AB} = 0.5$ nm; (b) activation controlled reaction.

$$\rho = \overleftarrow{\mathcal{J}}\,[A] \qquad\qquad (\text{E 9})$$

and

$$k = \rho/[A][B] = 4\pi r_{AB} D_{AB}. \qquad\qquad (\text{E 10})$$

Since the reaction is purely diffusion-controlled, k is also equal to k_d. For r_{AB} and D_{AB} expressed in SI units, k has units $m^3\,s^{-1}$ (with molecule^{-1} understood). It may be converted into the more normal units, $dm^3\,mol^{-1}\,s^{-1}$, by

$$k = 1000 \times 4\pi r_{AB} D_{AB} \times N_A \qquad\qquad (\text{E 11})$$

Table 6.1 lists diffusion coefficients for ions in aqueous solution at 298 K. Inserting typical values ($D_{AB} = 2 \times 10^{-9}\,m^2\,s^{-1}$, $r_{AB} = 0.5$ nm) into eqn (11)) we find $k \approx 8 \times 10^9\,dm^3\,mol^{-1}\,s^{-1}$; $k > 10^9\,dm^3\,mol^{-1}\,s^{-1}$ is usually indicative of a diffusion-controlled reaction. Diffusion-controlled reactions include:

1. Recombination reactions of atoms and radicals, which have small, zero or even negative activation energies in the gas phase. For iodine atoms in hexane at 298 K, the rate coefficient is $1.3 \times 10^{10}\,dm^3\,mol^{-1}\,s^{-1}$. Note how much larger the rate coefficient is than for the gas phase reaction discussed in Chapter 5.

2. Acid–base reactions which involve transport of H^+ or OH^- and are often very fast. The reaction

$$H^+ + OH^- \rightarrow H_2O \qquad\qquad (\text{R 3})$$

has a rate coefficient at 298 K of $1.4 \times 10^{11}\,dm^3\,mol^{-1}\,s^{-1}$, and is one of the fastest solution reactions known, reflecting the high diffusion coefficients of H^+ and OH^-. We shall return to this interesting class of reactions later.

3. Quenching of electronically excited molecules. Photoexcitation generally leads to formation of an electronically excited singlet state (S_1) which has a short lifetime, 10^{-8} s. These S_1 states have their own absorption spectra, and may be studied by laser flash photolysis or by fluorescence. Efficient transfer of electronic energy takes place to molecules with S_1 states of lower energy, e.g. naphthalene (S_1) + biacetyl → naphthalene + biacetyl (S_1), $k = 2.2 \times 10^{10}\,dm^3\,mol^{-1}\,s^{-1}$ (see also Chapter 12, p. 276).

Study note
Although the derivation of eqn (10) is mathematically quite straightforward, the model itself causes some conceptual difficulties and is a frequent cause of tutorial discussion. The objection that is usually raised is that we evaluate the rate of reaction by considering the flux of B molecules into a central 'sink', yet that central sink disappears as soon as a B molecule enters it. The important point to remember is that the distance dependent concentration we consider in Fig. 6.3 is obtained by averaging the distributions of B molecules about all the remaining A molecules. As

Table 6.1 Diffusion coefficients for ions in aqueous
solution at 298 K

Ion	$D/m^2\,s^{-1}$	Ion	$D/m^2\,s^{-1}$
H^+	9.1×10^{-9}	OH^-	5.2×10^{-9}
Li^+	1.0×10^{-9}	Cl^-	2.0×10^{-9}
Na^+	1.3×10^{-9}	Br^-	2.1×10^{-9}

soon as an A molecule reacts, that cell and the B distribution in it are removed from our consideration. Thus, reaction depletes the distribution in the remaining cells at small r and diffusion from the unaffected region of the distribution (at large r) tries to make good this depletion. We evaluate the rate of reaction from this diffusive flux.

We know to our cost that this argument does not convince everyone! At the risk of some increase in mathematical complexity, an alternative and much more satisfying derivation can be made and a discussion can be found in Clegg, R. M. (1986). *Journal of Chemical Education*, **63**, 571–4. This approach is very flexible and the paper describes several applications which we shall address, in a different form, in the next few sections. The approximations inherent in the present model are also more clearly indicated in this alternative approach.

Box 6.1 Time-dependent rate 'constants'

In the treatment given above, we have assumed that the concentration profile given by eqn (8) always applies. If we produced the A molecules 'instantaneously' (for example if we generated S_1 naphthalene molecules by pulsed laser excitation) then the unreactive ground state naphthalene/biacetyl distribution would apply i.e. $[B]_r$ would be constant. Reaction then occurs rapidly for those molecules separated by short distances and the concentration gradient changes with time until, eventually, the distribution described by eqn (8) is established. Since the concentration gradient is changing, we may no longer employ eqn (8) but must use instead the time-dependent form:

$$\frac{d[B]_r}{dt} = D \left(\frac{d^2[B]_r}{dr^2} + \frac{2}{r} \frac{d[B]_r}{dr} \right). \tag{E 12}$$

This eqn is solved, subject to the boundary conditions $[B]_{r_{AB}} = 0$, $[B]_\infty = [B]$ and the rate of reaction once more evaluated from the flux across the reaction boundary i.e. across the spherical shell, radius r_{AB}, centred on A, using eqn (10). (Note that J is not now independent of r and so must be evaluated at $r = r_{AB}$). If we follow normal practice, and set the rate of reaction equal to $k[A][B]$, then k is time dependent! This arises because $d[B]_{r_{AB}}/dr$ is larger at short times, as can be seen from Fig. 6.4, and it is this gradient that drives the flux across the reaction boundary at $r = r_{AB}$. Eventually, the steady-state concentration profile is established, when the consumption of B molecules at $r = r_{AB}$ is quantitatively balanced by their diffusion from larger distances and the time-independent treatment becomes valid. The time–dependent form of k is:

$$k(t) = 4\pi r_{AB} D_{AB} (1 + r_{AB}/\sqrt{(\pi D_{AB}t)}) \tag{E 13}$$

Which shows the necessary asymptotic behaviour as t tends to infinity. A rather disturbing feature is that as $t \to 0$, $k \to \infty$, but this need not concern us too much, since the validity of the whole diffusion model breaks down for times less than ≈ 1 ps and more complex models must be employed. Experimental examples of time–dependent rate coefficients are discussed by Rice (Rice, S. A. (1984). Diffusion controlled reactions. In *Diffusion limited reactions* (Ed. C. H. Bamford, C. F. H. Tipper, and R. G. Compton). pp. 30–45. Elsevier, Amsterdam.) A relevant example is the fluorescence lifetime of excited naphthalene in 1,2-propandiol which can only be characterized by a time–dependent rate coefficient of the form of eqn (13).

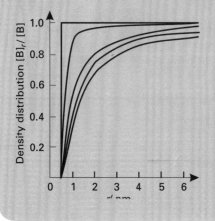

Fig. 6.4 Density distribution for a diffusion coefficient of $D = 10^{-9}$ m^2 s^{-1} and encounter distance $r = 0.5$ nm. The five curves progressively shift to smaller densities at any radial distance and are for times $t = 0$, 10^{-9}, 10^{-8}, 10^{-7} s, and an infinite time, respectively.

6.3.1 Dependence of the rate coefficient for diffusion-controlled reactions on viscosity and temperature

The diffusion coefficient is related to solvent viscosity by the Stokes-Einstein equation:

$$D = k_B T / \pi \beta \eta r_s \qquad \text{(E 14)}$$

where η is the viscosity, r_s the hydrodynamic radius of the diffusing species, and β is a numerical constant. The equation was derived for a continuum solvent ($r_s \gg$ radius of the solvent molecules) in which case $\beta = 6$. For molecular diffusion, β is usually given the value 4, but the validity of the equation is somewhat questionable. Diffusion-controlled reactions are certainly slower in high viscosity solvents, showing that there is an inverse relationship of some sort between D and η. The viscosities of some common solvents are shown in Table 6.2

The hydrodynamic radius is governed by solvent–reactant interactions, and so is different from the encounter radius, which depends on reactant-reactant interactions. There is rarely much information available on either, and the two radii are often assumed equal (i.e $r_{SA} = r_A$; $r_{SB} = r_B$; $r_{AB} = r_A + r_B$). If r_A and r_B are also approximately equal, eqns (10) and (14) may be combined to give:

$$k = 16 \, k_B T / \beta \eta \qquad \text{(E 15)}$$

Table 6.2 Viscosities of some common solvents at 293 K

Solvent	$10^3 \eta / \text{kg m}^{-1} \text{ s}^{-1}$	Solvent	$10^3 \eta / \text{kg m}^{-1} \text{ s}^{-1}$
Hexane	0.326	Propan–1–ol	2.26
Water	1.00	Ethan–1,2-diol	19.9
Ethanol	1.20	Propan–1,2,3-triol	1490

which is independent of r_{AB}. A large encounter distance increases the rate coefficient, but is also associated with a greater viscous drag (equation (14)) which decreases D. These two effects cancel in eqn (15), however, the approximate nature of this equation must be stressed.

Viscosity decreases with increasing temperature, and in consequence D and k increase. This is often expressed in terms of an Arrhenius equation:

$$k = A \exp(-E/RT) \qquad \text{(E 16)}$$

where $E \approx 15 \text{ kJ mol}^{-1}$ and $A \approx 10^{12} \text{ dm}^3 \text{ mol}^{-1} \text{ s}^{-1}$ for reactions in water. The origin of this 'activation energy' is much more complex than that of gas phase reactions, and indeed some experimental data do not conform to the Arrhenius equation. Motion in liquids is co-operative and depends on the associated movement of several molecules. As these molecules move, they must squeeze past each other, and in so doing must overcome potential energy barriers arising from intermolecular repulsion. Their movement is facilitated by an increase in temperature, leading to the observed increase in D. The complexity of the process, involving motion of several adjacent molecules, renders a discussion of the functional form of the temperature dependence of k much more difficult than we found for gas phase reactions. The pre-exponential factor for a diffusion-controlled reaction in solution is a little larger than the collision frequency factor Z in the gas phase.

6.3.2 Diffusion controlled reactions between ions

The diffusion equations (eqn (8) and the time-dependent form, eqn (13), which may be found in Box 6.1) may be derived by considering the motion of the particles to correspond to Brownian motion where small, random jumps are made at high frequency. Other things being equal, a particle has an equal probability of jumping in any direction. The net flux arises because of the presence of the concentration gradient. When the reactants are ions, however, the Coulomb force between them biases their relative motion i.e. similarly charged ions prefer to jump away from each other and oppositely charged ions tend to jump towards one another. The importance of this bias increases as r decreases, because the Coulomb force depends on r^{-2}. The eqn for the flux, $\bar{\mathscr{J}}$, must therefore be modified by the introduction of a *drift* term

$$\bar{\mathscr{J}} = \underbrace{4\pi r^2 D_{AB} \text{d}[B]_r/\text{d}r}_{\text{diffusion term}} + \underbrace{(4\pi r^2 D_{AB}[B]_r/k_B T) \, \text{d}V(r)/\text{d}r}_{\text{drift term}} \qquad \text{(E 17)}$$

where the drift term accounts for the biased motion introduced by the interaction. In eqn (17), $V(r)$ is the potential energy of interaction of the reactants. For ions,

$$V(r) = z_A z_B e^2/(4\pi\varepsilon_0\varepsilon_r) \qquad \text{(E 18)}$$

where $z_A e$ and $z_B e$ are the charges on the ions, ε_r is the relative permittivity of the medium and ε_0 is the permittivity of vacuum. A useful shorthand form for $V(r)$ is:

$$V(r) = (r_c/r)k_B T \qquad \text{(E 19)}$$

where $r_c = z_A z_B e^2/(4\pi\varepsilon_0\varepsilon_r k_B T)$. r_c is termed the Onsager escape distance and $|r_c|$ is the separation at which $|V(r_c)| = k_B T$ i.e. the distance at which the Coulomb interaction energy is roughly equal to the thermal energy.

Equation (17) may be rearranged to give a <u>linear differential equation</u>, whose solution is:

$$[B]_r = \exp\left\{\frac{-V(r)}{k_B T}\right\} \left[[B] - \frac{\overleftarrow{J}}{4\pi D_{AB}} \int_r^\infty \exp\left\{\frac{-V(r)}{k_B T}\right\} \frac{dr}{r^2}\right] \qquad (E\ 20)$$

Setting $[B]_{r_{AB}} = 0$, as before, $\overleftarrow{J}$ may be evaluated and hence the rate coefficient by setting $k = \overleftarrow{J}/[B]$. This gives

↳ see E9 & 10

$$\boxed{k = 4\pi r_{eff} D_{AB}} \qquad (E\ 21)$$

where $r_{eff} = \{\int_r^\infty \exp(V(r)/k_B T)\, dr/r^2\}^{-1}$ or, for a Coulomb interaction,

$$r_{eff} = r_c/\{\exp(r_c/r_{AB}) - 1\}. \qquad (E\ 22)$$

Equation (21) is directly comparable to that for neutral species (E 10), except that r_{AB} has been replaced by r_{eff}. The <u>Coulomb field modifies the probability that the reactant ions will encounter one another, increasing this probability for unlike charges and decreasing it for like charges.</u> Table 6.3 lists values for r_c and r_{eff} in solvents of different relative permittivity.

For <u>oppositely charged ions</u>, $r_{eff} \approx r_c > r_{AB}$; as one would expect the Coulomb forces increase the effective encounter distance. This does not mean that the ions react at a distance larger than r_{AB}, but simply that the attractive Coulomb forces draw ions together from distances greater than r_{AB}. Their relative motion can no longer be described as a random walk, since they show a propensity, which increases with decreasing interionic distance, to jump *towards* rather than away from one another. The effect is most marked in solvents of low permittivity, where the forces extend over many molecular diameters, and hence ensure that encounter takes place even for widely separated reactants (see Fig. 6.5). For like charges, the Coulomb forces reduce the probability of a reactive encounter, and $r_{eff} < r_{AB}$. The most extreme case cited here is hexane, where r_{eff} is only 10^{-23} nm. This means that the repulsive Coulomb forces reduce the probability that the ions will approach within 0.5 nm of each other by a factor of 3×10^{24}. In water, which has a high permittivity, the solvent molecules arrange themselves around the ion, providing an efficient screen and reducing the Coulomb forces. The probability of encounter is reduced by a factor of only 3.5.

Table 6.3 The influence of solvent permittivity on effective encounter distance at 298 K ($r_{AB} = 0.5$ nm, $|z_A| = |z_B| = 1$)

Solvent	ε_r	$|r_c|$/nm	r_{eff}/nm	
			$z_A = z_B$	$z_A = -z_B$
Water	78.5	0.7	0.2	0.9
Methanol	32.6	1.7	0.06	1.8
Ethanol	24.3	2.3	0.02	2.3
Diethyl ether	4.27	13	10^{-10}	13
Hexane	1.89	29	10^{-23}	29

$|r_c|$ is the distance at which the Coulomb interaction energy between two approaching ions is equal to kT; this interaction alters the probability of encounter, and effectively changes the encounter distance to r_{eff} (see text).

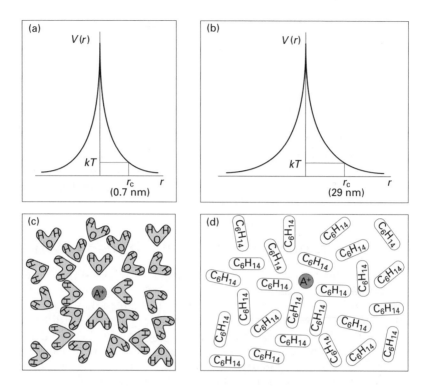

Fig. 6.5 (a) and (b) $1/r$ decay of the Coulombic potential of an ion in water and hexane. r_c is defined as the radius at which this potential is of equal magnitude to the thermal energy (kT). (c) and (d) show how the solvent molecules are ordered in water and hexane.

It might be argued that hexane is an unlikely solvent for the study of ionic reactions, but charged transient species can readily be generated in alkanes by the technique of pulse radiolysis. Ionization is a primary process which occurs when a liquid is exposed to high energy electrons.

$$A \rightarrow A^+ + e^-. \tag{R 4}$$

The ejected electron has a substantial kinetic energy and travels several nanometres before it is slowed down and eventually solvated. In water, the Coulomb forces extend only short distances, and a high yield of solvated electrons is found. In hexane, however, only an electron which travels more than 29 nm (see Table ~~force~~ 6.3) from its parent ion will have a high probability of escape from the attractive ~~longer~~ force. Most of the ejected electrons will return to A^+, and the yield of solvated ~~range.~~ electrons is low.

The technique of pulse radiolysis has been used to measure the rate coefficient for diffusion-controlled reactions between solvated electrons and cations in ~~$A^+ + e^-$~~ alkanes. Values of 10^{14} dm^3 mol^{-1} s^{-1} have been obtained. This is much greater than the typical value of $\approx 10^{10}$ dm^3 mol^{-1} s^{-1} which we discussed earlier for uncharged reactants, reflecting the high value for r_{eff} shown in Table 6.3 and the high diffusion coefficient for the electron. In water, diffusion-controlled reactions of the solvated electron show more 'normal' behaviour since the Coulomb interaction introduces a factor of only 2–3 into the rate coefficient.

A complication arises, particularly in high permittivity solvents such as water, because of interactions between reactive ions and the adjacent solvent molecules. The first solvation shell of an ion is highly ordered, because of the strong ion-dipole forces, and the solvent molecules are unable to re-orient, and reduce the interionic field, on the approach of the other reactant ion. The second and subsequent solvation shells are also affected, but much less so. The net result is that the permittivity of water decreases as r decreases, and eventually reaches a value more characteristic of non-polar solvents ($\varepsilon_r \approx 6$), thus increasing $V(r)$, and modifying r_{eff}. This phenomenon is known as dielectric saturation, and illustrates a frequent problem in solution kinetics, that of applying bulk measurements, in this case of ε_r, to molecular processes.

since on small scale, ε_r varies.

6.4 The transition to activation-controlled reactions

In the treatment of diffusion-controlled reactions, discussed above, we set $k_r = \infty$ so that reaction occurred instantaneously once A and B encountered. We were, in consequence, able to set $[B]_{r_{AB}} = 0$. This gives us one of the boundary conditions—the inner boundary condition—which is necessary to solve the differential diffusion eqn (5). The other necessary (outer) boundary condition is that as $r \to \infty$, $[B]_r \to [B]$, the bulk concentration. The use of this particular inner boundary condition (called the Smoluchowski boundary condition after the first person to use it) makes life comparatively easy, but it is unrealistic and also inflexible—it would be useful to develop an expression for k which was applicable for all values of k_r i.e. to diffusion-controlled, intermediate and activation-controlled reactions. We proceed as before, but instead of setting $[B]_{r_{AB}} = 0$, we allow non zero values to occur. We can envisage a diffusive flux, $\vec{J}$ given by eqns (E 5) and (E 7) and also a reactive flux $\vec{J}_{react}$, which describes the flow of B molecules across the reaction boundary. This latter flux is equal to the concentration of B at the reaction boundary, $[B]_{r_{AB}}$, multiplied by a rate coefficient k_R (we shall examine k_R in greater detail later). In the steady state $\vec{J} = \vec{J}_{react}$ (otherwise B would build up or be depleted at r_{AB}), so that

$$\boxed{\vec{J} = k_R[B]_{r_{AB}}.}$$

(E 23)

We now insert this new boundary condition into eqn (6) to obtain an expression for the concentration of B at the encounter distance in terms of k_R:

$$[B]_{r_{AB}}, = [B]/[1 + \{k_R/(4\pi r_{AB}D_{AB})\}]$$

(E 24)

whilst the rate coefficient may be determined as before

$$k = \vec{J}/[B] = k_R[B]_{r_{AB}}/[B]$$

$$= 4\pi r_{AB}D_{AB}k_R/(k_R + 4\pi r_{AB}D_{AB})$$

(E 25)

Finally, the distance dependent B concentration can also be obtained from eqn (26)

$$[B]_r = [B][1 - r_{AB}/\{r(1 + 4\pi r_{AB}D_{AB}/k_R)\}].$$

(E 26)

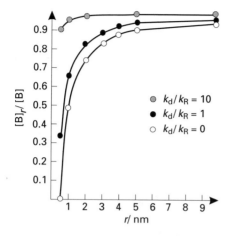

Fig. 6.6 Variation of $[B]_r/[B]$ for varying ratios of diffusive and reactive rate coefficients. As the rate coefficient for reaction decreases (tending towards activation-controlled reactions) the graph of $[B]_r/[B]$ tends towards that of (b) in Fig 6.3.

Legend in figure:
$k_d/k_R = 10$
$k_d/k_R = 1$
$k_d/k_R = 0$

y-axis: $[B]_r/[B]$
x-axis: $r/$ nm

Our understanding of the interaction between activation and diffusion control may be best improved by examining eqns (24) to (26) in turn.

1. The magnitude of $[B]_{r_{AB}}$ depends on the ratio $k_R/4\pi r_{AB}D_{AB}$, or, using our previous terminology, k_R/k_d. When this ratio is very large, $[B]_{r_{AB}} \rightarrow 0$ and the Smoluchowski boundary condition becomes applicable. On the other hand, when $k_d \gg k_R$ $[B]_{r_{AB}} \rightarrow [B]$, reaction is so slow that diffusion is able to rectify any reactive depletion of B, so that the equilibrium distribution applies at all distances and the reaction is activation controlled. At intermediate values of k_R, $[B]_{r_{AB}}$ falls below the equilibrium value, but is non-zero.

2. Substituting k_d for $4\pi r_{AB}D_{AB}$, and comparing eqns (3), (4), and (26), we find that

$$k_R = k_r k_d/k_{-d} \qquad (E\ 27)$$

whilst

$$k = k_R k_d/(k_R + k_d) \qquad (E\ 28)$$

Thus k_R is the rate coefficient which would apply if the equilibrium distribution of B were maintained. For activation-controlled reactions, this is so, and $k = k_R$. As k_r (or k_R) increases, however, diffusion is unable fully to make good the reactive depletion so that $[B]_{r_{AB}}$ falls below $[B]$ and $k < k_R$. — since less B at r_{AB} now.

3. Figure 6.6 shows $[B]_r$ vs. r for several values of k_d/k_R. The differing gradients of the plots, at a given value of r, reflect the relative fluxes and hence the relative rates of reaction. The concentration gradient is greatest for a diffusion-controlled reaction where k tends to its maximum value, k_d. For an activation controlled reaction, the rate of reaction and the diffusive flux are very much smaller.

6.5 Thermodynamic formulation of the rate coefficient

In a rapid reaction, which is diffusion-controlled or nearly so, diffusion is unable to maintain the equilibrium spatial distribution, so that thermodynamic arguments, which require equilibrium distributions in space as well as in energy, cannot be applied. Conversely in activation controlled reactions the concentration of encounter pairs, $\{AB\}$, is maintained at its equilibrium value, determined by K_{AB}. We can easily obtain a rough estimate of K_{AB} if we assume that A, B and the solvent S are all similar (i.e. that together they obey Raoult's law). Since they will be distributed randomly we can calculate the concentration of encounter pairs on a simple statistical basis:

$$[\{AB\}] = [A] \times \text{(probability of finding B next to A)}$$

$$= [A] \times (n[B]/[S]) \qquad (E\ 29)$$

where n is the coordination number. Thus

$$K_{AB} = [\{AB\}]/[A][B] = n/[S]. \qquad (E\ 30)$$

For water ($[S] = 55.5$ mol dm^{-3} at 298 K), $K_{AB} \approx 0.14$ dm^3 mol^{-1}, taking n to be 8.

The rate coefficient is given by

$$k = k_r K_{AB} \qquad (E\ 4)$$

and if transition state theory can be applied to k_r, then the rate coefficient can be written in the form:

$$k = \kappa k_B T/h \exp(-\Delta G^{\ddagger}/RT) \qquad (E\ 31)$$

where $\Delta G^{\ddagger}$ is the *overall* free energy of activation, from separated reactants, A and B, to the activated complex. Alternatively we can divide eqn (4) into its component parts of encounter formation and reaction, setting

$$K_{AB} = \exp(-\Delta G^{\circ}_{AB}/RT) \qquad (E\ 32)$$

where ΔG°_{AB} is the free energy change on forming the encounter pair, and

$$k_r = \kappa k_B T/h \exp(-\Delta G^{*}/RT) \qquad (E\ 33)$$

where ΔG^{*} is the free energy of activation from the encounter pair, then:

$$k = \kappa k_B T/h \exp\{-(\Delta G^{\circ}_{AB} + \Delta G^{*})/RT\} \qquad (E\ 34)$$

Note that ΔG^{*} does not include the mode corresponding to the reaction coordinate (cf. pp. 70–1). Because it is more difficult to apply statistical mechanics in solution, this thermodynamic formulation is widely employed. In the following sections, we shall examine its application to reactions between ions and to the effects of pressure on solution phase rate coefficients.

6.5.1 Ionic strength effects

In solutions of non-zero ionic strength,[1] interactions occur between many ions, whilst we have only considered those between reactant pairs. This approximation is particularly inaccurate when comparatively high concentrations of non-reactive ions are present. Since many reactions are studied in buffer solutions, the problem requires more detailed examination. The thermodynamic formulation of the rate coefficient may be rearranged to give

$$k = (\kappa k_B T/h) K^{\ddagger} \qquad (E\ 35)$$

---

[1] The ionic strength, I, is defined by $I = \frac{1}{2}\sum c_i z_i^2$ where c_i is the concentration (mol dm^{-3}) of the ith ion of charge z_i. The summation is made over all ions present in solution.

where $K^{\ddagger}$ is the equilibrium constant for formation of the activated complex, given by $RT \ln K^{\ddagger} = -\Delta G^{\ddagger}$. $K^{\ddagger}$ is expressed in terms of activities whilst it is concentrations that are significant in chemical kinetics. Equation (35) is only valid, therefore, for reactions of ions at infinite dilution and needs modification at higher ionic strengths. We start by considering the equilibrium between the separated reactants and the encounter complex. For non-ideal solutions,

$$K_{AB} = ([\{AB\}]/[A][B])(\gamma_{AB}/(\gamma_A\gamma_B)) \qquad (\text{E } 36)$$

where γ_A is the activity coefficient of ion A etc. The rate of reaction is given by

$$\rho = k[A][B] = k_r[\{AB\}] \qquad (\text{E } 37)$$

and

to remove

$$k = k_r K_{AB}(\gamma_A\gamma_B/\gamma_{AB}). \qquad (\text{E } 38)$$

Thus eqns (31) and (35) must be modified by inclusion of the activity coefficient term:

$$k = (\kappa k_B T/h)K^{\ddagger} (\gamma_A\gamma_B/\gamma_{AB}) \qquad (\text{E } 39)$$

At low ionic strength, the activity coefficient of an ion may be calculated from the limiting Debye–Hückel equation:

$$\log \gamma_A = -A'z_A^{2}\sqrt{I}. \qquad (\text{E } 40)$$

In aqueous solution at 298 K, $A' = 0.509$ dm$^{\frac{3}{2}}$mol$^{-\frac{1}{2}}$. Substituting this into eqn (38), and remembering that the encounter pair has a charge $(z_A + z_B)$, we find

$$\log(k/k_0) = 1.018z_Az_B\sqrt{I} \qquad (\text{E } 41)$$

where k_0 is equal to $k_r K_{AB}$, and is the rate coefficient at zero ionic strength. Equation (41) has had ample experimental verification provided the ionic strength is not too high ($\leq10^{-2}$ mol dm^{-3}). A plot of log k vs. $\sqrt{I}$ is a straight line with slope $\approx z_Az_B$, and so the rate increases with I for reactant ions of like charge, and decreases for oppositely charged ions. Debye–Hückel theory uses a model in which each ion is surrounded by an ionic atmosphere of equal but opposite charge, which lowers the chemical potential of the central ion by partial neutralization of its charge. This effect is more pronounced the higher the ionic charge. If A and B have like charges, high ionic strength will favour formation of the more highly charged encounter pair, and increase the reaction rate.

(+ve slope)

(+ve slope)

more stabilis

The kinetic salt effect can give useful information on transient ions, for which conventional equilibrium techniques are inapplicable. The carbonate radical ion, CO_3^{-}, is formed by photolysis of $[Co(NH_3)_4CO_3]^{+}ClO_4^{-}$. It is short-lived ($t_{\frac{1}{2}} \approx 10^{-2}$ s) but it may be studied by flash photolysis combined with absorption spectroscopy, since it has an absorption band at 600 nm. It reacts with indole-3-propionic acid (IPA), which exists as the conjugate base ($z_B = -1$) at pH 6. The rate of reaction between the carbonate radical and IPA has been measured as a function of ionic strength. At pH 12, the normal salt effect for $z_Az_B = +1$ was observed, but at lower pH the slope of log k versus $\sqrt{I}$ was reduced until it became zero at pH 7. Presumably protonation of CO_3^{-} is occurring, and at the lowest pH values the reaction is between HCO_3 and IPA for which $z_Az_B = 0$. At intermediate pH, the reaction mixture contains both CO_3 and HCO_3, and the mean value of z_Az_B lies between 0 and 1. Figure 6.7 shows a plot of the rate

more H+

∴ slope = 0

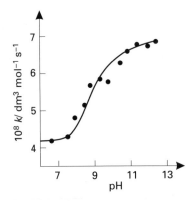

Fig. 6.7 The dependence of the rate coefficient for CO_3^- + IPA on pH at constant ionic strength.

coefficient versus pH at constant ionic strength. At high pH the rate is accelerated by the salt effect; at low pH the normal rate is observed. The plot has the form of a neutralization curve, and gives a pK_a of 9.6 for the carbonate radical. CO_3^- and HCO_3 react with IPA at the same rate in the limit of zero ionic strength.

At higher ionic strengths, eqn (40) is no longer valid, and should be replaced by

$$\log \gamma_A = -A' z_A^2 \sqrt{I}/(1 + B'a\sqrt{I}) \tag{E 42}$$

where B' is a constant and a is the radius of the ion. This equation has been used to determine the sizes of transient ions.

Box 6.2 The effect of ionic strength on diffusion-controlled reactions

The above treatment is only strictly applicable to activation-controlled reactions, since we have assumed an equilibrium concentration for [{AB}], the encounter complex. It is possible to examine the effect of ionic strength on diffusion-controlled reactions between ions. The strategy is to note that the potential between the reactive ions is no longer Coulomb, but is now screened by the presence of the other ions.

$$V(r) = \left(\frac{z_A z_B e^2}{4\pi\varepsilon_0\varepsilon_r r}\right) e^{-r/r_D} \tag{E 43}$$

where r_D is the Debye length, which decreases with increasing I. We see that the effect of increasing ionic strength will have qualitatively the same effect as for activation-controlled reactions: for similarly charged ions the ionic atmosphere reduces the effect of the repulsive potential and so enhances the rate, whilst for oppositely charged ions the potential is attractive and screening reduces the rate. A quantitative treatment (Logan, S. R. (1967). Effects of temperature on the rates of diffusion controlled reactions. *Transactions of the Faraday Society*, 1712–19) leads to an identical expression to eqn (E 41).

6.5.2 The effect of pressure on the rate coefficient

It is much more difficult to study the effect of pressure on a solution phase reaction than it is to study the effect of temperature. The reaction rate is comparatively insensitive to changes in pressure and very substantial pressures (up to several kilobars) must be applied before reasonably large changes can be observed. There are two major reasons for studying such changes. Firstly, as we shall see, they provide a useful insight into the mechanism of the reaction. Secondly, we often wish to understand processes which occur under conditions which are difficult to reproduce in the laboratory. Detonation provides a good example of such a process, since it involves the generation of very high pressures indeed, in the range 10–100 kbar. The energy providing this massive compression derives from chemical reactions and an important parameter is the rate at which this energy is generated. Many of the component reactions involve free radicals whose rate coefficients can be measured, using the techniques described in Chapter 2, but it is difficult to make such measurements at pressures greater than a few atmospheres. Thus, to understand and model detonation processes, we must be able to assess what will be the effects of large increases in pressure on these reaction rates.

We rewrite eqn (31) as

$$\ln k = \ln(\kappa k_B T/h) - \Delta G^{\ddagger}/RT. \tag{E 44}$$

Differentiating with respect to pressure, at constant temperature, and recognising that $(\kappa k_B T/h)$ is independent of pressure,

$$(\partial \ln k/\partial p)_T = -\Delta V^{\ddagger}/RT \tag{E 45}$$

where $\Delta V^{\ddagger}$ is the volume change of activation, i.e. it is the difference between the partial molar volume of the transition state and that of the reactants. The derivation of eqn (45) is discussed in somewhat greater detail in Box 6.3.

Box 6.3 Derivation of eqn (45)

The derivation of eqn (45), and the precise meaning of $\Delta V^{\ddagger}$ deserve more detailed consideration. The chemical potential, μ, for a solute A, molality m_A, is given by:

$$\mu_A = \mu'_A + RT\ln(m_A \gamma_A) \tag{E 46}$$

where γ_A is the activity coefficient and μ' the chemical potential in the reference state, which is the state with unit activity, at the pressure p of the system. Note that the solute based standard state is employed so that $\gamma_A \to 1$ as $m_A \to 0$ and the standard state is hypothetical (see Atkins, P. W. (1994). *Physical Chemistry*, (5th edn), p. 233. Oxford University Press). For the equilibrium

$$A + B \rightleftarrows AB,$$

$$\Delta G = \mu_{AB} - \mu_A - \mu_B = \Delta G' + RT \ln K_m = 0 \tag{E 47}$$

where

$$K_m = \{m_{AB}\gamma_{AB}/m_A\gamma_A m_B\gamma_B\}_{eq}. \tag{E 48}$$

Thus

$$(\partial \ln K_m/\partial p)_T = -1/RT(\partial \Delta G'/\partial p)_T = -\Delta V'/RT \tag{E 49}$$

where

$$dG = VdP - SdT$$
$$\left(\frac{\partial G}{\partial P}\right)_T = V$$

$$\Delta V' = V'_{AB} - V'_A - V'_B$$

is the change in partial molar volumes under standard state conditions (i.e. at infinite dilution). This result may be contrasted with the usual derivation of the equilibrium constant at pressures close to atmospheric, where the μ'_A is replaced by u°_A, the standard state (at a pressure of 1 atm). Very high pressures (≈ 1000 atm) are required before the pressure dependence of K_m becomes significant.

Turning to kinetics and following the treatment on p. 156, we find, for the rate coefficient ignoring $\varkappa$ the transmission coefficient,

$$k = (k_B T/h) K^{\ddagger} \gamma_A \gamma_B / \gamma_{AB}^{\ddagger} \tag{E 39}$$

where strictly, $K^{\ddagger}$ is $K_c^{\ddagger}$, the equilibrium constant in terms of concentrations (moles per unit volume, rather than mass of solvent). Assuming that the solution is dilute

$$c_i = (m_i 1000/M_0 V_0) \tag{E 50}$$

where V_0 is the molar volume of the solvent and M_0 its molar mass, thus

$$K_c^{\ddagger} = K_m^{\ddagger} V_0 \tag{E 51}$$

and

$$k = (k_B T/h) K_m^{\ddagger} V_0 (\gamma_A \gamma_B / \gamma_{AB}^{\ddagger}) \tag{E52}$$

The pressure dependence of k is given by

$$(\partial \ln k/\partial p)_T = (\partial \ln K_m^{\ddagger}/\partial p)_T + (\partial \ln V_0/\partial p)_T + \{\partial \ln(\gamma_A \gamma_B/\gamma_{AB})/\partial p)_T\}. \tag{E 53}$$

By definition, $-\Delta V'/RT$ $-\varkappa_0$ $-\Delta V/RT + \Delta V'/R$

$$(\partial \ln V_0/\partial p)_T = -\varkappa_0 \tag{E 54}$$

where $\varkappa_0$ is the coefficient of compressibility of the solvent, whilst the activity coefficient term may be found from the above equations remembering that the molality is independent of pressure:

$$\{\partial \ln(\gamma_A \gamma_B/\gamma_{AB}^{\ddagger})/\partial p\}_T = -1/RT\{\partial[(\mu_{AB} - \mu_A - \mu_B) - (\mu'_{AB} - \mu'_A - \mu'_B)]/\partial p\}_T$$

$$= -1/RT\{(\partial(\Delta G/\partial p)_T - (\partial(\Delta G'/\partial p)_T\}$$

$$= -(\Delta V - \Delta V')/RT. \tag{E55}$$

where ΔV is the difference in molar volumes of the activated complex and the reactants under the experimental conditions. Combining the above eqns gives

$$(\partial \ln k/\partial p)_T = -\Delta V/RT - \varkappa_0 \tag{E 56}$$

For water at 298 K and 1 atm, $\varkappa_0$ is 1.1×10^{-6} m^3 mol^{-1}, whilst for organic solvents it may be higher by a factor of three or four. It forms, therefore, a significant term.

A more detailed discussion of the effect of pressure on equilibria and activity coefficients may be found in Pitzer, K. S., and Brewer, L. (1961). *Thermodynamics* Chapter 20. (McGraw–Hill, New York), whilst a discussion of the effect on rate coefficients may be found in Weall, K. E. (1967). *Chemical Reactions at High Pressures*, Chapters 5 and 6. Spon, London.

The implication of eqn (45) is that the rate coefficient increases with increasing pressure if there is a volume decrease on forming the transition state. We might have predicted such a finding, given the thermodynamic basis of our current arguments, from our knowledge of Le Chatelier's principle. We can see, therefore, that experimental determinations of the pressure variation of the rate coefficient will give us information about the structure of the transition state.

We can consider the volume of activation, $\Delta V^{\ddagger}$, to be derived from two contributions, $\Delta V_1^{\ddagger}$ the volume change which occurs when the reactants form the transition state, and $\Delta V_2^{\ddagger}$, the volume change in the solvent, consequent on the formation of the transition state. For example, consider the hydrolysis of 2-chloromethylpropane (*t*-butyl chloride) in a polar solvent (such as an ethanol/water mixture), which involves the unimolecular rate-determining step

$$(CH_3)_3CCl \rightarrow (CH_3)_3C^+ + Cl^-. \tag{R 5}$$

The transition state has a stretched C–Cl bond, with an increased polarity. If we assume that the bond length increases by ≈ 0.1 nm, then a reasonable estimate of $\Delta V_1^{\ddagger}$ is $10^{-10}(\pi d^2/4)$, where d is the van der Waals diameter of the chlorine atom, giving $\Delta V_1^{\ddagger} \approx 6 \times 10^{-6}$ m^3 mol^{-1} and on that consideration alone the rate would be expected to decrease with pressure. However, the opposite is observed with k_5 increasing by a factor of 2.5 on increasing the pressure to 1500 bar, corresponding to $\Delta V^{\ddagger} \approx -1.5 \times 10^{-5}$ m^3 mol^{-1}. This discrepancy must be ascribed to a substantial decrease in the volume of the solvent, with $\Delta V_2 \approx -2.1 \times 10^{-5}$ m^3 mol^{-1}. The increased polarity in the C–Cl bond leads to stronger intermolecular bonding between the reactant and the polar solvent molecules and an overall decrease in volume on activation.

Solvent effects are particularly marked for bimolecular reactions involving ions. A well studied example is the Menschutkin reaction, which is a substitution reaction occurring between a tertiary amine and an organic halide, and which may be studied in both the forward (f) and reverse (r) directions:

$$CH_3I + C_6H_5N(CH_3)-C_2H_5 \rightleftharpoons [C_6H_5N(CH_3)_2-C_2H_5]^+I^- \tag{R 6, -6}$$

where $\Delta V_f^{\ddagger} \approx -2.0 \times 10^{-5}$ m^3 mol^{-1}, $\Delta V_r^{\ddagger} \approx 3.2 \times 10^{-5}$ m^3 mol^{-1}. The data demonstrate the dominance of the dipole/ or ion/solvent effects. An increase in charge separation, which occurs on forming the transition state, leads to a decrease in volume and a consequent enhancement of the rate coefficient on increasing the pressure. In the reverse reaction, in which the ion pair forms a polar transition state, there is a decrease in charge separation, so that the solvent is less tightly held and $\Delta V^{\ddagger}$ is positive.

$\Delta V_1^{\ddagger}$, the change in volume associated with the geometrical modification of the reactant(s), plays a more dominant role in reactions in which there is little charge separation, and hence less ordering forces on the solvent. For example, the dissociation of pentaphenyl ethane:

$$(C_6H_5)_3C-CH(C_6H_5)_2 \rightarrow (C_6H_5)_3C + CH(C_6H_5)_2 \tag{R 7}$$

has an average $\Delta V^{\ddagger}$, over the pressure range 1–1500 bar, of 1.3×10^{-5} m^3 mol^{-1}, which corresponds to a bond extension in the transition state of approximately 10 per cent.

The decomposition of diazonium cations in water provides an example of the use of volumes of activation to elucidate the reaction mechanism. Two routes are possible:

$$RN_2^+ + H_2O \rightarrow ROH_2^+ + N_2 \tag{R 8}$$

or

$$RN_2^+ \rightarrow R^+ + N_2 \text{ (slow)} \tag{R 9}$$

$$R^+ + 2H_2O \rightarrow ROH + H_3O^+ \text{ (fast)} \tag{R 10}$$

The rate-determining step is bimolecular in the first mechanism and unimolecular in the second, but the two routes cannot be distinguished in this way, because the solvent is the second reactant in the initial mechanism, so that the kinetics are pseudo-first-order. Experimental measurements on the effect of pressure on substituted benzene diazonium ions show a large positive value of $\Delta V^{\ddagger}$ ($\approx 10^{-5} \text{ m}^3 \text{ mol}^{-1}$), suggesting that the second mechanism, which involves a bond extension, is operative.

Pressure effects also occur in diffusion-controlled reactions and sometimes $\Delta V^{\ddagger}$ values are quoted, but they are difficult to interpret. The major effects arise from the decrease in the diffusion coefficient D_{AB} as the pressure is increased, which in turn is associated with the increase in viscosity. An interesting example occurs in the recombination of iodine atoms, which has been studied in the vapour phase from pressures of ≈ 1 atm up to several kilobars. In the gas phase, the reaction is third-order overall, because of the need to remove energy from the newly formed molecule. An increase in pressure leads, in consequence, to an increase in the pseudo-second-order rate coefficient k'_{11}

$$\left(+M\right) I + I \rightarrow I_2. \tag{R 11}$$

As the pressure continues to increase, the fluid viscosity increases and eventually the approach of the iodine atoms becomes slower than their reaction on encounter. Thus the reaction becomes diffusion-controlled and further increase in pressure leads to a decrease in k_{11}.

6.6 The dynamics of solution reactions

Thus far, our more detailed discussions have centred on the effects of diffusion. Reaction itself has only appeared via the parameters k_r and K_{AB}, which have served to determine the inner boundary condition of the diffusion equation, i.e. how rapidly the A,B pair react once they encounter one another. This device conceals a central kinetic problem, which we discussed at great length for gas phase reactions in Chapters 3 and 4. For example, if we were studying the reaction:

$$NH_4^+ + OH^- \rightarrow NH_3 + H_2O \tag{R 12}$$

then we ought to consider, in detail, the factors which determine the rate of proton transfer which presumably involves passage over a potential energy surface not dissimilar to that shown in Fig. 3.9. We could try to apply transition state theory, which would require a knowledge of the structure and vibrational frequencies of the reactants and the activated complex. But what is the role of the solvent? How do we incorporate the effects of the constant buffetings by the solvent molecules of the reacting molecules, as they move over the reaction potential energy surface? It is not possible, in the space available, to discuss this problem in detail, but some consideration of the basic ideas is worthwhile.

We can imagine, for an A + BC atom transfer reaction, a potential energy surface of the type shown in Fig. 3.10. The solution phase surface differs from that for the gas phase because of the influence of the solvent; this would be especially important for reactions involving charge transfer. An obvious consequence of such an effect is that we cannot use simple gas phase partition functions as we did on pp. 70–72, and it is generally necessary to use free energies, as in our discussion in Section 6.5. Nevertheless, the reactants do have vibrational modes, even if the frequencies are modified by the presence of solvent molecules. As with our earlier discussion on transition state theory, we can split the vibrational modes into those along the reaction coordinate (the reactive modes) and what might be termed the non-reactive modes. Following an approach comparable to that adopted on pp. 76–7 (Chapter 3), the rate coefficient is given by

$$k_r = v_R \exp(-\Delta G/RT) \tag{E 57}$$

where v_R is the average velocity of the system in the reactive mode through the transition state and ΔG is the free energy change between the encounter complex {A, BC} and the transition state. Provided a Boltzmann distribution pertains, then v_R may be replaced by $(k_B T/2\pi\mu)^{\frac{1}{2}}$:

$$k_r = (k_B T/2\pi\mu)^{\frac{1}{2}} \exp(-\Delta G/RT) \tag{E 58}$$

where μ is the reduced mass of the system.

It is now necessary to consider any dynamic effects of the solvent. The constant buffetings donate energy to the encounter complex and extract energy from it so that, provided the interactions are strong enough, and the collisions frequent enough, we might expect them to induce a Boltzmann distribution in the complex at all stages of its passage over the potential energy barrier, so that transition state theory would be applicable because of the maintenance of this equilibrium distribution without the need to invoke the dynamical requirements we discussed in Chapter 4. There is a further important factor which arises from the way in which the solvent interacts with the reactive mode. In the gas phase, provided there is sufficient energy in this mode and there are no bizarre features on the surface, the system moves smoothly over the energy barrier. In the liquid phase, this motion can be interrupted by collision and energy transfer with the solvent molecules. The direction of motion could even be reversed leading, possibly, to a recrossing of the transition state and a consequent reduction in the rate coefficient. In the limit of very frequent collisions, we might think of the system *diffusing* over the reaction surface with the frequent changes of direction leading to a Brownian motion.

These arguments are very qualitative, but give the essence of some of the more important dynamic solvent effects. The theory lags some way behind that of gas phase reactions in terms of its quantitative application, but rapid advances are being made in the development of an overall theoretical framework.

Study notes

It is difficult to suggest further reading which is easily digestible. The papers generally invoke frictional effects to describe the the effect of the solvent and employ friction coefficients. In order to understand some of the concepts, it is necessary to do some background reading on liquid phase dynamics. Articles dealing with correlation functions and the Langevin eqn could be consulted with profit.

A readable account of solvent effects, and the paper on which the discussion outlined above is based, is by Grote, R. F., and Hynes, J. T. (1981). Reactive modes in condense phase solutions. *Journal of Chemical Physical Chemistry*, **74**, 4465–75. It is quite mathematical, but Sections I, II, and III should prove understandable, provided the general principles are addressed, rather than the detailed mathematics.

Hynes has discussed a number of dynamical problems in solution in a review article (Hynes, J. T. (1985). Chemical reaction dynamics in solution. *Annual Review of Physical Chemistry*, **36**, 573–98). A major section of the review deals with the problem of solvent induced recrossing of the potential barrier and applications to atom transfer and isomerization reactions are considered. Further sections discuss how the effect of tunnelling in solution reactions is beginning to be considered and, especially relevant for Section 6.7, cage effects are also examined.

6.7 Cage reactions

Although the encounter complex plays an important role in our phenomenological treatment (Section 6.2), it is difficult to obtain direct information on it from a study of the kinetics of reactions in solution. Our understanding of the overall reaction process is limited and the relative diffusional behaviour of the proximate pair of reactants is only one aspect of this overall process. Is it possible to obtain information more directly?

If molecular iodine vapour is exposed to light of wavelength less than 499 nm, the quantum yield for dissociation (i.e. the probability of dissociation on absorption of a photon, see p. 277) is unity. The molecule is excited to a repulsive electronic state from which it rapidly dissociates before the energy can be removed. In solution, the separation of the iodine atoms is hindered by the surrounding solvent molecules, and there is now a non-zero probability that they will reform the iodine molecule, undergoing a process termed *geminate recombination*. In an inert solvent, each of the remaining iodine atoms, which have diffusively separated and so escaped geminate recombination, will eventually encounter another iodine atom originating from a different iodine molecule, and undergo *random recombination*. If we are interested only in the iodine atoms which escape geminate recombination, then we see that the effective quantum yield Φ is reduced below unity. At 435.8 nm, Φ is 0.66 whilst decreasing the wavelength to 404.7 nm increases Φ to 0.83. These observations can be rationalized by considering the excess kinetic energy (photon energy − bond dissociation energy of I_2)/2, of the iodine atoms. As the photon energy increases, the atoms are produced with greater kinetic energy and are therefore more likely to break out of the reaction cage. If the experiment is performed in hexachlorobutadiene, which has a higher viscosity than hexane, the quantum yield at 435.8 nm decreases to 0.075 reflecting the inability of the iodine atoms to penetrate the walls of their stronger cage and hence the increased probability of geminate recombination.

These measurements were made by a classical 'scavenging' technique which relies on the differences in the timescales of geminate and random recombination. Geminate recombination involves diffusion and re-encounter over very short distances (a few tenths of a nanometre) and takes place on a timescale of 10–100 ps. The half-life $t_{\frac{1}{2}}$ for random recombination, a diffusion-controlled process which is second order in iodine, is equal to $\{k_d[\mathrm{I}]_0\}^{-1}$, where $[\mathrm{I}]_0$ is the initial concentration of iodine atoms and k_d the diffusion-controlled rate coefficient. Taking $[\mathrm{I}]_0 \approx 10^{-6}\,\mathrm{mol\,dm^{-3}}$, $t_{\frac{1}{2}}$ is $\approx 8 \times 10^{-5}\,\mathrm{s}$ in hexane. The two processes may be separated by adding a solute which reacts with I such as allyl iodide. Addition of sufficient quantities of the iodide intercepts all of the iodine atoms which have escaped from their initial cage, before they can undergo random recombination and a kinetic analysis enables the number escaping geminate recombination to be determined. Much larger concentrations are needed to compete with geminate recombination, since a pseudo-first-order rate coefficient k_s of 10^{10}–$10^{11}\,\mathrm{s}^{-1}$ is needed in order to compete with the rapid timescale of geminate re-encounter.

The advent of picosecond lasers has enabled more direct experiments to be made using flash photolysis. Figure 6.8 shows a decay profile for the iodine atom concentration following the photolysis of iodine in carbon tetrachloride by a 6 ps pulse of light of wavelength 530 nm. The molecular iodine concentration was followed by absorption spectroscopy, also at 530 nm, and the atom concentration found by difference. Following the dissociation pulse, the atom concentration rises and then falls rapidly following geminate recombination. It does not fall to zero, however, because a substantial fraction escape and subsequently undergo random recombination on a much longer timescale.

Study notes
1 The geminate recombination process may be described mathematically using a time–dependent diffusion eqn directly analogous to that given on p. 149

$$\mathrm{d}p/\mathrm{d}t = D(\mathrm{d}^2 p/\mathrm{d}r^2 + (2/r)\mathrm{d}p/\mathrm{d}r) \tag{E 59}$$

where p is the probability of finding the pair of iodine atoms at separation r at time t. The eqn is solved subject to the boundary conditions $p(r_0,0) = 1$, i.e. the pair have

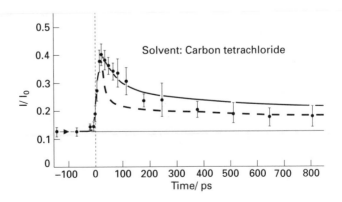

Fig. 6.8 Decay of iodine atoms as a function of time following excitation of iodine in CCl_4. The broken line represents the expected decay of the iodine atoms for simple geminate recombination.

an initial separation r_0, $p(r_{AB},t) = 0$ at all times i.e. the particles react instantaneously on encounter at the separation r_{AB} and $p(\infty,t) = 0$ at all times, i.e. the particles cannot attain an infinite separation in a finite period of time. The equation can be solved to obtain the probability, $\Omega(t)$, that the species have not reacted by time t:

$$\Omega(r_0,t) = \text{erf}\{(r_0-r_{AB})/\sqrt{(4Dt)}\} \tag{E 60}$$

where erf is the error function given by

$$\text{erf}(x) = (2/\pi) \int_0^x \exp(-y^2)\,dy.$$

The values of the error functions can be found in standard mathematical tables. The probability of escape, Ω, may be determined by letting t tend to infinity, which gives the simple expression

$$\Omega(r_0) = 1 - (r_{AB}/r_0). \tag{E 61}$$

Forces between the diffusing particles require the introduction of an additional term to the right hand side of eqn (59), which incorporates $V(r)$, the potential energy of interaction. For ions, where $V(r)$ is the Coulomb potential the escape probability is

$$\Omega(r_0) = 1 - (r_{eff}/r_{0,eff}) \tag{E 62}$$

where r_{eff} is defined in eqn (22) and $r_{0,eff}$ is defined similarly, replacing r_{AB} with r_0 (Fig. 6.9). When we remember that $r_{eff} \approx r_c$ for oppositely charged ions (Table 6.3), the origin of the term 'escape distance' for r_c becomes apparent.

2   The picosecond experiments on iodine were first reported in 1974 (Chuang, T. J., Hoffman, G. W., and Eisenthal, K. B. (1974). Picosecond studies of the cage effect and collision induced predissociation of iodine in liquids. *Chemical Physics Letters*, **25**, 201–5). Ten years later, the interpretation of the results was still the subject of some controversy. The problem is that the observed decay is too slow for geminate recombination, as may be seen from Fig. 6.8. The broken line represents the theoretical prediction of the recombination process, matched to the longer time atom concentration (where the theory should be most accurate. The experimental

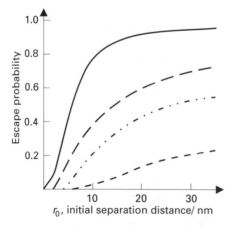

Fig. 6.9 Escape probability of a pair of unlike unit charges initially formed with a distance of separation r_0 in a solvent of relative permittivity: $\varepsilon = 10$ (solid line), $\varepsilon = 7$ (dashed line), $\varepsilon = 4$ (dashed and dotted line), $\varepsilon = 2$ (short-dashed line), for $r_{AB} = 0.5$ nm at room temperature.

data show a significantly slower decay than this theoretical prediction. Problems arise because the recombination process is not just a simple recombination along a potential energy surface like that shown in Fig. 5.11, p. 138. Instead, there are many interacting excited states of I_2 which can become involved. A detailed discussion of the electronic states of I_2 may be found in *Photochemistry of Small Molecules* by Okabe, H. (1978). (Wiley, New York) p. 187. Several experimental investigations of the iodine recombination have been taken and a description of one of them, which includes a discussion of previous experimental and theoretical work, may be found in the paper by Kelley, D. F., Abul–Haj, N. A., and Jang, D.-J. (1984). Geminate recombination of molecular iodine. *Journal of Chemical Physics,* **80**, 4105–11). The paper is reasonably easy to follow and contains a useful 'simplified' diagram of the relevant potential energy curves for the ground and excited electronic states of I_2. A few comments here may prove helpful. The initial excitation pulse at 533 nm raises I_2 to an excited, bound electronic state (the B state) which then undergoes predissociation to yield iodine atoms. Kelley *et al.* observed a reduction in light absorption over the wavelength range 450–575 nm, corresponding to the dissociation of ground state iodine and an increase in absorption over the wavelength range 575–800 nm which they assigned to formation of the bound excited A or A' states of I_2. The formation of the A or A' states was very rapid (taking $\approx$ 30 ps) and they suggest that this corresponds to geminate recombination into these particular states. Formation of the ground state is slower and they were unable to decide whether the rate-determining step is geminate recombination or a transition from one electronic state to another. Iodine recombination is not a simple process!

3   A more modern approach to the description of the diffusion kinetics of recombination in solution into a single electronic state may be found in a paper by Hynes, J. T., Kaprai, R., and Torrie, G. M. (1980). Stochastic trajectory simulation of iodine recombination. *Journal of Chemical Physics,* **72**, 177–88. It specifically discusses the iodine case, but concentrates simply on recombination into the ground electronic state, without taking into account the interaction with excited electronic states. The paper contains some detailed arguments and some new (to us) mathematical development, but it is clearly written and at least sections of it can be read with profit, particularly I and II, which discuss the background and explain how numerical simulations based on the Langevin eqn may be employed with a realistic potential between the iodine atoms, which incorporates the effects of the solvent structure.

4   A short review article by Wang, Y. and Eisenthal, K. B. (1982). Picosecond laser studies of ultrafast processes in chemistry. *Journal of Chemical Education,* **59**, 482–9 describes the methods which may be employed to study processes on a picosecond timescale and includes an outline of the 1974 iodine recombination experiment and of a study of diffusion-controlled reaction in the time regime in which the rate coefficient is time dependent.

6.7.1 Cluster reactions

We have spent quite some time on cage reactions. This emphasis reflects the recent interest in the area, which is likely to lead to a significant improvement in our fundamental understanding of liquid phase rate processes. In many ways, the experimental investigation and theoretical description of this type of correlated process has the same potential for exciting and revealing developments as does the field of gas phase reaction dynamics. An important new and rapidly growing field which attempts to bridge the gap between the gas and liquid phases is that of cluster studies. Under certain conditions, small clusters of molecules may be generated in the throat of a molecular beam. Molecules surrounded by a known number of solvent molecules can then be selected and studies made under carefully defined and readily variable conditions. The sizes of clusters which can be generated is enormous, ranging from simple Ar–HCl triatomic clusters, to several hundred atom rare gas clusters. A number of different examples of cluster sizes and geometries are shown in Fig. 6.10.

Study notes
A relevant example for our discussion involves the work of Lineberger and co-workers (1991). (Papanikolas, J. M., Gord, J. R., Levinger, N. E., Ray, D., Vorsa, V., and Lineberger, W. C. (1991). Photodissociation and geminate recombination dynamics. *Journal of Physical Chemistry,* **95**, 8028–40). They were able to make a

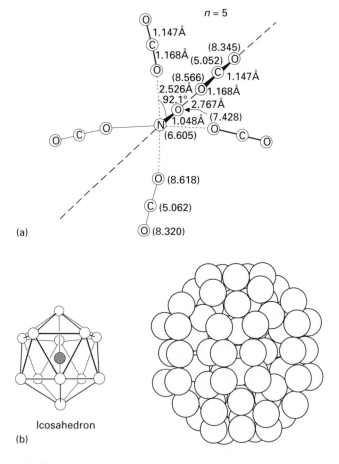

(a)

Icosahedron

(b)

Fig. 6.10 Schematic diagram showing some of the varieties and sizes of clusters which have been studied (a) possible structure of $(CO_2)_5NO^+$; (b) potential 'magic' (very stable) structures of $(Ar)_{12}NO^+$ and Kr_{111}^+.

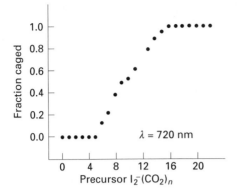

Fig. 6.11 Branching ratio for the production of caged photofragment ions as a function of precursor cluster size.

variety of clusters of the form $I_2^-(CO_2)_n$. The generation of ionic clusters has considerable practical advantages in that different sized clusters can be accurately 'mass selected' for isolated study. The I_2^- ions were then dissociated with a picosecond laser pulse and the fraction of recombination measured. The gradual build up of the first solvation shell of the ion can be seen from Fig. 6.11 which shows the fraction of iodine atoms that are caged following photolysis. For $n < 6$ complete dissociation occurs, however, for $n > 6$ the quantum yield for atom production decreases and by $n = 16$, all the dissociated I_2^- ions recombine within the solvent cage, indicating that this point marks the completion of the first 'solvation shell'. Obviously the initial energy given to the cluster by the absorption of the dissociation photon must be dissipated in some way. Analysis of the clusters following recombination shows that a majority of this energy is used to evaporate several of the CO_2 molecules from the cluster. The paper gives a very readable account of complex formation, mass selection and the use of pico-second techniques.

Castleman, A. (1986). *Accounts of Chemical Research,* **19**, 413–15 focuses on the use of clusters to provide an ideal link between the gas and condensed phase.

6.8 Reactions of hydrogen and hydroxyl ions

The rate coefficients for a few representative reactions of H^+ and OH^- are shown in Table 6.4. Most of them correspond to diffusion-controlled reactions, and the high H^+ and OH^- diffusion coefficients (see Table 6.1) lead to very large rate coefficients. The magnitude of these diffusion coefficients are a consequence of

Table 6.4 Rates of Reaction for H^+ + base and OH^- + acid

Acid	Conjugate base	$k_{H+}/dm^3\,mol^{-1}\,s^{-1}$	$k_{OH-}/dm^3\,mol^{-1}\,s^{-1}$	T/K
H_2O	OH^-	1.4×10^{11}		298
H_2O (ice)	OH^-	8.6×10^{12}		263
D_2O	OD^-	8.4×10^{10}		298
H_2CO_3	HCO_3^-	4.7×10^{10}		298
NH_4^+	NH_3	4.3×10^{10}	3.4×10^{10}	293
Dimethylanthranilic acid	—	—	1.1×10^7	285

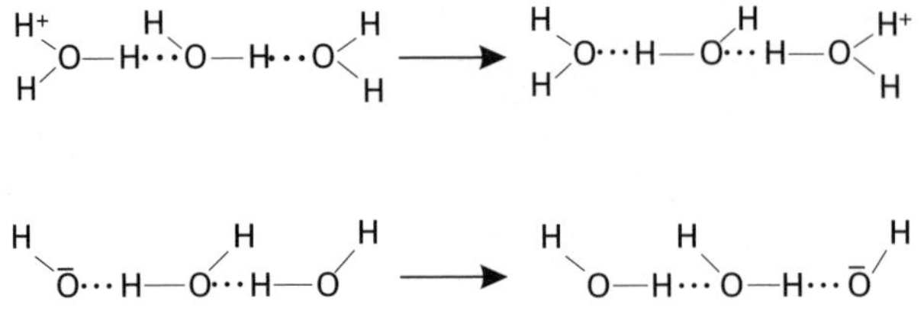

Fig. 6.12 Molecular structures for the rapid transfer of H^+ and OH^- ions.

the strong hydrogen bonding present in water. There is well–documented evidence that the proton associates with a water molecule to form the hyroxonium ion, H_3O^+. The positive charge is distributed throughout the ion, and this leads to strong hydrogen bonding with three more water molecules, which form a secondary hydration shell. The positive charge is delocalized, to some extent, over the three outer molecules, and there is some tertiary hydration, consisting of labile H_2O bridges. Figure 6.12 shows how a structure of this sort can lead to rapid H^+ and OH^- transport (the Grotthus mechanism). The mean lifetime of an H_3O^+ ion is thought to be $\approx 10^{-12}$ s.

It is pertinent to note that the proton mobility in ice is greater than that in water, and that the rate of the first reaction in Table 6.4 is faster in ice. The rigid hydrogen-bonded structure in ice is optimized for facile proton transfer from secondary to tertiary water molecules, and the rate–determining step is proton transfer from H_3O^+ to the secondary hydration shell. In water, the structure is less ideal, and the rate–determining step is now re-orientation of the water molecules in the tertiary structure.

Since the diffusion coefficients of H^+ and OH^- are known, the rate coefficient may be used to determine the encounter distance r_{AB}, from eqn (10). The validity of this approach is questionable, since we are using bulk values of D and ε_r, but it is interesting that a large value of 0.8 nm is obtained, which is compatible with the proposal that $H_9O_4^+$ and $H_7O_4^-$ are the reacting species, and not H_3O^+ and OH^-.

Other proton transfer reactions are generally diffusion-controlled, provided the free energy change for the reaction is negative. They are slower than the H^+, OH^- reactions, because of the smaller diffusion coefficient of the large reactant partner. There are some exceptions. For example N,N-dimethylanthranilic acid reacts slowly with OH^- because the proton is involved in intramolecular hydrogen bonding, and reaction can only take place if this bond is broken.

6.9 The solvated electron

Solvated electrons may be formed by the interaction of high energy radiation with liquids, the initial step being the production of ionized and of excited solvent molecules. For water the sequence of events shown in Fig. 6.13 occurs. The electron is initially free, but is rapidly trapped (-10^{-12} s) in solvent cavities, and is further stabilized following orientation of the solvent dipoles. In very pure water it has a half-life of $\leqslant 5 \times 10^{-4}$ s, and decays by reaction with water:

$$e_{aq}^- + H_2O \rightarrow OH^- + H. \qquad (R\ 13)$$

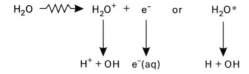

Fig. 6.13 Possible processes following pulsed radiolysis of water.

At first it was thought that the electron was too short lived $(t \approx 10^{-12} \text{ s})$ to be chemically significant, and only H and OH were presumed to be responsible for the reactions observed in radiolysed water. The presence of a second reducing agent R soon became apparent; for example R reacts with N_2O forming N_2, whilst H does not. R was thought to be e_{aq}^-, H_2^+, or some modified form of H. The effect of increasing the ionic strength on the relative rates of reactions such as R + N_2O and R + $Fe(CN)_6^{4-}$ showed that the charge on R is -1, providing strong evidence for the existence of the solvated electron. Later a broad absorption spectrum, peaking around 700 nm, was observed in the pulsed radiolysis of water, and assigned to e_{aq}^-. Since then the majority of rate measurements have been performed by pulse radiolysis.

The electron is a strong reducing agent, the rate coefficients for some of its reactions are shown in Table 6.5. Its rapid reaction with H^+ causes it to be efficiently converted to H at low pH. The diffusion coefficient of the hydrated electron is only $4.25 \times 10^{-9} \text{ m}^2 \text{ s}^{-1}$, i.e. it is slower than both H^+ and OH^-, and evidently there is no mechanism for electron transport comparable with the Grotthus mechanism. Diffusion possibly involves movement of a trap molecule, permitting the electron to jump out and hop to another trap, although in view of the low mass of the electron, tunnelling may play some role.

Table 6.5 Rate coefficients for reactions of $e^-(aq)$

Reactant	$k/\text{dm}^3 \text{ mol}^{-1} \text{ s}^{-1}$
H_3O^+	2.2×10^{10}
Ag^+	3.6×10^{10}
MnO_4^-	2.6×10^{10}
Naphthalene	5.4×10^9
F^-	$<2 \times 10^4$
H_2O	16.0

6.10 Electron transfer reactions

In many inorganic reactions, an electron is transferred directly from one molecule to another, without the participation of an intermediate solvated electron. Such reactions necessarily involve both oxidation and reduction, and are frequently termed redox reactions. They may follow one of two mechanisms.

6.10.1 Inner sphere

A ligand of one reactant displaces a ligand of the other, thus forming a bridge along which electron transfer can take place, e.g. Fig. 6.14. Red and Ox refer to

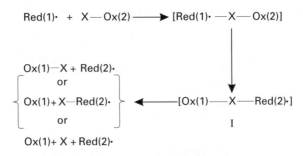

Fig. 6.14 Possible redox reaction mechanisms.

the reduced and oxidized forms of species (1) and (2) and the dot shows the position of the electron being transferred. The initial step involves ligand substitution, and frequently the only evidence supporting an inner sphere mechanism is the observation of a rate coefficient typical of ligand substitution. For example

$$V_{aq}^{2+} + Cu_{aq}^{2+} \rightarrow V_{aq}^{3+} + Cu_{aq}^{+} \tag{R 14}$$

where k_{14} is 26.6 dm^3 mol^{-1} s^{-1}, which is similar to that found for water exchange at V_{aq}^{2+}.

It is often very difficult to prove that the ligand is directly involved in the reaction, although some convincing experiments have been conducted on the oxidation of Cr^{2+}, e.g.

$$(NH_3)_5Co^{III}Cl^{2+} + Cr^{II}(H_2O)_6^{2+} \rightarrow [(NH_3)_5Co \cdots Cl \cdots Cr(H_2O)_5OH_2]^{4+}$$
$$\underset{I}{}$$
$$\rightarrow \rightarrow Co^{2+} + 5NH_3 + H_2O + Cr^{III}(H_2O)_5Cl^{2+} \tag{R 15}$$

The chromium ion has been oxidized, and at the same time one water ligand has been replaced by chloride. This suggests the intermediate I, which implicates Cl$^-$ as a bridging ligand in the redox process. The success of this model reaction depends on the relative kinetic stabilities of the various oxidation states involved. Cr(III) is stable with respect to ligand substitution, and the product ion is sufficiently long-lived to be detected. Cr(II) on the other hand, is labile and readily undergoes the incipient ligand substitution reaction to give I. The same is true for cobalt—the Co(III) complex is inert, but when electron transfer takes place, Cl$^-$ is readily displaced by water, and transferred to Cr(III).

6.10.2 Outer sphere

An electron is transferred rapidly (in 10^{-14}–10^{-15} s) with the coordination shells intact. For thermoneutral reactions, e.g.

$$Fe(H_2O)_6^{2+} + Fe(H_2O)_6^{3+} \rightarrow Fe(H_2O)_6^{3+} + Fe(H_2O)_6^{2+} \tag{R 16}$$

the rapidity of electron transfer imposes strict geometrical requirements. The Fe-O distances are 0.221 nm and 0.205 nm in Fe(II) and Fe(III) respectively. Electron transfer is much faster than nuclear vibration, and the molecules have no time to adjust to their new geometries during the transfer process (the Franck Condon principle). The reaction takes place more efficiently if the Fe^{II}–O bond contracts and the Fe^{III}–O bond lengthens beforehand. Such distortions are effected during the normal vibrational motion of the molecules, but the amplitudes are greater for molecules in the higher vibrational levels. As a result, the reaction is faster for vibrationally excited molecules, and it has been estimated that in the present case ≈ 23 kJ mol^{-1} of vibrational energy, out of total activation energy of 44 kJ mol^{-1}, are needed to achieve the required geometry changes.

Similar reactions include

$$\text{Fe(phen)}_3^{2+} + \text{Fe(phen)}_3^{3+}; \quad k = 10^5 \text{ dm}^3 \text{ mol}^{-1} \text{ s}^{-1} \qquad (\text{R } 17)$$

and

$$\text{IrCl}_6^{3-} + \text{IrCl}_6^{2-}; k = 10^3 \text{ dm}^3 \text{ mol}^{-1} \text{ s}^{-1} \qquad (\text{R } 18)$$

where phen = 1, 10-phenanthroline. The reactions may be studied by isotopic labelling.

The requirements of the Franck–Condon principle are less rigid for exothermic reactions, since some of the excess of energy of the reaction may be channelled into product vibrational motion. This leads to a smaller activation energy and a faster reaction, e.g.

$$\text{Fe(CN)}_6^{4-} + \text{Fe(phen)}^{3+} \rightarrow \text{Fe(CN)}_6^{3-} + \text{Fe(phen)}^{2+} \qquad (\text{R } 19)$$

$$k_{19} = 10^8 \text{ dm}^3 \text{ mol}^{-1} \text{ s}^{-1}$$

The 1993 Nobel prize for chemistry was awarded to R. A. Marcus for the development of the theory of electron transfer reactions. Marcus was also responsible for key developments in the theory of unimolecular reactions—he is the 'M' in RRKM theory (see p. 131).

6.11 Questions

6.1 Account for the relative magnitudes of the diffusion coefficients for the positive ions listed in Table 6.1. If the rate coefficient for the reaction of H$^+$ and OH$^-$ ions is 1.4×10^{11} dm^3 mol^{-1} s^{-1} what is the value of r_{AB}? Compare your answer to typical ionic radii.

6.2 Estimate the rate of recombination of iodine atoms (diameter 0.4 nm) in hexane and in propan−1,2,3,-triol at 300 K. You will need to use the data in Table 6.2 and eqn (15).

6.3 For the reaction H$^+$ + e$^-$ estimate the rate coefficient in water, methanol, and hexane. Use the data in Tables 6.1 and 6.3. Take the mean diffusion coefficient of the solvated electron to be 4.3×10^{-9} m^{2}s^{-1}.

6.4 The rate of reaction of solvated electrons with a potassium iron cyanate complex of unknown charge has been measured relative to that of e$_{aq}^-$ + N$_2$O as a function of ionic strength.

log (relative rate)	0.0	0.21	0.28	0.32	0.51
$\sqrt{I}/(1 + \sqrt{I})$	0.0	0.9	0.15	0.16	0.27

Determine the charge on the complex.

6.5 The first hydrolysis constant for Fe(H$_2$O)$^{3+}$ can be described by the equation

$$\log K_h = \log K_o - (4AI^{\frac{1}{2}})/(1 + aBI^{\frac{1}{2}}) \tag{E 63}$$

Show that the data given below are consistent with this eqn and not with the simplified version of (E 41)

I/M	0.0	0.1	1.0	2.67	3.0
$10^3 K_h/M$	6.7	2.75	1.60	1.16	1.28

The constants A and B are 0.509 and 3.29, calculate the distance of closest approach using the results of your graph. (see Birus, M. *et al.* (1993).

6.6 Determine the volume of activation at 300 K and at 1 atm and 3000 atm for the reaction

$$C_2H_5Br + CH_3O^- \rightarrow C_2H_5OCH_3 + Br^- \tag{R 20}$$

from the following data:

$P/$atm	1	3000	15000
$10^5 k/dm^3 mol^{-1} s^{-1}$	3.8	10	33.2

6.7 (i) Account for the observed volumes of activation for the reactions tabulated below. What sign would you expect for the entropy of activation for each reaction?

Reaction	Solvent	$\Delta V^{\ddagger}/cm^3 mol^{-1}$
$Co(NH_3)_5Br^{2+} + OH^- \rightarrow Co(NH_3)_5OH^{2+} + Br^-$	H_2O	9
$CH_2ClCO_2^- + OH^- \rightarrow CH_2OHCO_2^- + Cl^-$	H_2O	-6
$(C_6H_5CO)_2O_2 \rightarrow 2C_6H_5CO_2$	CCl_4	10
$(CH_3)_3CCl \rightarrow (CH_3)_3C^+ + Cl^-$	$EtOH/H_2O$	-16

(ii) For each reaction calculate the pressure required to change the rate coefficient by 50 per cent.

6.8 When $(CO_2)_{16}I_2^-$ clusters are irradiated with 720 nm light the average number of CO_2 molecules evaporated after geminate recombination is seven. Estimate the binding energy per CO_2 molecule and list any assumptions you have made in the calculation. Will this be an upper or lower bound to the true value? What other information would you require to make a more accurate-determination?

References

Atkins, P. W. (1994). *Physical Chemistry*, (5th edn), pp. 846–56. Oxford University Press.
Birus, M., Kujundz, N., and Pribanic, M. (1983). Complexation of iron (III) in aqueous solution. *Progress in Reaction Kinetics*, **18**, 171–271.

7 Surface reactions

7.1 Introduction

Many reactions are catalysed by solid materials, the reaction taking place more readily on the surface than in the homogeneous (gas or liquid) phase. We can rationalize this observation in the following way: in general the atoms or molecules which constitute the bulk solid have a high coordination number which must necessarily be lowered at the surface of the solid. Therefore the surface molecules have a high propensity for reaction, and in turn these reactions may act as steps for a catalytic cycle. Surface catalysis is of enormous industrial importance (e.g. the synthesis of ammonia via the Haber process) and makes a significant contribution to a number of atmospheric processes including some mechanisms for the formation of acid rain and, by reactions occurring on polar stratospheric clouds, to the production of the ozone hole.

Although we shall limit our discussion to reactions on solid surfaces and more specifically metal surfaces, we should note that a number of important processes, especially in atmospheric chemistry, can also occur on liquid surfaces. With liquids and droplets we face the further complication of separating reactions occurring on the surface from those taking place in the bulk solution. However, even in the latter case surface effects are important as the gas phase molecules must first be adsorbed onto the surface before diffusing into the bulk solution, and the gas–liquid interface has recently seen increased research activity.

Both academic and industrial research in surface science have recently been stimulated by the revolution in information technology. The ever shrinking sizes of electronic devices means that more and more information has to be put onto each silicon chip. These chips are made in a variety of ways, one of which involves the etching of the surface by reactive species such as fluorine atoms. Obviously gas-surface reactions are of crucial interest in this lucrative field in investigation.

Laboratory investigation of the nature of adsorbed species and the kinetics and dynamics of surface processes represents a major research area. We shall examine, very briefly, a few of the important aspects and make use of examples and references to further reading in an attempt to demonstrate the wide-ranging academic and industrial interest.

7.2 Adsorption on metal surfaces

A molecule can interact with a surface via a physical process (physisorption) in which the attractive force is van der Waals in nature and the enthalpy of adsorption, ΔH_{ad}, comparatively small (typically $\approx -40 \text{ kJ mol}^{-1}$). Strong chemical bonds can also be made between the surface and the adsorbate (chemisorption), for example ΔH_{ad} for CO on Pd is $\approx -150 \text{ kJ mol}^{-1}$, whilst dissociative adsorption can also occur. In the latter case the mechanism might involve initial physisorption of the molecule, followed by dissociative chemisorption to produce strongly bound atoms on the surface. Depending on the shapes of the potential energy surfaces for these processes (Fig. 7.1) dissociative adsorption may have an activation energy.

In a catalysed reaction, the adsorbed species are stable intermediates in a sequence of reactions and it is of advantage to characterize them in detail—to have information on their structure (the geometric position on the surface and the nature and geometry of the bond with the surface), their energetics (ΔH_{ad}) and dynamics. By the general term dynamics we mean the ease with which the species can migrate on the surface, the rate coefficient for desorption and its dependence on temperature, and the accommodation of the species on the

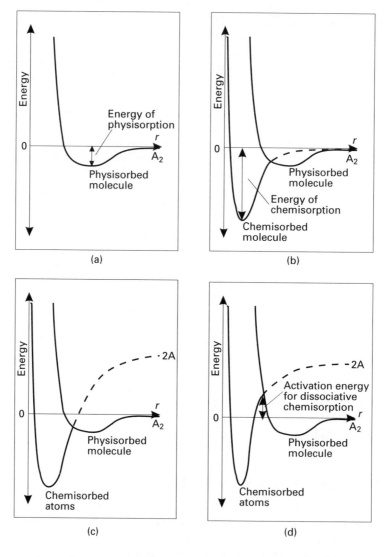

Fig. 7.1 (a) Physisorption, (b) non-activated chemisorption. It is possible for there to be an activation barrier between the physisorbed and chemisorbed states, (c) Non-activated dissociative chemisorption—the energy required to break the A–A bond is provided by the energy of physisorption. (d) Activated dissociative chemisorption.

surface—is the energy of the colliding species easily redistributed into the vibrational modes of the surface or does the species rapidly desorb before such a redistribution can occur? Space precludes a detailed discussion of most of these topics and we shall content ourselves by enumerating the techniques employed and listing references. The dynamical properties are more closely allied to kinetics and we shall spend a little longer on this area.

Catalytic activity not only involves the formation of so called chemisorbed species, but often the combination of two different chemisorbed species to form a new compound. Such reactions can only take place if the chemisorbed species have a degree of mobility on the catalyst surface. The catalytic ability of a series of transition metals often shows a maximum curve (Fig. 7.2). As one moves along the series of similar metals the strength of the bonds formed decreases, diminishing the rate of the absorption process, but also the lower bond strengths increase the mobility of the adsorbed species. The combination of these two opposing trends leads to a peak in catalytic activity of the transition metal species.

7.3 Structural characterization of surface species

Although industrial catalysts are often 'dirty', i.e. they are complex systems, often developed by empirical means, most laboratory studies that are of interest to us here, employ specific surface planes of single crystals, for the simple reason that the different planes show different behaviour. Ideally we want to be able to attribute any observed reactivity to a particular molecular process on the surface and this is only possible on a clean and well characterized metal surface.

The first problem confronting the surface scientist is the production and maintenance of a clean, well characterized surface. Surfaces can be cleaned by sputtering (bombarding with high energy inert ions) or heating. Once produced, the clean surface must be maintained, (which is a non-trivial process, even at

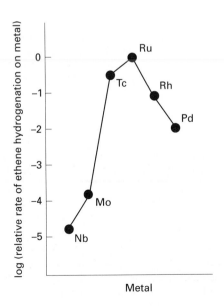

Fig. 7.2 Varying catalytic activity for the hydrogenation of ethene.

pressures in the region of 10^{-6} Torr the metal surface would be oxidized in minutes—see question 7.1) and this requires ultrahigh vacuums and associated
technology.

Secondly, we need to be able to characterize both the structure of the surface
and the molecular properties of the absorbent molecules. Low energy electron
diffraction (LEED) is the most commonly employed technique for characterizing both the structure of the surface and the arrangement of the adsorbate
molecules (or atoms) on it—the patterns resulting from the diffraction of low
energy (20–500 eV) electrons are determined in the presence and absence of the
adsorbate. The structure is further defined by determining the vibrational frequencies of the adsorbate on the surface using IR spectroscopy and electron
energy loss spectroscopy. UV photoelectron spectroscopy, which probes the
valence shell electrons in the adsorbate, has been used to study the modification
of the molecular orbitals on adsorption, whilst higher energy techniques, such as
X-ray photoelectron spectroscopy and Auger electron spectroscopy have been
used to identify surface species. Figure 7.3 shows the structure of CO adsorbed in
the (100) surface of palladium at a particular coverage—the CO is adsorbed
between the metal atoms, at right angles to the surface plane. The Pd–CO bond,
which has a frequency of 236 cm^{-1}, involves the carbon lone pair electrons and
back donation of electron density from the populated bands of the metal into the
$2\pi*$ antibonding orbitals of CO. The nature of the bonding can be seen from the
UV-PES of CO which shows the stabilization of the lone pair electrons and the
other experiments which show the lengthening of the CO bond in comparison to
its gas phase value. We conclude that the intermediates involved in surface reactions are capable of detailed structural characterization.

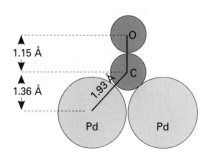

Fig. 7.3 The structure of CO adsorbed on
a Pd(100) surface at a coverage of $\theta = 0.5$.
Note the slight lengthening of the CO
bond over its gas phase value (1.128 Å).

Study notes
Surface characterization is a branch of spectroscopy in its own right and the interested reader will no doubt be able to find many texts (many of them from physics)
giving detailed accounts of the techniques introduced above. Good examples are
suggested below:
1   pp. 978–984 of Atkins, P. W. *Physical Chemistry,* (5th edn). Oxford University
Press provides more detail than that given in Section 7.3 above but the interested
reader is encouraged also to consult the more specific texts detailed below.
2   Chapters 5, 6, and 7 of *An Introduction to Chemisorption and Catalysis by Metals*
(Gasser, R. P. H. (1985). Oxford Science Publications) provide detailed accounts
of LEED, electron emission techniques, and electron energy loss spectroscopy. The
book also provides more background for other parts of this chapter.

3 Chapter 4 of *Catalysis at Surfaces* (Campbell, I. M. (1988). Chapman and Hall, London) provides a very readable account of both destructive (e.g. thermal desorption, secondary ion mass spectroscopy) and non-destructive techniques for investigating the properties of surfaces.

4 A more detailed and comprehensive account of surface investigation and analysis may be obtained from Chapter 3 of *The Chemical Physics of Surfaces* (Morrison, S. R. (1990). Plenum, New York) which contains many references to the original literature.

We shall see an example of the importance of surface characterization in Chapter 11 on oscillating reactions. In this example the nature of the surface changes as CO is adsorbed. The new surface is now activated for oxygen adsorption. A rapid surface oxidation of CO takes place and, as the CO is removed, the surface changes back to its original structure. Surface studies have demonstrated the correlations between a particular surface structure and CO oxidation.

7.4 The Langmuir adsorption isotherm

The treatment of the equilibrium aspects of adsorption, which we shall extend to include kinetics in a later section, proceeds via an examination of the adsorption and desorption processes and the equating of their rates at equilibrium:

$$X + site \rightleftarrows X\text{–site.} \tag{R 1,-1}$$

The surface is considered to consist of a number of adsorption sites and the rate of desorption, ρ_{-1}, is assumed to be proportional to the fraction, θ, of the surface covered. The rate of adsorption, ρ_1, is assumed proportional to the fraction of vacant sites, $(1-\theta)$ and to the pressure, p, of the gas X. Thus, at equilibrium

$$k_1 p(1-\theta) = k_{-1}\theta \tag{E 1}$$

$$\theta = bp/(1 + bp). \tag{E 2}$$

where

$$b = k_1/k_{-1}.$$

Equation (2) is known as the Langmuir adsorption isotherm. As p increases from zero, θ rises, at first linearly in p ($bp \ll 1$) and then tends to unity (full coverage) as bp becomes much larger than unity. If the heat of adsorption can be assumed independent of θ, then eqn (2) can be rearranged and k_1/k_{-1} determined from a study of the volume of gas taken up versus pressure—a series of experiments over a range of temperatures then enables ΔH_{ad} to be obtained. In practice, ΔH_{ad} is not independent of θ (not all sites are equivalent and interactions can occur between adsorbed species) and ΔH_{ad} is determined for a specific θ by examining the temperature dependence of the equilibrium pressure of X for that fractional surface coverage.

Example 7.1
The data below are for the adsorption of CO on charcoal at 273 K. Confirm that they fit the Langmuir isotherm, find the constant b and the volume corresponding to complete surface coverage.

p/Torr	100	200	300	400	500	600	700
V/cm^3	10.2	18.6	25.5	31.4	36.9	41.6	46.1

Firstly we need to relate θ, the fractional coverage to the volume of CO adsorbed. If V_∞ is the volume of gas corresponding to complete monolayer coverage then $\theta = V/V_\infty$. Equation (2) can be rearranged to

$$p/V = p/V_\infty + 1/bV_\infty \qquad \text{(E 3)}$$

and hence a plot of p/V vs. p will have gradient $1/V_\infty$ and intercept $1/bV_\infty$ (Fig. 7.4). The gradient is 0.0090 cm^{-3}, and so $V_\infty = 111$ cm^3. The intercept is 9.0 torr cm^{-3} and so

$$b = 1/(111 \text{ cm}^3) \times (9.0 \text{ torr cm}^{-3}) = 1.0 \times 10^{-3} \text{ torr}^{-1}$$

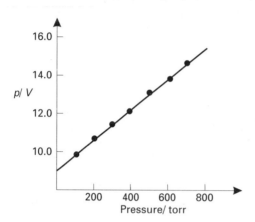

Fig. 7.4 Test of the Langmuir isotherm for CO on charcoal.

Box 7.1 Langmuir isotherm for dissociative adsorption

Many important reactions are initiated by dissociative adsorption. The significance of this step lies in the fact that an unreactive molecule, such as N_2, is converted into two reactive radicals. The same principles of equilibrium apply, but we need to account for the fact that adsorption requires two available sites (hence the $(1-\theta_A)^2$ term in the rate of adsorption) and desorption is a bimolecular step involving the combination of two surface atoms.

$$A_2(g) \rightleftarrows 2A(ads). \qquad \text{(R 2, -2)}$$

The rate of absorption $= k_2 p_{A_2}(1-\theta_A)^2$. The rate of desorption $= k_{-2}\theta_A^2$. Equating the two rates gives the following expression for the isotherm for dissociative absorption:

$$\theta_A = (p_{A_2}b)^{\frac{1}{2}}/\{1 + (p_{A_2}b)^{\frac{1}{2}}\} \qquad \text{(E 4)}$$

The Langmuir isotherm (E 2) is just the simplest of a number of eqns designed to model the equilibrium partitioning between surface and gas phases. However some of its simple assumptions are not valid, especially at high surface coverages. On real surfaces some sites will be preferentially filled and as the surface coverage increases adsorbate–adsorbate interactions will become significant. The final assumption of the Langmuir isotherm is that only a monolayer of gas can be absorbed but further layers of gas may be physisorbed. Other isotherms have been developed to account for these simplifications and further information on these more complex isotherms, which attempt to account for the varying ΔH_{ad} with θ, can be found in Gasser (1985) pp.14–17.

7.5 Dynamics of surface adsorption

As with gas phase collisions, molecular beam experiments provide a considerable insight into the dynamics of gas/surface collisions. A beam (effusive or nozzle) of molecules strikes a crystal surface (Fig. 7.5) and the intensity of scattering is measured mass-spectrometrically as a function of angle, γ, and coverage, θ. The translational energy of the scattered molecule can also be measured by, for example, time of flight mass spectrometry, whilst in some cases, its internal energy distribution can be detected by laser induced fluorescence. Figure 7.6 shows data for the scattering of NO from Pt (111). The scattering intensity is low at zero coverage, because most of the molecules are adsorbed on to the surface– the *sticking coefficient*, s (s = number of molecules adsorbed/number striking surface) being ≈ 0.85. If there were no transfer of translational energy with the surface, then the scattering would be specular with the angle of reflection equal to the angle of incidence. The diagram shows a distribution of angles, demonstrating some energy transfer for the scattered molecules, but it is peaked at the specular angle. At a higher surface coverage, the scattered intensity increases substantially, demonstrating a reduced sticking coefficient, and the angular distribution is much broader and approaches the cosine distribution which would apply if the energy were accommodated on the surface. Time of flight measurements show that the translational energy of the scattered molecules is close to that expected from equilibration with the solid and LIF measurements show a

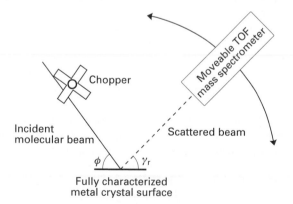

Fig. 7.5 Schematic diagram of apparatus to study gas–surface interactions. The composition and energy of the molecular beam may be varied (see Chapter 4 p. 102). The use of the chopper and time of flight (TOF) mass spectrometer allows for time resolution and a determination of the residence time.

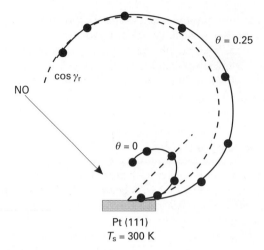

$\theta = 0.25$

$\cos \gamma_r$

NO

$\theta = 0$

Pt (111)

$T_s = 300$ K

Fig. 7.6 Polar diagram for the angular distribution of NO molecules scattered from a clean ($\theta = 0$) and a partially NO-saturated Pt(111) surface. The broken lines show the expected distributions for molecules scattering elastically off the surface (angle of reflection = angle of incidence) and for molecules accommodated on the surface (cosine distribution). (Modification of Ertl (1982).)

Boltzmann rotational distribution at the surface temperature. Thus energy accommodation is rapid and efficient and occurs, it is thought, by transfer to the vibrational modes of the solid (phonon excitation). The increase in energy accommodation with surface coverage may be due at least in part, to weak physisorption at occupied sites, from which surface migration may occur, leading, in some cases, to eventual chemisorption.

In view of this efficient energy transfer, the desorption process can be described using transition state theory, following arguments similar to those adopted for unimolecular reactions at the high pressure limit (p. 133). Thus

$$k_{-1} = \frac{k_B T}{h} \cdot \frac{Q^{\ddagger}}{Q} \exp\left(\frac{-E_d}{RT}\right) \tag{E 5}$$

where Q and $Q^{\ddagger}$ are the partition functions for the adsorbed species (adsorbate + surface site) and for the activated complex respectively. Since the complex will be looser than the adsorbed species, $Q^{\ddagger} > Q$ and desorption typically has an A factor of $\approx 10^{15}$ s^{-1}. For physisorption, $E_d \approx -\Delta H_{ad}$, whilst chemisorption may be activated with $E_d > -\Delta H_{ad}$.

Study notes
Collisions of gas phase molecules with surfaces is the first step in any surface catalysed reaction and not surprisingly these events have been the subject of ever more detailed dynamical investigation. An example of the molecular information which can be obtained from such studies can be found in an article by Roth *et al.* (1992) (Roth, C. Häger, J., and Walther, H. (1992). Scattering of NO molecules from a diamond surface. *Journal of Chemical Physics*, **97**, 6880–9). In this a molecular beam was used to create a well defined, rotationally cold beam of NO molecules, whose translational energy could be varied by altering the ratio of xenon and helium in the beam (p. 102). The scattering angle of the NO beam was measured by a mass spectrometer, but in addition, information on the internal energy (rotational and vibrational) was obtained by resonance enhanced multi-photon ionization. This allowed the authors to investigate the efficiency of trans-

lational to rotational and translational to vibrational excitation which in turn gives some insight into the molecular processes occurring during the collision.

Sections I and II (introduction and experimental) should be relatively easy to follow and contain a number of useful references to earlier studies. The results section (III) is long and rather complex. It should probably only be dipped into after studying the discussion and conclusions sections of the paper and the reader should consult to see what kind of detailed molecular information can be obtained rather than to understand all the experimental results.

An appreciation of the current status of the dynamics of gas/surface interactions, including reaction, may be found in *Faraday Discussions,* 1993, **96**, which gives the proceedings of a conference on dynamics at the gas/solid interface. Most of the articles are of interest but the following can be read with particular profit: pp. 1–16 (Harris, J., *Some remarks on surface reactions*, which discusses surface reactions from a molecular viewpoint and examines the links between theory and experiment); pp. 17–31 (Rettner, C. T., Michelsen, H. A., and Averback, D. J., *From quantum-state-specific dynamics to reaction rates*, which develops the relationship between rate coefficients and molecular adsorption probabilities); pp. 129–149 (Barclay, V. J., Hung, W.-H., Polanyi, J. C., Zhang, G., and Zeiri, Y., *Photochemistry of adsorbed molecules*, discuss an experiment which probes the photolysis of HI on a LiF surface and compares the results with gas phase photolysis); pp. 255-264 (Hopkinson, A. and King, D. A., *Modelling the CO induced surface phase transition on Pt {111};* this paper examines the detailed kinetics of CO adsorption on Pt and discusses oscillations (see Chapter 11) in CO/NO adsorption) and pp. 325–336 (Oakes, D. J., McCoustra, M. R. S., and Chesters, M. A., *Dissociative adsorption of methane on Pt{111}*, which describes modern techniques for determining sticking probabilities). An important feature of *Faraday Discussions* is, as the title suggests, the discussion following presentation of the papers, which is reported in full in Volume 96. The discussion is lively and sometimes controversial. The proceedings demonstrate clearly that surface dynamics is entering the period of rapid development enjoyed by gas phase dynamics some years ago and described briefly in Chapter 4.

7.6 Kinetics of surface catalysed reactions

A wide variety of rate laws and an even wider variety of reaction mechanisms can apply to surface catalysed reactions. We cannot undertake an exhaustive survey and will examine, instead, two typical, simple cases, whilst further examples can be found in the problems. The two systems we shall take are both examples of the Langmuir–Hinshelwood mechanism, in which the surface reaction is sufficiently slow that the adsorption equilibrium is not disturbed so that the Langmuir adsorption isotherm can be applied to the adsorbed gases.

7.6.1 Unimolecular decomposition

Some unimolecular decompositions, such as that of phosphine on tungsten, may be described by the mechanism:

$$A(g) \underset{k_{-3}}{\overset{k_3}{\rightleftarrows}} A(ads) \overset{k_4}{\rightarrow} products \qquad (R\ 3,\ -3,\ and\ 4)$$

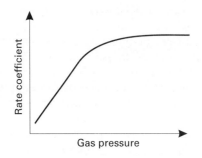

Fig. 7.7 Variation of the rate coefficient with gas pressure showing the transition from first-order to zeroth-order behaviour.

where (g) and (ads) refer to gas phase and surface adsorbed species. The rate of product formation, ρ, is $k_4\theta_A$ and, provided $k_4 \ll k_{-3}$, the fractional coverage, θ_A may be determined from the Langmuir adsorption isotherm (E 2), thus

$$\rho = k_4\theta_A = k_4 b_A p_A/(1 + b_A p_A). \qquad (E\ 6)$$

The reaction shows two limiting rate laws corresponding to the two extremes of behaviour of the Langmuir isotherm (Fig. 7.7): (i) At low pressures (<1 Pa for PH_3 at 1000 K), $k_3 p_A \ll k_{-3}$ and θ_A is very small and proportional to pressure, so that the reaction is first-order in A(g) with

$$\rho = k_4 b_A p_A; \qquad (E\ 7)$$

(ii) At high pressures (> 100 Pa for PH_3 at 1000 K), $k_3 p_A \gg k_{-3}$, the surface is saturated ($\theta_A = 1$) and the reaction is zeroth order with

$$\rho = k_4. \qquad (E\ 8)$$

7.6.2 Bimolecular reactions

Consider the reaction (R 5) mechanism shown in Fig. 7.8, where A and B compete for the same surface sites. Following the isotherm treatment detailed above

$$k_A(1 - \theta_A - \theta_B)p_A = k_{-A}\theta_A \qquad (E\ 9)$$

$$k_B(1 - \theta_A - \theta_B)p_B = k_{-B}\theta_B. \qquad (E\ 10)$$

Solving for θ_A and θ_B and setting $\rho = k_5\theta_A\theta_B$ we find

$$\rho = k_5 b_A b_B\, p_A p_B\,/(1 + b_A\, p_A + b_B\, p_B)^2. \qquad (E\ 11)$$

A variety of expressions follow, depending on the magnitude of b_A and b_B (which in turn are dependent on the nature and strength of the surface bond). A frequent

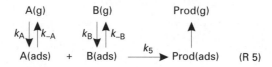

Fig. 7.8 Mechanism for reaction (5).

case has one of the species (say A) weakly adsorbed (e.g. via a van der Waals complex), so $b_A \ll b_B$, so that for $p_A \approx p_B$,

$$\rho = k_5 b_A b_B \, p_A \, p_B / (1 + b_B \, p_B)^2 \tag{E 12}$$

which reduces, at high pressures, to

$$\rho = k_5 b_A \, p_A / b_B \, p_B. \tag{E 13}$$

High pressures of B inhibit the reaction, because they lead to saturation of the surface by B, to the exclusion of A, whilst at low pressures, where $b_B \, p_B \ll 1$ the rate increases with p_B

$$\rho = k_5 b_A b_B \, p_A \, p_B. \tag{E 14}$$

A maximum therefore occurs in the rate as p_B increases and ρ_{max} may be determined by differentiating eqn (12) with respect to p_B.

Once again we have only been able to describe a small fraction of the possible reaction mechanisms. The texts referenced in the previous study notes will provide further examples. We complete this chapter by considering two specific systems which illustrate some of the principles and techniques we have discussed.

7.7 Oxidation of carbon monoxide

Oxygen undergoes dissociative adsorption on a Pt(111) surface, whilst CO is strongly chemisorbed. If oxygen is adsorbed first, its surface distribution is quite open, so that subsequent chemisorption of CO is feasible and either a Langmuir–Hinshelwood mechanism (dissociative adsorption):

$$O_2(g) \rightarrow 2O(ads) \tag{R 6}$$

$$CO(g) \rightarrow CO(ads) \tag{R 7}$$

$$O(ads) + CO(ads) \rightarrow CO_2(ads) \tag{R 8}$$

$$CO_2(ads) \rightarrow CO_2(g) \tag{R 9}$$

or an Eley–Rideal mechanism (reaction of adsorbed O with gaseous CO)

$$O_2(g) \rightarrow 2O(ads) \tag{R 6}$$

$$O(ads) + CO(g) \rightarrow CO_2(g) \tag{R 10}$$

could apply. Given our knowledge that CO is strongly adsorbed we might expect the former mechanism to apply. Convincing evidence in favour of the Langmuir–Hinshelwood mechanism was obtained using molecular beam techniques, in which it was demonstrated that the average time lag between the CO arriving at the surface and the CO_2 leaving it is about a millisecond at a temperature of ≈ 440 K. CO_2 itself is only weakly adsorbed on the surface and such a long lifetime can only be compatible with reaction via a strongly adsorbed CO species (Langmuir–Hinshelwood mechanism). Varying the temperature of the surface in the molecular beam experiments permits the activation energy of the surface reaction to be determined and Fig. 7.9 shows a potential energy profile and a potential energy surface for the overall reaction. To illustrate, finally, the

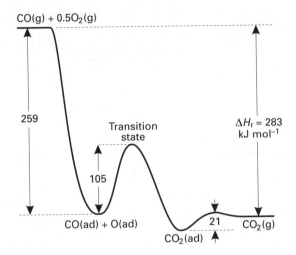

Fig. 7.9 Potential energy diagram illustrating the catalytic formation of CO_2 on a Pt(111) surface. Energies in kJ mol^{-1}. (Modification of Ertl (1982).

amount of detail now being obtained on surface reactions, it is of interest to note that time of flight and infrared emission measurements have shown that the CO_2 is both translationally and vibrationally hot as it leaves the surface.

A more detailed discussion of this reaction may be found in Gasser (1985) (Chapter 9) and in Ertl (1982).

7.8 The synthesis of ammonia

Ammonia is synthesized industrially from nitrogen and hydrogen by the Haber–Bosch process, which involves a modified Fe_3O_4 catalyst. Both reactants are adsorbed via a dissociative mechanism, the adsorption of nitrogen being slow, even though the activation energy is comparatively small. The pre-exponential factor is small primarily because the initially adsorbed N_2, which, like CO, is probably adsorbed end on, is much more ready to undergo desorption. Hydrogen, on the other hand, undergoes much more facile dissociative adsorption; under the conditions which apply synthetically, the hydrogen coverage is small, but the atoms are very mobile.

For a long time there was uncertainty as to whether the mechanism involved reaction between H and adsorbed N_2 or H and adsorbed N. It was known that nitrogen undergoes dissociative adsorption, but it was not clear whether ammonia synthesis involved such a step, or proceeded via a parallel route involving adsorbed N_2. i.e. the two possible mechanisms are:

$$N_2(g) \rightarrow N_2(ads) \tag{R 11}$$

$$N_2(ads) + 6H(ads) \rightarrow 2NH_3 \tag{R 12}$$

or

$$N_2(g) \rightarrow N_2(ads) \rightarrow 2N(ads) + 6H(ads) \rightarrow 2NH_3. \tag{R 13}$$

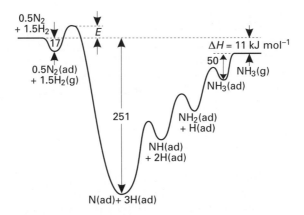

Fig. 7.10 Potential energy diagram illustrating the
production of ammonia. The activation energy E depends on
the structure and coverage of the surface. Energies in kJ
mol^{-1}. (Modification of Ertl (1982).)

The latter mechanism was confirmed by Auger spectroscopy at temperatures
below 450 °C, where desorption of N(ads) is very slow. The hydrogen pressure
was varied and it was found that [N(ads)] fell rapidly at the higher hydrogen
pressures, demonstrating not only that the adsorbed atoms are involved in the
reaction, but that the dissociative chemisorption of nitrogen is the rate determining step.

The overall mechanism involves the sequential reaction with adsorbed H
atoms:

$$H_2(g) \rightarrow 2H(ads) \quad \text{(dissociative chemisorption)} \qquad \text{(R 14)}$$

$$N_2(g) \rightarrow N_2(ads) \quad \text{(physisorption)} \qquad \text{(R 15)}$$

$$N_2(ads) \rightarrow 2N(ads) \quad \begin{array}{l}\text{(physisorbed molecules} \rightarrow \\ \text{chemisorbed atoms)}\end{array} \qquad \text{(R 16)}$$

$$N(ads) + H(ads) \rightarrow NH(ads) \quad \text{(reaction of chemisorbed species)} \qquad \text{(R 17)}$$

$$NH(ads) + H(ads) \rightarrow NH_2(ads) \qquad \text{(R 18)}$$

$$NH_2(ads) + H(ads) \rightarrow NH_3(ads) \qquad \text{(R 19)}$$

$$NH_3(ads) \rightarrow NH_3(g) \quad \text{(desorption of ammonia)} \qquad \text{(R 20)}$$

and an energy profile is shown in Fig. 7.10.

Study notes
Interesting accounts of ammonia synthesis, which illustrate the complexity of the industrial catalyst and which justify laboratory investigations on 'cleaner' surfaces, may be found in two papers: Ertl, G. (1980) Surface science and catalysis studies on the mechanism of ammonia synthesis. *Catalysis Reviews,* **21**, 201–23; Ertl, G. (1983) Primary steps in catalytic synthesis. *Journal of Vacuum Science and Technology,* **A1**, 1247–53).

7.9 Questions

7.1 Calculate the collision frequency per unit area ($\mathcal{Z}_w = p/(2\pi m k_B T)^{\frac{1}{2}}$) at 300 K for the following pressures of air; 760, 1, 10^{-3}, 10^{-6}, 10^{-9} torr (760 torr = 1 atm $\approx 1 \times 10^5$ Pa) on a metal surface. Estimate the number of metal atoms per unit area and hence the average number of collisions each atom will receive per second.

7.2 A clean surface is exposed to a diatomic gas. What factors will affect the degree of surface coverage?

7.3 The following rotational state data (Fig. 7.11) were obtained for the scattering of an NO beam from a diamond (110) surface for different translational energies of the incident NO beam which is rotationally cold. Calculate the rotational temperatures for the linear portions of the Boltzmann plots. Discuss possible collision mechanisms in the light of the observed plots and the surface temperature of the diamond sample.

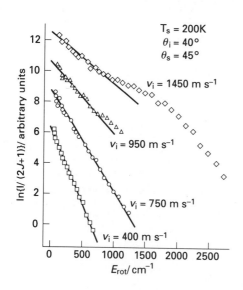

$T_s = 200K$
$\theta_i = 40°$
$\theta_s = 45°$

$v_i = 1450$ m s^{-1}

$v_i = 950$ m s^{-1}

$v_i = 750$ m s^{-1}

$v_i = 400$ m s^{-1}

Fig. 7.11 Rotational state distributions for different incoming velocities of NO. (NB the vertical distance between the different distributions has no significance.) The surface temperature of the diamond is 200 K.

7.4 The following data were obtained for the volume of nitrogen (converted to 273 K and 1 atm) adsorbed per gram of charcoal at 196 K.

pN_2/atm	3.5	10.0	16.7	25.7	33.5	39.2
volume N$_2$/cm^{3}g^{-1}	101	136	153	162	165	166

Test the applicability of the Langmuir isotherm and find values for b and V_∞. Describe the expected variation of these parameters with temperature.

7.5   The rate of propene oxidation on a transition metal catalyst obeys the following rate law:

$$\text{Rate} = kpO_2^{\frac{1}{2}}$$

What can one say about the relative rates and mechanisms of propene and oxygen adsorption and the surface reaction?

7.6   Devise kinetic schemes to explain why the decomposition of N_2O on gold shows first-order kinetics, whilst the decomposition of NH_3 on Pt follows the rate expression:

$$-d[NH_3]/dt = k[NH_3]/[H_2].$$

7.7   The bond dissociation energies (in kJ mol^{-1}) for metal hydrides, carbides and oxides are tabulated below. The dissociation energies of the following gas phase molecules are: $D(H_2) = 434$ kJ mol^{-1}, $D(O_2) = 494$ kJ mol^{-1}, and $D(CO) = 1071$ kJ mol^{-1}.

Metal (M)	Fe	Ni	Pt
$D(M–H)$	287	280	266
$D(M–C)$	705	666	600
$D(M–O)$	438	415	320

Use these data to interpret the following observations: (i) H_2 and O_2 may be dissociatively chemisorbed on all of these metals. (ii) When mixtures of CO and H_2 are passed over the metals at temperatures of around 500 K, methanol (CH_3OH) is produced with Pt (partial reduction of CO) whereas with Ni and Fe the products are hydrocarbon species (C–H and C–C bonds only) and water (complete reduction of CO).

7.8   At temperatures of around 550 K nitric oxide is reduced by hydrogen on a rhodium metal catalyst. The mechanism takes the form

$$NO(g) \rightleftarrows NO(ads)$$

$$H_2(g) \rightleftarrows 2H(ads)$$

$$H(ads) + NO(ads) \rightarrow \text{products.}$$

The last step is the rate-determining step leading to the formation after several rapid steps, of N_2 and H_2O. Devise the form of the rate law in terms of the partial pressures of nitric oxide and hydrogen. As subsequent steps, including desorption, are so rapid, the surface coverage of species other than NO or H can be ignored.

References

Ertl, G. (1982). Chemical dynamics in surface reactions. *Berichte Bunsengesellschaft für Physikalische Chemie*, **86,** 425–32.

Gasser, R. P. H. (1985). *An introduction to chemisorption and catalysis by metals*. Oxford University Press.

8 Complex reactions

8.1 Introduction

The preceding chapters of this book have concentrated extensively on the study of isolated elementary reactions in various media. In many, though not all, cases the goal has been to elucidate detailed molecular mechanisms. Another, more practical, aspect of reaction kinetics is to measure reaction rates so that we can understand the time evolution of complex systems such as enzyme catalysed reactions, stratospheric ozone depletion, ethene polymerization or hydrocarbon combustion. These more complex systems are still, nevertheless, composed of the same types of elementary reactions. The aim of this chapter (and succeeding chapters) is to see how elementary reactions couple together in these more complex environments and how the comparatively simple time-dependencies of several elementary reactions can lead to either steady-state behaviour (chain reactions), rapidly accelerating reactions (explosions), or even oscillations.

In Section 8.2 we investigate how elementary reactions combine together in *parallel*, *opposing*, and *consecutive* reactions. A detailed analysis of the latter case leads us to a fuller understanding of the *steady-state hypothesis* that we have already come across in several of the preceding chapters. The steady-state treatment allows us to simplify some complex looking reaction mechanisms and obtain analytical solutions for the time-dependence of reactants and products. In other cases such simplifications cannot be made and we are left with the problem of solving a number of simultaneous differential rate eqns which control the concentrations of reactants and products. Table 8.1 details *part* of a chemical mechanism for methane combustion. Solution of the complete set of coupled differential equations arising from the 213 reactions of the full mechanism would be a daunting task to say the least! Fortunately a number of computer programs are now available for the solution of such problems and in Section 8.3 we discuss some of the principles underlying these *numerical integration* techniques.

In the final sections of Chapter 8 we shall briefly consider applications of complex reactions in atmospheric chemistry and enzyme kinetics. This chapter provides important background for the succeeding chapters in which many more applications of complex reactions will be detailed.

8.2 The interaction of elementary reactions

8.2.1 Parallel reactions

Besides the brief introduction in Section 1.8, we have already come across one example of parallel reactions in Chapter 2 (p. 32) when we discussed the measurement of elementary reactions using pseudo-first-order methods. For example, consider the reaction

$$CH + CH_4 \rightarrow C_2H_4 + H \tag{R 1}$$

Table 8.1 Part of a reaction scheme for methane combustion (rate coefficients are expressed in the form $k = AT^n \exp(-E/RT)$)

	Reaction	A/cm^3 molecule^{-1}s^{-1}	n	E/kJ mol^{-1}
1	$CH_4 + M \rightarrow CH_3 + H + M$	1.7×10^{-7}	0	368.2
2	$CH_4 + H \rightarrow H_2 + CH_3$	3.7×10^{-20}	3.00	36.6
3	$CH_4 + O \rightarrow CH_3 + OH$	2.0×10^{-17}	2.08	31.9
4	$CH_4 + OH \rightarrow CH_3 + H_2O$	5.8×10^{-21}	3.08	8.4
5	$CH_4 + CH_2 \rightarrow 2CH_3$	2.2×10^{-11}	0	39.7
6	$CH_3 + M \rightarrow CH_2 + H + M$	3.2×10^{-8}	0	383.2
7	$CH_3 + H \rightarrow CH_2 + H_2$	1.5×10^{-10}	0	63.2
8	$CH_3 + O \rightarrow CH_2O + H$	1.1×10^{-10}	0	0
9	$CH_3 + OH \rightarrow CH_2 + H_2O$	2.5×10^{-11}	0	20.9
10	$CH_3 + O_2 \rightarrow CH_2O + OH$	8.6×10^{-11}	0	144.6
11	$CH_3 + O_2 \rightarrow CH_3O + O$	1.2×10^{-11}	0	107.3
12	$CH_3O + M \rightarrow CH_2O + H + M$	1.7×10^{-10}	0	104.6
13	$CH_3O + H \rightarrow CH_2O + H_2$	3.3×10^{-11}	0	0
14	$CH_3O + O \rightarrow CH_2O + OH$	1.7×10^{-11}	0	0
15	$CH_3O + OH \rightarrow CH_2O + H_2O$	1.7×10^{-11}	0	0
16	$CH_3O + O_2 \rightarrow CH_2O + HO_2$	1.1×10^{-13}	0	10.8
17	$CH_2O + M \rightarrow HCO + H + M$	5.5×10^{-8}	0	338.9
18	$CH_2O + H \rightarrow HCO + H_2$	3.7×10^{-16}	1.77	43.9
19	$CH_2O + O \rightarrow HCO + OH$	3.0×10^{-11}	0	12.9

conducted under pseudo-first-order conditions ($[CH_4] \gg [CH]$). We generate CH from the laser flash photolysis of CHBr$_3$ and so the parallel reaction

$$CH + CHBr_3 \rightarrow products \qquad (R\ 2)$$

will also be taking place (again under pseudo-first-order conditions). The rate law for the disappearance of CH will therefore be:

$$-d[CH]/dt = (k_1' + k_2')[CH] \qquad (E\ 1)$$

where $k_1' = k_1[CH_4]$ and $k_2' = k_2[CHBr_3]$. Equation (1) is a simple first-order differential equation and can be integrated (see p. 9) to give:

$$[CH]_t = [CH]_0 \exp\{-(k_1' + k_2')t\}. \qquad (E\ 2)$$

The rate of disappearance of CH is determined by the *sum* of the pseudo-first-order rate coefficients for the individual processes. In order to extract the bimolecular rate coefficients we need to measure $(k_1' + k_2')$ as a function of $[CH_4]$ at a constant $[CHBr_3]$.

$$(k_1' + k_2') = k_2[CHBr_3] + k_1[CH_4] \qquad (E\ 3)$$

thus a plot of $(k_1' + k_2')$ vs. $[CH_4]$ (Fig. 8.1) will have gradient k_1 and intercept $k_2[CHBr_3]$.

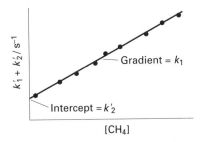

Fig. 8.1 Bimolecular plot of $k_1' + k_2'$ vs. $[CH_4]$ used to separate the rate coefficients of the two parallel reactions (R 1) and (R 2).

For a generalized first or pseudo-first-order system e.g.

$$A \rightarrow B \qquad \qquad (R\ 3)$$

$$A \rightarrow C \qquad \qquad (R\ 4)$$

one might intuitively imagine that the products would be formed with different time constants, dependent on the relative values of k_3 and k_4, however, *this is not the case*. Both products are controlled by the same overall rate coefficient, but the relative yields of B and C will be determined by the relative magnitudes of k_3 and k_4.

$$d[B]/dt = k_3[A]_t \qquad \qquad (E\ 4)$$

but

$$[A]_t = [A]_0 \exp\{-(k_3 + k_4)t\} \qquad \qquad (E\ 5)$$

therefore

$$d[B]/dt = k_3[A]_0 \exp\{-(k_3 + k_4)t\}. \qquad \qquad (E\ 6a)$$

This can be integrated to give

$$[B]_t = \{k_3[A]_0/(k_3 + k_4)\}[1 - \exp\{-(k_3 + k_4)t\}]. \qquad (E\ 6b)$$

A similar expression can be obtained for $[C]_t$. The exponential growths of both [B] and [C] have the same time constant which is identical to that for the decay of [A] and are the sum of the time constants for the individual reactions. If we allow $t \rightarrow \infty$ then:

$$[B]_\infty = k_3[A]_0/(k_3 + k_4) \qquad \qquad (E\ 7a)$$

$$[C]_\infty = k_4[A]_0/(k_3 + k_4) \qquad \qquad (E\ 7b)$$

i.e. the relative yield of each component is dependent on the relative magnitudes of the rate coefficients, and this argument can be extended to systems of more than two parallel reactions.

Example 8.1 Acid–base catalysis
The iodination of propanone is catalysed by both acid and base:

$$(R\,5)$$

By varying the pH, the pseudo-first-order rate coefficient for reaction may be shown to be given by:

$$k = k_0 + k_H[H_3O^+] + k_{OH}[OH^-] \qquad (E\,8)$$

where $k_0 = 2.8 \times 10^{-8}\,s^{-1}$, $k_H = 1.64 \times 10^{-3}\,dm^3\,mol^{-1}\,s^{-1}$, and $k_{OH} = 15\,dm^3\,mol^{-1}\,s^{-1}$.

Study notes
Acid and base catalysis depend on proton transfer and H_3O^+ and OH^- are not unique in their ability to act as proton donors or acceptors, but are members of a general class. Thus the type of catalysis to which we have referred above is termed specific acid–base catalysis and forms a subset of general acid–base catalysis. There are distinct and obvious problems in unravelling the specific and general forms and determining individual rate coefficients, but a lucid early account may be found in a paper by Bell,R. P., and Lidwell, O. M., (1940) (Acid–base catalysis *Proceedings of the Royal Society*, **176A**, 88) which shows that the above expression should be modified by the addition of two extra terms, $k_B[B]$ and $k_A[A]$ where B is a base and A the conjugate acid. For B = trimethylacetate, $k_B = 2.45 \times 10^{-5}\,dm^3\,mol^{-1}\,s^{-1}$ and $k_A = 4.4 \times 10^{-6}\,dm^3\,mol^{-1}\,s^{-1}$; indeed, on the presumption that the 'spontaneous' reaction arises from proton transfer involving the solvent, water, k_0 can be written $k_{H_2O}[H_2O]$, whence, $k_{H_2O} = 5 \times 10^{-10}\,dm^3\,mol^{-1}\,s^{-1}$, since $[H_2O] = 55.5\,mol\,dm^{-3}$ at 298 K. A thorough discussion of acid base catalysis, which includes many examples, may be found in Bell, R. P. (1973). *The Proton in Chemistry* (2nd edn), Chapters 8 and 9. Chapman and Hall, London. Further important examples in the field of photochemistry may be found in Section 12.3 of this book.

8.2.2 Competing first- and second-order reactions

It is important to recognize that the simple analysis outlined above only applies if the parallel reactions are first-order or pseudo-first-order. If the reactions involve more than one species whose concentrations change with time, then the differential equations become more complex and must be treated on their own merits—sometimes it may not be possible to solve them analytically. A common, and important example involves parallel first- and second-order reactions:

$$A \rightarrow B \tag{R 6}$$

$$A + A \rightarrow C \tag{R 7}$$

$$d[A]/dt = -(k_6[A] + 2k_7[A]^2). \tag{E 9}$$

This equation can be solved analytically and gives the solution:

$$[A]_t = [\{1/a_0 + (2k_7/k_6)\}\exp(k_6 t) - (2k_7/k_6)]^{-1} \tag{E 10}$$

where $a_0 = [A]_{t=0}$. An example of this occurs in the determination of the rate coefficient for reaction (8):

$$CH_3 + O_2 + M \rightarrow CH_3O_2 + M \tag{R 8}$$

(an important step in the atmospheric oxidation of methane) by laser flash photolysis coupled with UV absorption spectroscopy to follow the methyl radical concentration. Owing to the relative insensitivity of the monitoring technique, high (10^{13} molecules cm^{-3}) concentrations of methyl radicals are required and the recombination reaction:

$$CH_3 + CH_3 + M \rightarrow C_2H_6 + M \tag{R 9}$$

plays a significant role in determining the overall loss rate for methyl. Under the experimental conditions $[M] \gg [O_2] \gg [CH_3]$ so that reactions (8) and (9) are in first- and second-order regimes respectively. In order to extract k_8 the methyl decay traces had to be analysed according to eqn (10). More sensitive detection techniques would allow the initial methyl radical concentration to be lowered to such an extent that reaction (9) becomes insignificant and methyl decay is controlled solely by pseudo-first-order kinetics.

8.2.3 Opposing reactions

All chemical reactions are potentially reversible, although in many cases the reverse reaction is so slow that it may be neglected (as we have done in the majority of cases to date). For systems close to equilibrium, however, forward and backward rates are comparable, and the analysis of opposing reactions becomes important: as the product concentrations build up the *overall* forward rate must be reduced. The simplest cases occur with opposing first-order (or pseudo-first-order) reactions. For our analysis of such a system let us consider a generalized isomerization reaction,

$$A \rightleftharpoons B \tag{R 10, -10}$$

Let the equilibrium concentrations of A and B be a and b, and let the concentration differences from these equilibrium values at any time t be x and y. The time-dependent behaviour of B is given by:

$$d(b-y)/dt = k_{10}(a - x) - k_{-10}(b - y). \tag{E 11}$$

From the stoichiometry, $x = -y$, and

$$db/dt - dy/dt = k_{10}a - k_{-10}b + (k_{10} + k_{-10})y \tag{E 12}$$

At equilibrium there is no net change and

$$db/dt = k_{10}a - k_{-10}b = 0. \tag{E 13}$$

Thus

$$-dy/dt = (k_{10} + k_{-10})y. \tag{E 14}$$

Equation (14) has the form of a first-order differential equation with rate coefficient $(k_{10} + k_{-10})$, i.e. the equilibrium concentrations are approached exponentially with a rate coefficient equal to the sum of the forward and reverse rate coefficients.

$$y = y_0 \exp\{- (k_{10} + k_{-10})t\}. \tag{E 15}$$

The system is said to *relax* towards its equilibrium state (Fig 8.2), and the reciprocal of the rate coefficient for approach, $(k_{10} + k_{-10})^{-1}$, is termed the relaxation time (τ). If the equilibrium coefficient (k_{10}/k_{-10}) is known, both k_{10} and k_{-10} can be determined from τ.

The use of forward and reverse rate coefficients in determining equilibrium constants has already been discussed in Section 1.7. In that example the individual reactions were determined in separate experiments; however, the original paper (Seakins and Pilling, 1991) describes relaxation measurements on the system, with both reactions being studied under pseudo-first-order conditions.

$$Br + i\text{-}C_4H_{10} \rightleftharpoons HBr + t\text{-}C_4H_9. \tag{R 11, -11}$$

Section 2.10 of Chapter 2 discussed examples of relaxation experiments which are based on the principles outlined in this section. In general the analysis is complex for systems involving reaction of overall order greater than unity. However, if the perturbations from equilibrium are small, simplifying assumptions may be made. An analysis, similar to that given above, for the reaction

$$A + B \rightleftharpoons C \tag{R 12, -12}$$

leads to the equation

$$-dz/dt = \{k_{12}(a + b) - k_{-12}\}z + k_{12}z^2 \tag{E 16}$$

where z is the difference in the concentration of C from its equilibrium value and a and b are the equilibrium concentrations of A and B. Since the disturbance is slight, z is small, and the quadratic term, $k_{12}z^2$ may be neglected. The equation now shows first-order form, with a relaxation time $\{k_{12}(a + b) - k_{-12}\}^{-1}$. Both k_{12} and k_{-12} may be evaluated by measuring τ as a function of a and b.

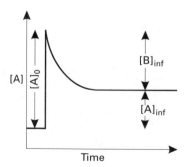

Fig. 8.2 Approach to equilibrium for the reaction A $\rightleftharpoons$ B starting from $[A]_0$.

Study notes

An excellent recent example of the use of opposing reactions in thermochemical studies can be found in a paper by Russell, J. J., Seetula, J. A., Gutman, D., Davis, F., Caralp, F., Lightfoot, P. D. (1990) (Kinetics and thermochemistry of the equilibrium $CCl_3 + O_2 \rightleftharpoons CCl_3O_2$. *Journal of Physical Chemistry,* **94**, 3277–83), which details the experimental study of the equilibrium

$$CCl_3 + O_2 \rightleftharpoons CCl_3O_2 \qquad\qquad (R\,13, -13)$$

There has been a recent upsurge in experimental studies on chlorinated radicals for two reasons, firstly their significance in stratospheric ozone depletion (see Section 8.4) and secondly, incineration has been proposed as one of the cleanest methods of disposing of chlorinated solvents and other compounds. In order to comprehend fully both of these complex systems we need to understand the underlying kinetics and thermochemistry. The paper is a model example of how to perform a thorough investigation; two different experimental techniques were used to study the reaction and the experimental results were backed up by theoretical calculations. The reader is strongly recommended to study the original paper in depth.

The first method uses the technique of laser photolysis coupled to photoionization mass spectrometry (PIMS). CCl_3 radicals are generated by a laser pulse and the temporal profile of $[CCl_3]$ is monitored in real time with an excess of oxygen (pseudo-first-order conditions) using mass spectrometry. Figure 8.3 shows a typical experimental trace, and the approach to equilibrium can clearly be seen. The

CCl_3^+

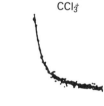

0	10	20

Time/ ms

Fig. 8.3 Approach to equilibrium of $[CCl_3]$ following its generation by flash photolysis of CCl_4 and reaction with O_2.

figure differs somewhat from our idealized profile of Fig. 8.2 as even at long times the $[CCl_3]$ continues to decay. This slow decay arises from the loss of CCl_3 and CCl_3O_2 onto the reactor wall. The resulting decay is biexponential, with the faster decay constant corresponding approximately to the approach to equilibrium ($k_{13} + k_{-13}$). The sensitivity of the PIMS technique means that very low concentrations of CCl_3 can be used. Under these conditions reactions which are second order in radical concentration such as

$$CCl_3 + CCl_3 \rightarrow \text{products} \qquad\qquad (R\,14)$$

can be ignored.

In the second experimental technique, CCl_3 is again generated by flash photolysis and radical concentrations are followed in real time by UV absorption spectroscopy. This technique is less sensitive and hence higher radical concentrations are required. Now secondary reactions need to be considered and the equations governing $[CCl_3]$ as a function of time can no longer be solved analytically. However, k_{13} and k_{-13} can still be extracted by numerical integration techniques (see Section 8.3).

In Section 8.3 k_{13} and k_{-13} are used to calculate ΔH and ΔS for the reaction and the results have a number of interesting implications. Firstly the bond strength $D(CX_3-O_2)$ decreases as X changes from H to Cl. Presumably this is due to the strong electron withdrawing properties of the electronegative halogen atoms. In turn this means that the stability of the chlorinated peroxy species is significantly lower than that of the alkyl species. We shall see in Chapter 11 that the stability of peroxy species is crucial in determining the mechanism of hydrocarbon oxidation. For alkyl systems there is a change in oxidation mechanism at about 700–800 K, the lower bond strength of the chlorinated species means that this change will occur at much lower temperatures, possibly around 400 K. The results are of great importance for modelling chlorinated hydrocarbon incineration.

8.2.4 Consecutive reactions

Consecutive reactions form a large and important group of kinetic systems. The simplest case is

$$A \rightarrow B \rightarrow C. \qquad \text{(R 15, 16)}$$

Mother-and-daughter radioactive decay is a good example of this type of behaviour, e.g.

$$^{218}\text{Po} \rightarrow {}^{214}\text{Pb} \rightarrow {}^{214}\text{Bi} \qquad \text{(R 17, 18)}$$

where $k_{17} = 5 \times 10^{-3}\,\text{s}^{-1}$ and $k_{18} = 6 \times 10^{-4}\,\text{s}^{-1}$. The kinetics may be analysed as follows:

$$da/dt = -k_{15}a \qquad \text{(E 17)}$$

$$db/dt = k_{15}a - k_{16}b \qquad \text{(E 18)}$$

$$dc/dt = k_{16}b. \qquad \text{(E 19)}$$

We notice that $(da/dt + db/dt + dc/dt) = 0$ as is required by material balance. Equation (17) may be solved straightforwardly as it is a simple first-order differential equation (see Section 1.4) to give

$$a = a_0 \exp(-k_{15}t) \qquad \text{(E 20)}$$

This expression can now be substituted for a in eqn (18). This equation is not so straightforward to solve, requiring the use of an integrating factor. Assuming that only A is present initially, the solution is:

$$b = a_0 k_{15}/(k_{16} - k_{15})\{\exp(-k_{15}t) - \exp(-k_{16}t)\} \qquad \text{(E 21)}$$

where a_0 is the initial concentration of A. Because of the material balance requirements we do not need to solve the final differential equation. The concentration of C is given by:

$$c = a_0 - a - b \qquad \text{(E 22)}$$

$$= a_0 - \{a_0 k_{16}/(k_{16} - k_{15})\exp(-k_{15}t)\} + \{a_0 k_{15}/(k_{16} - k_{15})\exp(-k_{16}t)\} \qquad \text{(E 23)}$$

Figure 8.4 shows the time-dependence of a, b, and c for two cases, $k_{15} = 10k_{16}$ and $k_{15} = 0.1k_{16}$. In the former case, B builds up to an appreciable concentration before decaying, and there is a significant time lag between the decay of the initial reactant A and the formation of the ultimate product C. This type of behaviour is

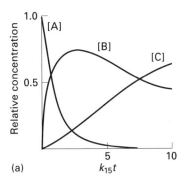
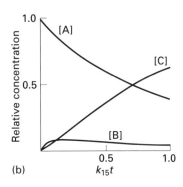

Fig. 8.4 The time-dependence of reactant (a), intermediate (b), and product (c) concentrations: (a) $k_{15} = 10k_{16}$, (b) $k_{15} = 0.1k_{16}$.

shown by the ^{218}Po sequence. Essentially the initial reactant is more reactive than the intermediate. When $k_{15} = 0.1k_{16}$ the situation is reversed and the intermediate is comparatively reactive. In this case the concentration of the intermediate is small at all times. After initiation of the reaction, b at first increases, since the rate of removal of B, $k_{16}b$, is initially zero. At longer times, when $\exp(-k_{16}t)$ is effectively zero, eqn (21) becomes

$$b = a_0 k_{15}/(k_{16} - k_{15}) \exp(-k_{15}t) \qquad \text{(E 24)}$$

and the decay of B parallels that of A. Differentiating eqn (24) gives

$$db/dt = -a_0 k_{15} \exp(-k_{15}t)\{k_{15}/(k_{16} - k_{15})\} \qquad \text{(E 25)}$$

$$= -k_{15}a\{k_{15}/(k_{16} - k_{15})\}$$

$$= (da/dt)\{k_{15}/(k_{16} - k_{15})\}. \qquad \text{(E 26)}$$

If $k_{16} \gg k_{15}$, the rate of decay of B is very much less than that of A, and we may assume that $db/dt \approx 0$, and that B has attained a steady concentration. This is the basis of the *steady-state approximation*. The period during which b increases is known as the induction period; the steady-state approximation is invalid in this time zone.

In Fig 8.4 the growth of C is mainly determined by the slower of the two reactions. In Fig. 8.4(a) c is still small when A has decayed considerably, and its growth parallels the decay of B, i.e. reaction (16) is the *rate-determining step*. In 8.4(b) reaction (15) is the rate-determining step; b remains small and the growth of C parallels the decay of A. Note that if we only follow the concentration of C it is not possible to say which of the fast or slow steps is first or second.

From our examples and discussion, we see that the steady-state approximation may be applied to an intermediate if the first-order rate coefficient for its production is very much less than that for its removal. Under these circumstances, after an initial build-up period, the intermediate reaches a very small and effectively constant concentration (b). The slope of the graph of b vs. time is approximately zero, so that $db/dt \approx 0$. However, as A is still being consumed and C is still being produced, B must also be produced and then react. These two observations can only be accounted for if the rate of production of B is equal to its rate of consumption,

$$k_{15} a = k_{16}b. \qquad \text{(E 27)}$$

These limitations are somewhat stricter than they need be—the steady-state approximation can also be applied under some other circumstances but they are sufficient conditions that will serve us well in the remainder of this, and following chapters.

Study note

A fuller discussion of the steady-state approximation may be found in Turanyi, T., Tomlin, A., and Pilling, M. J. (1993) On the error of the quasi-steady-state approximation. *Journal of Physical Chemistry*, **97**, 163–72. The paper examines the fractional error introduced by using the approximation ($\Delta b/b$ in the present context, where Δb is the difference between the true concentration and that calculated using the steady-state approximation). It shows that $\Delta b/b \approx (\tau/b)(\mathrm{d}b/\mathrm{d}t)$, where τ is the lifetime of the species, k_{16}^{-1} in the above example. Thus the steady-state approximation can be used to good effect if either $\mathrm{d}b/\mathrm{d}t$ or τ is small. The latter condition means that little error is introduced, even if $\mathrm{d}b/\mathrm{d}t$ is quite large, provided τ is very small.

Box 8.1 The interplay between parallel, opposing and consecutive reactions

It is instructive to consider the following simple reaction scheme.

$$A \rightleftarrows B \rightarrow C \qquad\qquad (R\ 19, -19, 20)$$

which is reminiscent of those we encountered on p. 121 and p. 145, and in which B is in a quasi-steady-state.

$$\mathrm{d}b/\mathrm{d}t = k_{19}a - (k_{-19} + k_{20})b = 0 \qquad (E\ 28)$$

$$b = k_{19}a/(k_{-19} + k_{20}) \qquad (E\ 29)$$

$$-\mathrm{d}a/\mathrm{d}t = \mathrm{d}c/\mathrm{d}t = k_{20}b = k_{19}k_{20}a/(k_{-19} + k_{20}) \qquad (E\ 30)$$

Thus, if k_0 is the overall first-order rate coefficient for the decay of A, $k_0 = k_{19}k_{20}/(k_{-19} + k_{20})$ and

$$1/k_0 = 1/k_{19} + (k_{-19}/k_{19})(1/k_{20}). \qquad (E\ 31)$$

k_0^{-1} has units of time and corresponds to the time for a to fall to $1/e$ of its initial value. If we define the decay constant, τ, as the reciprocal of the rate coefficient for any first-order (or pseudo-first-order) reaction, then,

$$\tau_0 = \tau_{19} + K\tau_{20} \qquad (E\ 32)$$

where $K = k_{-19}/k_{19}$. We thus have an interesting comparison between parallel and consecutive reactions. In the former, the total rate coefficient is the sum of the individual rate coefficients and is dominated by the largest one. In the latter, the decay time is the sum of the individual decay times (suitably modified by the constant K, which arises because reaction (19) is reversible. We can think of the overall time it takes for A to reach C as being made up of two times, one of which is the time it spends in reaction (19), the other in reaction (20). Hence the concept of the *rate-determining step*, since the reaction with the larger τ (or smaller k) holds up the reactants the most.

8.3 Numerical integration

Thus far we have primarily considered elementary chemical reactions, in which the time-dependence of the concentration of the reactants or products may be obtained by an analytic integration of the rate equations. This is also possible for some more complex reactions and in other cases the steady-state approximation is frequently employed, but the steady-state treatment is not of universal validity. Before the advent of fast computers, approximate methods had to be employed but these had severe limitations and the solution of very complex mechanisms, such as those occurring in biological environments or in atmospheric chemistry, was not feasible. Now mechanisms involving tens of species and hundreds of elementary reactions can be rapidly integrated, giving the concentration of all the reactants, products and intermediates as a function of time. We illustrate the technique by reference to a simple first-order reaction

$$A \rightarrow B. \tag{R 21}$$

Of course there is an analytical solution for this system $a = a_0 \exp(-k_{21}t)$ and this will enable us to assess the accuracy of the numerical technique. The strategy employed in a numerical integration is to evaluate the concentration of a at some time $t + \delta t$ given its concentration at time t:

$$a(t + \delta t) = a(t) + \Phi \delta t \tag{E 33}$$

where Φ is a function which depends on the rate coefficients (and possibly, in some reaction schemes, the concentrations of the reactants and intermediates at time t) which are known.

In the present case we know that

$$a(t + \delta t) = a(t)\exp(-k_{21}\delta t)$$
$$= a(t) \left\{ 1 - k_{21}\delta t + \tfrac{1}{2}(k_{21}\delta t)^2 - \ldots \right\} \tag{E 34}$$

Provided we make δt sufficiently small (e.g. such that $k_{21}\delta t < 0.1$) then

$$a(t + \delta t) \approx a(t) - a(t)k_{21}\delta t \tag{E 35}$$

or

$$\Phi = -a(t)k_{21}. \tag{E 36}$$

Thus we can evaluate $a(t + \delta t)$ simply by subtracting $a(t)k_{21}\delta t$ from $a(t)$, a very simple operation. Figure 8.5 compares the approximate numerical estimates of a with the exact values for two values of δt. For $k_{21}\delta t = 0.1$, the exponential decay is reproduced quite well. This type of integration is known as an explicit method. Φ is expressed as a polynomial and the form of the polynomial depends on the explicit method used. The most usual form is the fourth-order Runge–Kutta method.

Box 8.2 The Runge-Kutta Method

The technique we used above was obtained by explicit consideration of the exponential form of $a(t)$ and the derivation of a simple approximation. For more complex kinetic systems, more general approximations are needed. The differential equation describing a can be expressed in the form:

$$da/dt = f(a) \tag{E 37}$$

where $f(a)$ is a function of a and, usually, of other species. The fourth order Runge-Kutta method uses the following approximation:

$$a(t + \delta t) = a(t) + \tfrac{1}{6}(\alpha_0 + 2\alpha_1 + 2\alpha_2 + \alpha_3) \tag{E 38}$$

where

$$\alpha_0 = f(a(t))\delta t$$
$$\alpha_1 = f(a(t) + \tfrac{1}{2}\alpha_0)\delta t$$
$$\alpha_2 = f(a(t) + \tfrac{1}{2}\alpha_1)\delta t$$
$$\alpha_3 = f(a(t) + \alpha_2)\delta t.$$

Thus, provided $a(t)$ and the form of $f(a)$ are known, then $a(t + \delta t)$ may be evaluated. Starting at zero time, $a(0)$, the initial concentration may then be used to calculate sequentially the concentration of a at all subsequent times. The Runge–Kutta method provides an efficient numerical integrator, which differs from the normal numerical integration techniques such as the trapezium rule in that it predicts the forward behaviour of the dependent variable. A derivation of the method may be found in Gear, G. W. (1971) *Numerical Initial Value Problems in Ordinary Differential Equations*, Prentice–Hall, New Jersey. The number of repetitive numerical calculations mean that, for all but the simplest examples, computer programs are required to evaluate the data. Many program libraries (e.g. NAG, IMSL) contain ready made subroutines which can be incorporated into a user defined program. The series of books *Numerical Recipes for FORTRAN, PASCAL, etc* by (Press, W. H., Flannery, B. P., Teukolsky, S. A., and Vetterling, W. T. (1988). Cambridge University Press) contain examples of the coding required.

8.3.1 Stiffness

Figure 8.5 illustrates that even for simple kinetic systems we need to keep the stepsize short in order to calculate accurately the time-dependent concentrations. By short we mean short compared with the overall decay time, at least as

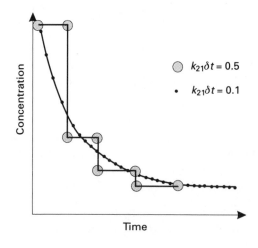

$k_{21}\delta t = 0.5$

$k_{21}\delta t = 0.1$

Fig. 8.5 Exponential decay of [A] (solid line) and numerical integration simulations. The small time steps are able to faithfully reproduce the actual decay but when the time interval is increased significantly large oscillations occur.

far as the simple exponential decay is concerned. A problem arises with more complex systems which can be appreciated by considering the A → B → C system discussed above, with $k_{15} \ll k_{16}$, i.e. the type of problem in which the steady state approximation applies. Consider a situation where $k_{15} = 10^{-3}$ s^{-1} and $k_{16} = 10^{3}$ s^{-1}. During the induction period we need to use a δt value of not greater than 10^{-4} s, but once the $\exp(-k_{16}t)$ term has decayed to very small values, say after 10^{-2} s, we might expect to be able to move on to much larger time steps typical of the decay determined by k_{15}. However, such a procedure introduces very large errors and the integration becomes unstable, the solutions may even show oscillatory behaviour. We need to continue to use small stepsizes over the very long time scale of the overall reaction, which clearly requires a great deal of computer time. This type of behaviour, in which the kinetics are determined by processes occurring with widely disparate timescales, produces what are known as stiff coupled differential eqns which are difficult to solve efficiently using techniques like the Runge–Kutta method. Fortunately, more efficient, so-called implicit techniques such as the 'backward' Euler method, have been devised and a discussion of one of these may be found in *An Introduction to Numerical Analysis* by Stoer and Bulirsh (1980).

8.3.2 Sensitivity analysis

The availability of rapid and efficient computer packages has enabled the kineticist to attempt to model complex systems such as combustion or atmospheric chemistry, possibly solving differential equations of mechanisms of several hundred chemical equations involving tens of reactants, intermediates and products. The temptation, when devising such models, is to write down every conceivable reaction and then perform the integration. However, what does this tell us? Whilst the numerical integration may be performed correctly, the quality of the end result is only as good as the input data, in this case the individual rate coefficients for each elementary reaction. The end results may match the experiments, but with several hundred elementary reactions being considered this may be as much due to chance as to good chemistry! We need to develop some way of analysing our mechanism to see which reactions are the most important. Once these reactions have been identified, the modeller will investigate the corresponding rate coefficients. Are they reliable? If not, then they need to be remeasured, or in some cases, measured for the first time.

The method most often used to investigate a reaction mechanism is sensitivity analysis. We shall illustrate our example by considering the pyrolysis of propane to form the industrial feed stocks ethene, propene, methane and hydrogen (see Section 9.4). Firstly we consider the overall rate of consumption of propane, and then vary each of the individual rate coefficients by a set amount (e.g. ±10 per cent), in each case noting the effect on the consumption of propane after a certain period of time. The most important, or sensitive, reactions in the model will induce the largest changes in the rate of propane consumption. We can be more specific than this and look at the effect of each rate coefficient on the production of one particular product. The sensitivity coefficient (S_{ij}) of species Y_i with respect to the rate coefficient k_j is defined by:

$$S_{ij} = \partial y_i(t)/\partial k_j(t_0).$$
(E 39)

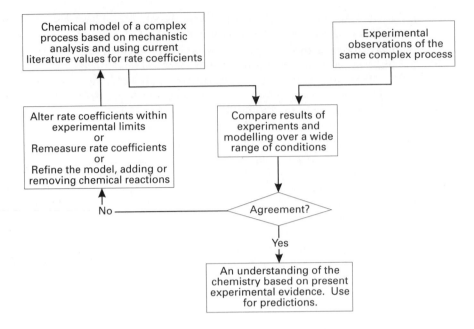

Fig. 8.6 Flow diagram showing the interplay between experiment and modelling studies.

Such analysis often provides important clues as to the dominant mechanisms occurring in complex systems. Note that the sensitivity coefficient is time-dependent. For example if we vary the rate coefficient of a reaction involving the combination of two product species, then this can only have a significant effect on the reaction when product concentrations have reached appreciable levels, i.e. late on in the reaction.

The method we have described relies more on 'brute force and ignorance' rather than any more elegant methods and can only be applied to simple systems or where a limited number of specific sensitivity coefficients are required. More efficient methods, such as the Decoupled Direct Method (Dunker, 1984) have been implemented in commercially available packages.

The relationship between modelling and experiment is a symbiotic one and is illustrated in the above schematic diagram (Fig. 8.6). Initially the modeller creates a reaction mechanism and performs the numerical integration. The results are compared with experiment, generally the agreement is less than perfect. A sensitivity analysis is performed and the crucial reactions are selected. The experimentalist will then examine these reactions and new values of the rate coefficients will be input into the model. This process is repeated until a convergence between experimental observations and numerical predictions is (hopefully!) obtained.

8.4 Atmospheric kinetics

On the face of it, the Earth's atmosphere looks singularly uninteresting from a kinetic viewpoint, consisting primarily of two very stable gases, nitrogen and oxygen. However, the interaction between these gases, sunlight and additional materials present in, or entering the atmosphere from natural or man-made sources, provides a system of great kinetic complexity, especially when coupled with the hydrodynamic processes occurring in the atmosphere. Interest in, and concern for, the effects of anthropogenic materials released into the atmosphere led, in the 1970s, to an unprecedented concentration of effort on the measurement of the rates of elementary reactions and to the development or refinement of many of the techniques discussed in Chapter 2.

8.4.1 The structure of the atmosphere

The variation in the temperature of the atmosphere with altitude leads to its division into reasonably well-defined layers which contain significantly different chemical and kinetic processes. Much of the energy from the sun passes through the atmosphere and heats the surface of the planet. The temperature of the lower atmosphere derives primarily from heat transfer from the surface and so decreases with increasing altitude. Above an altitude of ≈ 12 km (a figure dependent on latitude and season) the temperature, having fallen to ≈ 220 K, begins to increase again because of the absorption of sunlight in the ozone layer, whose origin we shall discuss below. The lower region is termed the troposphere and the upper one the stratosphere. The decrease in temperature with altitude promotes vertical mixing in the troposphere but the temperature inversion which occurs on entering the stratosphere has the converse effect. The transfer of the material from the troposphere to the stratosphere is comparatively slow. Further regions of the atmosphere occur at higher altitudes but will not concern us here.

8.4.2 Stratospheric chemistry

The interest in stratospheric chemistry is primarily in the creation and destruction of the ozone layer, which not only governs the temperature profile of the stratosphere, but is also responsible for absorbing solar radiation in the 240–320 nm region which would otherwise reach the Earth's surface and damage the surface cells of plants and animals. There is concern, therefore, about any processes leading to a reduction in the concentration of stratospheric ozone and the consequent danger of increased incidence of skin cancer in humans or interruption of the lower levels of food chains.

Ozone is produced by photodissociation of oxygen at wavelengths below ≈ 242 nm, followed by a termolecular combination reaction with molecular oxygen.

$$O_2 + h\nu \rightarrow 2O \qquad \text{(R 22)}$$

$$O + O_2 + M \rightarrow O_3 + M \qquad \text{(R 23)}$$

These reactions account for the layer-type structure of ozone in the stratosphere—the requisite short wavelengths for oxygen photolysis are absent at lower altitudes, having already been absorbed by oxygen molecules higher up. A similar reduction in the rate of photodissociation occurs at still higher altitudes because of the decrease in $[O_2]$, associated with the general reduction in pressure with altitude, which will also reduce the efficiency of the termolecular recombination reaction.

The ozone concentration is limited by two further reactions:

$$O_3 + h\nu \rightarrow O + O_2 \qquad \text{(R 24)}$$

$$O + O_3 \rightarrow 2O_2. \tag{R 25}$$

Reactions (22–25) constitute the Chapman mechanism for the creation and destruction of ozone in the unpolluted stratosphere. Reactions (23) and (24) are much faster than (R 22) and (R 25) and rapidly interconvert O and O_3. It has become customary for atmospheric chemists to refer to 'odd oxygen', where [odd oxygen] = [O] + [O_3]. The relative concentrations of the two species may be found by equating ρ_{23} and ρ_{24}, since their rates must be equal in the steady-state regime:

$$k_{23}[O][O_2][M] = P_{24}[O_3] \tag{E 40}$$

where $P_{24}[O_3]$ is the rate of ozone photolysis at a particular altitude. P_{24} is a complex function which contains wavelength dependent solar light intensity and extinction coefficient terms. Thus;

$$[O]/[O_3] = P_{24}/(k_{23}[O_2][M]) \tag{E 41}$$

In the lower stratosphere, where [O_2] and [M] are comparatively large and, because of absorption of the relevant wavelengths at higher altitudes, P_{24} is comparatively small, ozone accounts predominantly for the odd oxygen concentration. At higher altitudes, P_{24} is larger, and [M] smaller, so that the [O]/[O_3] ratio increases with altitude in the upper stratosphere.

Reactions (22 and 25) respectively produce and consume odd oxygen. For reaction (22) we have

$$d[O]/dt = d[\text{odd oxygen}]/dt = 2P_{22}[O_2] \tag{E 42}$$

where P_{22} is an equivalent term to P_{24} and refers to oxygen photolysis. For reaction (25)

$$-d[\text{odd oxygen}]/dt = 2k_{25}[O][O_3] \tag{E 43}$$

where the factor of two is included because both [O] and [O_3] are removed. Equating eqns (42) and (43), and recognizing that, in the majority of the stratosphere, [odd oxygen] $\approx$ [O_3], then

$$2P_{22}[O_2] = 2k_{25}[O_3]^2 P_{24}/(k_{23}[O_2][M]) \tag{E 44}$$

and

$$[O_3] = \{P_{22}[O_2]/k_{25}\}^{\frac{1}{2}}\{k_{23}[O_2][M]/P_{24}\}^{\frac{1}{2}}. \tag{E 45}$$

Equation (45) is shown in this unreduced form to demonstrate the dependence of [O_3] on a balance between the rate of two pairs of reactions: (a) reactions (22) and (25), which respectively produce and consume odd oxygen and (b) reactions (23) and (24), which determine the ratio of concentration of the odd oxygen species O and O_3.

This dependence arises because the rate of reaction (25) is proportional to the product of [O] and [O_3] and, for a given odd oxygen concentration, is maximized when the two concentrations are equal. The fact that, in the stratosphere, $k_{23}[O_2][M] \gg P_{24}$ not only determines that ozone is the primary odd oxygen species present, but also enhances the overall concentration of odd oxygen. Note that our simple treatment makes the assumption that the rates of reactions (23) and (24) are much faster than ρ_{22} or ρ_{25}, so that the [O]/[O_3] ratio, established through eqn (41), is undisturbed by reactions (22) and (25).

The substantial stratospheric ozone concentration can be perturbed by comparatively small traces of other substances, via catalytic cycles of the type

$$X + O_3 \rightarrow XO + O_2 \qquad \text{(R 26)}$$

$$XO + O \rightarrow X + O_2 \qquad \text{(R 27)}$$

$$\text{net: } O + O_3 \rightarrow 2O_2$$

where X is an atom or free radical. The cycle, in effect, increases the rate of reaction (25) in the simple Chapman mechanism and does not consume the perturbing species X.

Interest initially centred around NO and NO_2 as X and XO respectively, at a time when it was thought that supersonic aircraft would eventually constitute a major component of civilian airlines. Such aircraft fly at high altitudes and nitrogen oxides constitute important combustion products, so the catalytic cycle

$$NO + O_3 \rightarrow NO_2 + O_2 \qquad \text{(R 28)}$$

$$NO_2 + O \rightarrow NO + O_2 \qquad \text{(R 29)}$$

was, in consequence, considered a potential threat to the ozone layer. The number of supersonic aircraft has yet to reach the projected levels but the cycle is still important. Another class of anthropogenic pollutant, chlorofluorocarbons (CFCs), represent a dangerous threat to the ozone layer. Until the advent of the Montreal Protocol in 1989, CFCs were widely used as refrigerants, aerosol propellants and blowing agents (e.g. for making expanded polystyrene), where a major attraction is their chemical inertness. Paradoxically, it is this chemical inertness which provides the greatest threat to stratospheric ozone. CFCs have a very long lifetime in the troposphere and eventually pass up into the stratosphere, where the higher ultraviolet intensities lead to photolysis, producing chlorine atoms e.g.

$$CF_2Cl_2 + h\nu \rightarrow CF_2Cl + Cl \qquad \text{(R 30)}$$

followed by the catalytic cycle

$$Cl + O_3 \rightarrow ClO + O_2 \qquad \text{(R 31)}$$

$$ClO + O \rightarrow Cl + O_2. \qquad \text{(R 32)}$$

The cycle can be repeated many times before the chlorine atoms or ClO radicals react with other species to generate so called reservoir species which hold the destructive chlorine atoms as unreactive species e.g.

$$Cl + CH_4 \rightarrow HCl + CH_3 \qquad \text{(R 33)}$$

$$ClO + NO_2 + M \rightarrow ClONO_2 + M. \qquad \text{(R 34)}$$

Reactive chlorine can be regenerated e.g.

$$ClONO_2 + h\nu \rightarrow Cl + NO_3. \qquad \text{(R 35)}$$

or the reservoir species can be removed from the stratosphere reducing the active halogen species.

The lives of stratospheric kineticists would be considerably simplified if all they had to consider was a series of non-interacting catalytic cycles. Needless to say, interactions of considerable complexity do take place and much more detailed calculations are required to assess the future of the ozone layer. Indeed, at one stage projections varied from year to year, ranging from predictions of decreases to increases in ozone abundance, not because of real changes in the atmosphere, but because of changes in our understanding or in measured values for the component reactions.

New replacements for CFCs contain C–H bonds, e.g. $CHFCl_2$. Whilst these compounds are still relatively inert, the C–H bond allows them to be attacked by the OH radicals present in the troposphere,

$$CHFCl_2 + OH \rightarrow H_2O + CFCl_2 \qquad (R\ 36)$$

the resulting radicals are oxidized and the products rained out before they reach the stratosphere.

Box 8.3 The Ozone Hole

During the Antarctic spring very large reductions have been observed in the ozone layer above the Antarctic continent and later in the spring at lower latitudes. This 'hole' in the ozone layer illustrates the interactions between meteorology and chemistry and the complexity of atmospheric chemistry. Gas phase catalytic cycles alone cannot explain the dramatic reductions of ozone and we need to consider heterogeneous chemistry in addition.

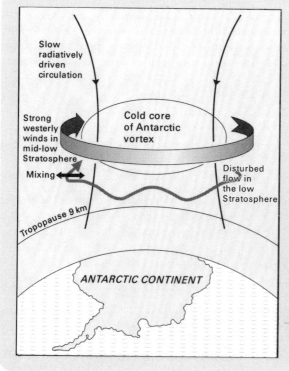

Fig. 8.7 The winter vortex over Antarctica. The cold core is almost isolated from the rest of the atmosphere, and acts as a reaction vessel in which the constituents may become chemically 'preconditioned' during the long polar night.

In winter a vortex of winds isolates the air above the continent (Fig. 8.7). The stratospheric air is cooled and polar stratospheric clouds (PSCs), consisting of solid ice crystals form. The surfaces of these ice crystals provide sites for the reaction of reservoir species HCl and chlorine nitrate:

$$HCl + ClONO_2 \rightarrow Cl_2 + HNO_3. \qquad (R\ 37)$$

Once the spring arrives the molecular chlorine is photolysed generating active chlorine atoms and the catalytic cycles can start up. The PSC have catalysed the conversion of reservoir species into photochemically labile chlorine atom precursors. The presence of the encircling vortex prevents the escape of the active chlorine and inward transport of fresh ozone bearing air and the ozone levels fall dramatically. Eventually the vortex breaks up and ozone deficient air spreads out over the inhabited areas of the south southern hemisphere.

That ClO is heavily implicated in ozone destruction can be seen from the results of an elegant experiment which monitored ClO and O_3 concentrations during a flight through the ozone hole (Fig. 8.8). The inverse correlation between the two molecules is all too apparent.

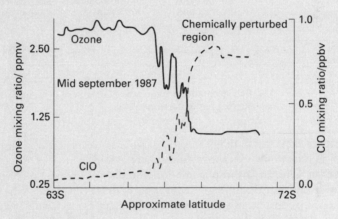

Fig. 8.8 Data recorded during an airborne experiment through the edge of the ozone hole monitoring both ozone and chlorine monoxide concentrations and showing the strong correlation between ozone depletion and ClO concentration. The experiment has been cited as conclusive evidence for enhanced ozone depletion by the products of CFC or HCFC degradation.

8.4.3 Tropospheric chemistry

The complexities of the reaction kinetics of the stratosphere are enhanced even further when the troposphere is considered and we illustrate this with two diagrams. The first (Fig. 8.9) is a schematic diagram of some of the processes and interactions present in the troposphere. Organic and inorganic material is released into the atmosphere from both natural (terrestrial and marine) and man-made sources. Reactions can occur in the gas phase, in cloud droplets or on particulate surfaces. Photochemical reaction rates will vary with latitude and season and reactant concentrations may depend on various meterological factors as much as chemistry.

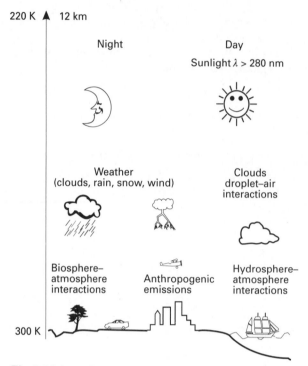

220 K ▲ 12 km

Night

Day

Sunlight λ > 280 nm

Weather
(clouds, rain, snow, wind)

Clouds
droplet–air
interactions

Biosphere–
atmosphere
interactions

Anthropogenic
emissions

Hydrosphere–
atmosphere
interactions

300 K

Fig. 8.9 Schematic cartoon indicating some of the complexities and interactions in the troposphere.

Figure 8.10 shows that even the gas-phase oxidation of a simple compound such as ethane, in a polluted atmosphere is fairly complex. The degradation of larger biogenic or anthropogenic compounds would take up several more pages. Despite this complexity, significant progress has been made in the understanding of many tropospheric processes such as urban smog formation and acid rain. However, there is much still to be done and this field remains a very active, fruitful, and interesting area of research. Aspects of photodissociation in atmospheric chemistry are briefly considered in Section 12.7.

Study notes
1   General coverage. A recent detailed account covering both terrestrial and extra-terrestrial atmospheres may be found in the book *Chemistry of Atmospheres* by Richard Wayne (2nd edn, Oxford University Press, 1992).
2   Stratospheric ozone. Chapter 4 of Wayne deals with the background to ozone formation and the interaction between the catalytic cycles. The chapter also contains a detailed account of the origin of the X and XO species.
3   The ozone hole. Recent articles relevant to ozone depletion are:
(a) Rowland, F. S. (1991). Stratospheric ozone depletion. *Annual Reviews of Physical Chemistry*, **42**, 731. Rowland was one of the pioneers of stratospheric ozone research and this account of the study of ozone depletion is in itself very readable and provides further access to the original literature.
(b) Johnston, H. S., (1992). Atmospheric ozone, *Annual Reviews of Physical Chemistry*, **43**, 1–32. A majority of the article is devoted to stratospheric ozone

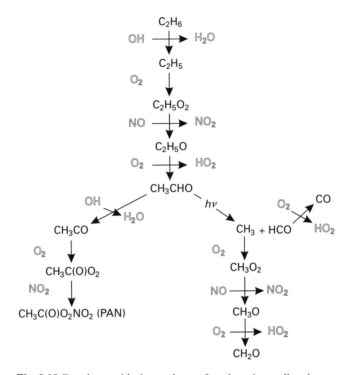

Fig. 8.10 Gas phase oxidation pathways for ethane in a polluted atmosphere (i.e. containing NO and NO_2). NO_2 is photolysed to produce O which reacts with O_2 to give O_3, an important component of smog. During the degradation there are several steps which involve NO oxidation and hence increase ozone concentrations. It is the combination of hydrocarbons and NO (both are primary automobile emissions).

and this highly personal account illustrates the complex interplay between science, economics and politics.

(c) Two articles in *Science* (Tolbert, M. A., Rossi, M. J., and Golden, D. M. (1988). Antarctic ozone depleting chemistry-reactions on ice surfaces. *Science*, **240**, 1018–21), and the *Journal of Physical Chemistry* (Quinlan, M. A., Reins, C. M., Golden, D. M., and Tolbert, M. A. (1990). Heterogeneous reactions on model polar stratospheric cloud surfaces. *Journal of Physical Chemistry*, **94**, 3255–60) discuss the relevance of heterogeneous chemistry in ozone destruction.

4   Tropospheric chemistry. Chapter 5 of Wayne, Chapter C 1.2 of *Modern Gas Kinetics* (ed. M. J. Pilling and I. W. M. Smith, Blackwell, Oxford, 1987) and the book *Atmospheric Chemistry* by Findlayson-Pitts, B. and Pitts, J. N. (John Wiley, Chichester, 1986) give extensive information about the various processes occurring in the troposphere.

5   The major environmental problems of the atmosphere not only involve chemistry but also meteorology, economics and politics. A more general perspective on these concerns may be found in a number of more broadly based books an example of which is *Atmospheric Pollution* by D. Elsom (Blackwell, Oxford, 1992).

8.5 Enzyme catalysis

Enzyme catalysis provides an excellent example of the analysis of complex reactions and the validity of the steady-state hypothesis. Enzymes (denoted as E in this section) are specific catalysts in biological reactions and they act by forming a complex (ES) with the reactant or substrate (S) which can then react to form the product (P). The overall reaction thus involves at least three steps:

$$\overset{k_{38}}{\underset{k_{-38}}{\rightleftarrows}} \quad \overset{k_{39}}{}$$
$$E + S \rightleftarrows ES \rightarrow P + E \qquad (R\ 38, -38, 39)$$

and regenerates the enzyme along with the product. The enzyme is a protein whose tertiary structure brings together certain functional groups, such as carboxyl or amino, into the active site, which provides the binding, specific to the substrate, for formation of the complex. The net result is an overall lowering of the barrier along the free energy surface linking S and P.

Much of the earlier work on enzyme catalysis employed the steady-state approximation for ES, which gives:

$$[ES] = k_{38}[E][S]/(k_{-38} + k_{39}) \qquad (E\ 46)$$

and

$$v = k_{38}k_{39}[E][S]/(k_{-38} + k_{39}) \qquad (E\ 47)$$

where v is the rate of the overall reaction (i.e. $v = d[P]/dt$ or $-d[S]/dt$). We use v, rather than our usual ρ to conform with the conventional practice in enzyme kinetics. At high substrate concentrations, the fraction of the enzyme in the form of the complex can be very high (i.e. $k_{38}[S] \gg (k_{-38} + k_{39})$) and the normal procedure is to evaluate v in terms of the total enzyme concentration, $[E_0]$, where

$$[E_0] = [E] + [ES] = [E][1 + \{k_{38}[S]/(k_{-38} + k_{39})\}] \qquad (E\ 48)$$

so that

$$[E] = [E_0](k_{-38} + k_{39})/(k_{38}[S] + k_{-38} + k_{39}) \qquad (E\ 49)$$

and

$$v = k_{38}k_{39}[E_0][S]/(k_{38}[S] + k_{-38} + k_{39}). \qquad (E\ 50)$$

These equations and the derivation are somewhat similar to those encountered in Chapter 5 for unimolecular reactions, or indeed in Chapter 6 for reactions in solution. In each case the mechanism of the reaction involves the formation of an intermediate complex which can either reform the reactants or go on to yield products. However, we shall see that the catalytic nature of the reaction and the small concentrations of enzymes result in some important differences and limit the applicability of the steady-state hypothesis.

The rate of reaction increases with $[S]$ at constant $[E_0]$, until it reaches a limiting maximum value $V = k_{39}[E_0]$, when the enzyme is saturated, $k_{38}[S] \gg (k_{-38} + k_{39})$ and $[ES] = [E_0]$. Inverting eqn (50)

$$1/v = 1/V + K_m/k_{39}[E_0][S] \qquad (E\ 51)$$

where K_m is the Michaelis constant, $(k_{-38} + k_{39})/k_{38}$: $K_m/[S]$ is equal to the ratio of free and complexed enzyme.

The classical method of studying enzyme reactions employs steady-state conditions and involves the determination of the initial rate of forming P (or consuming S). From a plot of $1/v$ vs. $1/[S]$, V and K_m may be found by extrapolation. Equation (51) is known as the Michaelis–Menten equation whilst the linear plot shown in Fig. 8.11 is a Lineweaver–Burk plot. Note that the rate law for [S] is quite complex (set $v = -d[S]/dt$ in eqn (50)), so that it is much easier to employ an initial rate method measuring the rate of reaction only over the first few percent where [S] is effectively constant.

The reaction scheme described above does not conform to the requirements we enumerated on p. 199 for the application of the steady-state approximation, since, ES, the intermediate, can be present at much higher concentrations than E, a reactant. The enzyme is different from the reactants discussed on pp. 121–2 and p. 145, however, since it is eventually regenerated, it is a true catalyst and $[E_0]$, the total enzyme pool, is constant throughout the reaction. Once the complex has been formed, any loss due to its decomposition is rapidly made good from the virtually inexhaustible supply of reactants and [ES] may justifiably be assumed constant. For example, for the fumarase/fumaric acid system, if $[E_0] = 5 \times 10^{-10}$ mol dm^{-3} and $[S] = 10^{-4}$ mol dm^{-3} only 10^{-3} s are required to bring [ES] within 10^{-3} per cent of its steady-state value, during which time the product P has attained only 10^{-3} per cent of its eventual value.

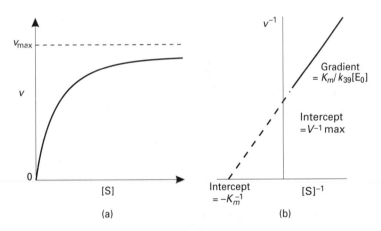

Fig. 8.11 Lineweaver–Burk plot of $1/v$ vs. $1/[S]$, showing the determination of V and K_m.

Box 8.4 The nature of enzymes

Enzymes are proteins made up of tens or hundreds of aminoacids (e.g. insulin, a small enzyme, contains a chain of 51 aminoacids). Their catalytic activity depends on the sequence of aminoacids in the chain and the functional groups present on the aminoacids. Aminoacids, as their name implies, have both basic and acidic functional groups and are of the general form:

$$R$$

$$H2N-C-COOH$$

$$H$$

where R can have various functional forms, e.g. alcoholic, basic, acidic etc. The amino acids are linked together via peptide linkages to form a chain structure of the form:

$$\cdots NH-CH(R)-CO-NH-CH(R)-CO \cdots$$

and in turn the chain wraps itself together to make a stable 3-D structure with various functional groups accessible on the substrate (Fig. 8.12). It is the complex 3-D structure of the enzyme that endows catalytic specificity. Only substrates of a precise shape and size can key into the enzyme surface and bind with the functional groups at the active site of the enzyme.

The specificity of enzyme catalysis can be seen by the reaction of the enzyme fumarase which catalyses the hydration of the double bond in fumaric acid, but not in the cis isomer, maleic acid (see (R 40)). See Fersht, A. *Enzyme Structure and Mechanism*, (2nd edn). Freeman, New York, (1985) for a wide-ranging discussion of enzymes, which sets the kinetic measurements in a broader context.

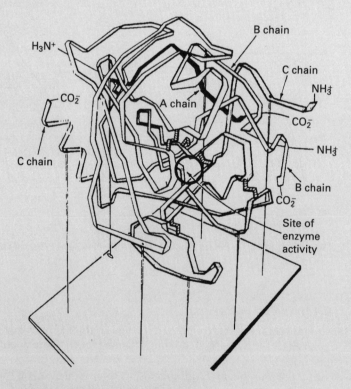

Fig. 8.12 Schematic drawing showing the conformation of the polypeptide chains in chymotrypsin.

$$\text{(R 40)}$$

8.5.1 Validity of the steady-state assumption

In view of the rather special circumstances of the steady-state approximation, it is worthwhile examining its establishment, and the conditions necessary for its application, in more detail. Equation (52b) describing the time-dependence of [ES] is obtained by substituting the expression $[E] = [E_0] - [ES]$ in the simple differential equation, eqn (52a)

$$d[ES]/dt = k_{38}[E][S] - (k_{-38} + k_{39})[ES] \tag{E 52a}$$

$$= k_{38}[E_0][S] - (k_{38}[S] + k_{-38} + k_{39})[ES]. \tag{E 52b}$$

Integration between $t = 0$ ($[ES] = 0$) and t and assuming [S] to be constant, gives

$$[ES] = k_{38}[S][E_0]/(k_{38}[S] + k_{-38} + k_{39}) [1 - \exp\{-(k_{38}[S] + k_{-38} + k_{39})t\}]$$

$$= [ES]_{ss} [1 - \exp\{-(k_{38}[S] + k_{-38} + k_{39})t\}] \tag{E 53}$$

where $[ES]_{ss}$ is the steady-state complex concentration obtained by setting eqn (52) equal to zero. Thus, the complex is in a steady-state for times $t_{ss} \gg (k_{38}[S] + k_{-38} + k_{39})^{-1}$, i.e. when the exponential term is small (e.g. $[ES] = 0.99[ES]_{ss}$ for $t_{ss} > 4.6(k_{38}[S] + k_{-38} + k_{39})^{-1}$). Note that all three reactions contribute towards establishment of the steady-state and, even for small [S], the steady-state will be established rapidly provided k_{-38} or k_{39} are large. Large substrate concentrations, which correspond to the circumstances under which most steady-state experiments are conducted, promote the rapid establishment of the steady-state via the term $k_{38}[S]$.

This discussion only provides us with part of the answer; the important criterion is that t_{ss} is much shorter than the decay time of S. The time-dependence of [S] is given by

$$K_m \ln([S_0]/[S]) + [S_0] - [S] = Vt. \tag{E 54}$$

From this equation, we can evaluate $t_{0.01}$, the time taken for consumption of 1 per cent of S (assuming that steady-state kinetics apply), by setting $[S] = 0.99[S_0]$

$$t_{0.01} = (0.99[S_0] - 0.01K_m)/V. \tag{E 55}$$

We can, therefore, get a reasonable idea of the overall applicability by requiring that $t_{ss} \ll t_{0.01}$. These arguments are somewhat qualitative, but are generally adequate. A more direct approach is to integrate the full, time-dependent eqn for [S]; this is generally difficult but can be simplified in the specific case of small [S].

8.5.2 Mechanism involving two intermediates

Sometimes more than one enzyme–substrate complex is involved in the catalytic reaction. For example, in the hydrolysis of p-nitrophenylacetate:

$$\text{(R 41)}$$

catalysed by chymotrypsin, the enzyme initially and rapidly forms a complex, ES_1, which does not involve the formation of any covalent bonds. Next, the enzyme is acylated to form a second complex, ES_2 with the release of nitrophenol as a product (P_1) and finally the enzyme is reformed and acetate (P_2) is released:

$$E + S \underset{k_{-42}}{\overset{k_{42}}{\rightleftarrows}} ES_1 \overset{k_{43}}{\rightarrow} ES_2 + P_1 \overset{k_{44}}{\rightarrow} E + P_2. \qquad \text{(R 42, −42, 43, 44)}$$

We may obtain an overall rate expression for such a reaction, by (i) applying the steady-state approximation to ES_1 and ES_2, (ii) setting the overall rate of reaction equal to $d[P_2]/dt = k_{44}[ES_2]$, which gives us an equation containing $[E]$ and (iii) substituting $[E_0]$ for $[E]$ using $[E_0] = [E] + [ES_1] + [ES_2]$.

(i)
$$d[ES_1]/dt = k_{42}[E][S] - (k_{-42} + k_{43})[ES_1] = 0$$

$$[ES_1] = \frac{k_{42}[E][S]}{(k_{-42} + k_{43})} \qquad \text{(E 56)}$$

similarly

$$[ES_2] = k_{43}[ES_1]/k_{44} = k_{42}k_{43}[E][S]/k_{44}(k_{-42} + k_{43}) \qquad \text{(E 57)}$$

(ii)
$$v = d[P_2]/dt = k_{44}[ES_2] = k_{42}k_{43}[E][S]/(k_{-42} + k_{43}) \qquad \text{(E 58)}$$

(iii)
$$[E_0] = [E] + [ES_1] + [ES_2]$$

$$[E] = \left\{ 1 + \frac{k_{42}[S]}{(k_{-42} + k_{43})} + \frac{k_{42}k_{43}[S]}{k_{44}(k_{-42} + k_{43})} \right\} \qquad \text{(E 59)}$$

$$[E] = \frac{[E_0]k_{44}(k_{-42} + k_{43})}{k_{44}(k_{-42} + k_{43}) + k_{42}[S](k_{44} + k_{43})} \qquad \text{(E 60)}$$

substituting into eqn (58) we find

$$v = \frac{k_{42}k_{43}k_{44}[E_0][S]}{k_{42}[S](k_{43} + k_{44}) + k_{44}(k_{-42} + k_{43})} \qquad \text{(E 61)}$$

or

$$v = \frac{V[S]}{([S] + K_m)} \qquad (E\,62)$$

where $V = k_{43}k_{44}[E_0]/(k_{43} + k_{44})$ and $K_m = k_{44}(k_{-42} + k_{43})/k_{42}(k_{43} + k_{44})$. Note:

1.   Equation (62) is identical in *form* to that obtained for a single intermediate, the only difference is that more complex expressions are obtained for V and K_m. Thus a double-intermediate mechanism follows the Michaelis–Menten equation. Conversely, by a straightforward study of the steady-state kinetics of an enzyme-substrate complex, it is not possible to determine the number of complexes involved.

2.   The following equations apply

$$d[P_2]/dt = d[P_1]/dt = -d[S]/dt$$

3.   In some circumstances it is possible to obtain further information on the mechanism, such as the rates of the individual processes, by adding a reagent which provides a competitive route for the generation of ES_2.

4.   A more complete discussion of multiple intermediate mechanisms may be found in Roberts (1977), Chapter 3. He uses a more complex vector diagram approach in his evaluation of v, which pays dividends for more complex mechanisms.

8.5.3 Enzyme inhibition

The efficiency of an enzyme may be reduced, or even destroyed, by the presence of an inhibitor (I), which competes with the substrate in forming a complex EI. A well-known, and possibly terminal, example is the inactivation of cytochrome oxidase (a central enzyme in aerobic oxidation) by cyanide. The following kinetic scheme now applies:

$$E + S \underset{k_{-45}}{\overset{k_{45}}{\rightleftarrows}} ES \overset{k_{46}}{\rightarrow} P + E \qquad (R\,45, -45, 46)$$

$$E + I \underset{k_{-47}}{\overset{k_{47}}{\rightleftarrows}} EI \qquad (R\,47)$$

We proceed as before, first evaluating [ES], which is given, once again, by eqn (56) and EI, which, according to the reaction scheme, is in equilibrium with E and I, i.e. $[EI] = K_I [E][I]$. We then obtain [E] as a function of $[E_0]$ and the overall rate, v, is given by the familiar expression

$$v = V[S]/([S] + K) \qquad (E\,62)$$

where $K = K_m(1 + [I]/K_I)$ whilst $V = k_{46}[E_0]$ and $K_m = (k_{-45} + k_{46})/k_{45}$. Thus the rate determinations (v vs. [S]) for different values of [I] show the same maximum limiting rate, V, (since if enough substrate is added, the enzyme becomes saturated) but the slope of the Lineweaver–Burk plot increases with [I] and the $v^{-1} = 0$ intercept moves to smaller values of $-[S]^{-1}$.

A more complete discussion of enzyme inhibition may be found in Roberts (1977), Chapter 3. In particular, he considers further possible reactions of EI and the case of inhibition in which I binds to ES but not to E.

8.6 Questions

8.1 The following relative concentrations of *cis*- and *trans*-cinnamonitrile ($C_6H_5CH = CHCN$) were determined as a function of time in a solution initially containing almost pure *cis*-($C_6H_5CH = CHCN$).

t/s	0	10^3	2×10^3	3×10^3	5×10^3	10^5
cis	0.98	0.92	0.81	0.77	0.67	0.42
trans	0.02	0.08	0.19	0.23	0.33	0.58

Evaluate the equilibrium constant and the rate coefficients of the forward and backward reactions.

8.2 Figure 8.13 shows the time-dependence of OH chemiluminescence from the first electronically excited state of OH produced from the reaction:

$$CH + O_2 \overset{k_1}{\to} OH(A) + CO. \tag{1}$$

carried out under pseudo-first-order conditions with $[O_2] = 5 \times 10^{14}$ molecule cm^{-3}. The electronically excited OH decays back to the ground state with the emission of light:

$$OH(A) \overset{k_2}{\to} OH(X) + h\nu \tag{2}$$

Derive an expression for the time-dependence of the OH emission and associate reactions (1) and (2) with the appropriate part of the profile if $k_2 \gg k_1[O_2]$. Estimate a value for the bimolecular rate coefficient k_1.

8.3 The Chapman mechanism for generating ozone in the stratosphere consists of the reactions:

$$O_2 + h\nu \to O + O \tag{3}$$

$$O + O_2 + M \to O_3 + M \tag{4}$$

$$O_3 + h\nu \to O_2 + O \tag{5}$$

$$O + O_3 \to 2O_2 \tag{6}$$

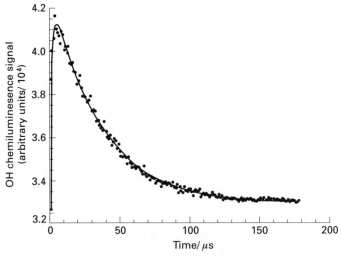

Fig. 8.13 Temporal profile of OH(A) spontaneous emission from the reaction of CH radicals with molecular oxygen.

(i) Explain why ozone is present in a well defined layer in the stratosphere. (ii) Explain why O and O_3 are strongly coupled together so that we can refer to them as odd oxygen. (iii) Why is the species M required in reaction (2) ? (see Chapter 5). (iv) Obtain an expression for the steady state concentration of odd oxygen. (v) Explain how small concentrations of Cl can lead to ozone depletion. (vi) From the following rate data, determine the ratio of [Cl]/[O] needed to increase the net rate of loss of odd oxygen by 10 per cent at 220 K.

Reaction	$A/cm^3\,molecule^{-1}\,s^{-1}$	$E/kJ\,mol^{-1}$
$O + O_3$	8×10^{-12}	19.1
$Cl + O_3$	2.8×10^{-11}	2.1

8.4 In polluted urban environments O_3, NO_2 and NO interact in the following way:

$$NO_2 + h\nu \rightarrow NO + O \tag{7}$$

$$O + O_2 + M \rightarrow O_3 + M \tag{8}$$

$$O_3 + NO \rightarrow NO_2 + O_2 \tag{9}$$

By assuming a steady state in [O] and [O_3] show that:

$$[O_3] = k_1[NO_2]/k_3[NO].$$

Measurements have been made of the mixing ratios of NO, NO_2 and O_3 in a trajectory across the plume from a power plant (22 km downwind from the source) as a function of distance (d) from the centre. At a particular time of day the values were as follows:

O_3/ppb	94.0	108.0	129.0
NO_2/ppb	74.5	49.1	30.2
NO/ppb	18.3	10.5	5.4
d/km	0	4.86	12.9

Show that these data are consistent with the establishment of the photostationary state.

8.5 In a temperature jump study of the lysozyme(E) $-$ N,N–biacetylglucosamine(S) interaction, the following first-order rate coefficients k for the approach to equilibrium were measured as a function of concentrations:

$10^4([E] + [S])/mol\,dm^{-3}$	2.0	4.0	6.0
$10^{-3}\,k/s^{-1}$	1.8	2.7	3.6

Evaluate the rate coefficients for formation and for dissociation of the E-S complex.

References

Durker, A. M. (1984). The decoupled direct method for calculating sensitivity coefficients in chemical kinetics. *Journal of Chemical Physics*, **81**, 2385–93.

Roberts, D. V. (1977). *Enzyme kinetics*. Cambridge University Press, Cambridge.

Seakins, P. W. and Pilling, M. J. (1991). A laser flash photolysis study of the Br + $i - C_4H_{10}$ $\rightarrow C_4H_9$ + HBr reaction. *Journal of Physical Chemistry*, **95**, 9874–8.

Stoer, J. and Bulirsh, R. (1980). *An introduction to numerical analysis*. Springer Verlag, Berlin.

9 Straight chain reactions

9.1 Introduction

Many reactions, several of them of industrial importance such as polymerization or alkane pyrolysis, proceed via a so-called *chain mechanism*, where the links of the chain, the elementary reactions, are repeated over and over again. A generalized scheme for such a reaction is shown in Fig. 9.1. Chain reactions normally involve elementary reactions of radical species, the unique feature being that the starting radical is regenerated either in the propagation reaction or in a subsequent reaction. The process can therefore be repeated and the chain cycle continues until all the reactants are consumed or the radical is removed in some other reaction.

In our initial description of chain reactions we shall draw on examples from the hydrogen/bromine system, where the overall reaction can be written as:

$$H_2 + Br_2 \rightarrow 2HBr. \tag{R 1}$$

Radicals are generated in an *initiation step* which is generally slow and usually involves thermal or photochemical dissociation of a relatively stable reactant to form free radicals. e.g.

$$Br_2 \rightarrow 2Br. \tag{R 2}$$

The thermal process has a high activation energy, so that its first-order rate coefficient is small. The radicals produced in this step are highly reactive and undergo rapid reaction with stable reactant molecules, which necessarily generate more radicals. e.g.

$$Br + H_2 \rightarrow HBr + H. \tag{R 3}$$

Further reaction ensues in an H_2/Br_2 mixture:

$$H + Br_2 \rightarrow HBr + Br \tag{R 4}$$

so that the original radical is regenerated. These are very rapid reactions so that the radical interconversion occurs on a very short timescale, and the process can repeat itself many times, each cycle representing a link in the overall chain process. The rapid steps are termed *propagation reactions* and the radicals are known as *chain centres* or *chain carriers*. Finally, the chain is broken when a chain centre reacts by an alternative route called a *termination reaction*, in which no new radicals are formed. Such a step may be linear (i.e. first order in radicals), a frequent example is the diffusion of the chain centre to the wall of the reaction vessel, or quadratic (i.e. second order in radicals) involving gas phase recombination e.g.

$$Br + Br + M \rightarrow Br_2 + M \tag{R 5}$$

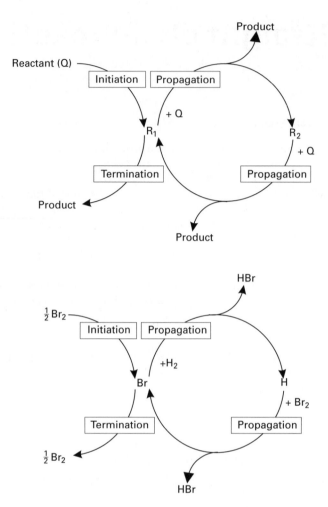

Fig. 9.1 Schematic diagram of a generalized chain reaction showing the relationship between initiation, propagation and termination and for a specific example, the $H_2 + Br_2$ reaction.

After a short induction period, a *steady-state* is established and the reaction cycles through the chain steps at a constant rate. These propagation steps do not affect the overall radical concentration, since each step involves a radical as both a reactant and product. The overall radical concentration is determined by a balance between the initiation reactions which form them, and the termination reactions which remove them. The average number of propagation steps, which occur between initiation and termination is called the *chain length*, which we shall discuss in more detail below. It is determined by the relative rates of the propagation and termination (or initiation) reactions.

There are a great many chain reactions but we shall examine their general features through a discussion of only three general classes, the hydrogen/halogen reactions, the pyrolysis of hydrocarbons, which is of importance in the cracking of petroleum feedstocks, and free-radical polymerization reactions.

9.2 The hydrogen/halogen reactions

The hydrogen/halogen reactions form an interesting class of reactions and we shall begin our discussion with a more detailed investigation of the hydrogen/bromine reaction. The reactions described in the previous section are summarized below in Table 9.1 where we have introduced a new class of reaction, *inhibition*, for the process in which the chain carriers can also attack the product, regenerating one of the reactants, but note that there is still no reduction in the number of chain carriers.

The kinetics may be deduced by applying the steady-state approximation (Section 8.2) to the reactive intermediates, i.e. to the chain carriers Br and H.

$$d[Br]/dt = 2k_{-5}[Br_2][M] - \mathbf{k_3[Br][H_2]} + \mathbf{k_4[H][Br_2]} +$$
$$\mathbf{k_{-3}[H][HBr]} - 2k_5[Br]^2[M] = 0 \tag{E 1}$$

$$d[H]/dt = \mathbf{k_3[Br][H_2]} - \mathbf{k_4[H][Br_2]} - \mathbf{k_{-3}[H][HBr]} = 0 \tag{E 2}$$

Note the similarity between the emboldened sections of eqn (1) and (2). These involve the propagation and inhibition steps, cycling Br to H and vice versa. Adding eqn (1) and (2) gives

$$2k_{-5}[Br_2][M] - 2k_5[Br]^2[M] = 0. \tag{E 3}$$

This simplification arises because radical formation and termination are solely controlled by reactions involving Br atoms. Equation (3) can be solved giving

$$[Br] = (k_{-5}[Br_2]/k_5)^{\frac{1}{2}} = (K_{-5}[Br_2])^{\frac{1}{2}} \tag{E 4}$$

where K_{-5} is the equilibrium constant for reaction $(-5,5)$; we see that Br and Br_2 are in equilibrium. Substituting eqn (4) into eqn (2) gives

$$[H] = \frac{k_3[H_2](K_{-5}[Br_2])^{\frac{1}{2}}}{k_{-3}[HBr] + k_4[Br_2]} \tag{E 5}$$

The rate of product formation is

$$d[HBr]/dt = k_3[Br][H_2] + k_4[H][Br_2] - k_{-3}[H][HBr] \tag{E 6}$$

Table 9.1 Reactions of the H_2/Br_2 system

Reaction	Class of reaction	$k_{500\,K}/dm^3\,mol^{-1}\,s^{-1}$	Reaction number
$Br_2 + M \rightarrow 2Br + M$	Initiation	$3.8 \times 10^{-8}[M]^a$	(R -5)
$Br + H_2 \rightarrow HBr + H$	Propagation	380	(R 3)
$H + Br_2 \rightarrow HBr + Br$	Propagation	9.6×10^{10}	(R 4)
$H + HBr \rightarrow H_2 + Br$	Inhibition	7.2×10^9	(R -3)
$Br + Br + M \rightarrow Br_2 + M$	Termination	$4.2 \times 10^{-13}[M]^b$	(R 5)

^{a} Unimolecular reaction–value quoted for low pressure region where $k_{uni}(s^{-1}) = k^0[M]$ (see section 5.8).
^{b} Termolecular reaction–value quoted for low pressure region where $k_{bi} = k^0[M]$. For 1 atm at 500 K, $[M] \approx 0.025$ mol dm^{-3} (see section 5.12).

$$= \frac{2k_3[H_2](K_{-5}[Br_2])^{\frac{1}{2}}}{1 + \left(\dfrac{k_{-3}[HBr]}{k_4[Br_2]}\right)} \tag{E 7}$$

The form of eqn (7) agrees well with that found by Bodenstein and Lind in their pioneering experiments on chain reactions, performed early in the twentieth century. The inhibition step introduces the term in [HBr] into the denominator. At low conversions, this reaction may be neglected, and the denominator becomes unity.

Study notes
The reaction scheme is characteristic of many chain mechanisms and therefore deserves a more detailed examination.
1 The dissociation of hydrogen is much slower than that of bromine, because the H–H bond is much stronger than the Br–Br bond and initiation proceeds exclusively via reaction (−5).
2 The rate coefficient for reaction (4) is larger than that for reaction (3), once again reflecting the relative bond energies of the diatomic molecules. In the steady-state, the reaction moves smoothly through the propagation steps and, at low conversions where we can ignore the inhibition term, the *rate* of reaction (3) must equal that of reaction (4) i.e.

$$k_3[Br][H_2] = k_4[H][Br_2] \tag{E 8}$$

Thus for a stoichiometric mixture of bromine and hydrogen, $[Br]/[H] = k_4/k_3$ and $[Br] \gg [H]$. An alternative way of understanding this inequality is through the average time the system takes to go through steps (3) and (4). As we discussed in Section 8.2. the average time for a reaction is equal to the reciprocal of the pseudo-first-order rate coefficient. Labeling these times τ_3 and τ_4, then it is clear that τ_3 $(=1/k_3[H_2])$ is much longer than τ_4 $(=1/k_4[Br_2])$ so that there is only a short time lag in between Br being consumed and its regeneration.
3 Because of the preponderance of bromine atoms, recombination via hydrogen atoms is slow and only termination via reaction (5) need be considered. The balance between initiation and termination and the exclusive identification of the latter with bromine recombination, results in an equilibrium concentration of bromine atoms in the system (eqn (4)). If k_3 and k_4 were closer in magnitude, and hence the concentrations of H and Br more comparable, this simplification and the resulting equilibrium would not apply, since some termination would then take place via H.
4 Neglecting the inhibition term, which will be a good approximation at short times, where [HBr] is very small,

$$d[HBr]/dt = 2k_3[H_2]\{K_{-5}[Br_2]\}^{\frac{1}{2}} \tag{E 9}$$

The overall reaction order is 1.5 and the effective activation energy is given by:

$$E_{act} = RT^2 d\ln(2K_{-5}^{\frac{1}{2}}k_3)/dT = E_3 + \tfrac{1}{2}\Delta H_5 \tag{E 10}$$

where E_3 is the activation energy for reaction (3) and ΔH_5 the enthalpy change for the equilibrium (−5, 5)

5 The chain length v, is equal to the average number of links in the chain, the mean number of propagation steps which occur before termination.

$$v = \text{(rate of propagation)}/\text{(rate of termination)} \qquad (E\,11)$$

$$= k_3[\text{Br}][\text{H}_2]/k_5[\text{Br}]^2[\text{M}]. \qquad (E\,12)$$

9.2.1 A comparison of the hydrogen/halogen reactions

It is instructive to compare the rates of reaction of hydrogen with the three halogens, chlorine, bromine, and iodine. As we have just seen, the principal factors that govern the rate of the $\text{H}_2 + \text{Br}_2$ reaction are the steady-state concentration of bromine atoms, and the rates of the propagation steps. Since bromine atoms and bromine molecules are in equilibrium, we shall, as a first step in our comparison, consider the equilibrium

$$\text{Hal}_2 \rightleftarrows 2\text{Hal} \qquad (R\,6,\,-6)$$

the position of which is largely determined by the dissociation energy. We see from Table 9.2 that the equilibrium concentration is high for iodine and low for chlorine, compared to bromine. The rate coefficients for the propagating steps are shown in Table 9.3.

The A factors are similar for all the reactions, but the activation energies vary considerably. For all three reactions (see Table 9.3) step (**3**) ($\text{Hal} + \text{H}_2$) is the rate-determining propagation step. This means that, since step (**4**) is much faster than step (**3**), the halogen atoms removed in (**3**) are replaced almost immediately. The propagating steps do not materially affect [Hal], which depends solely on (**5**) and (**−5**), and thus is equal to the equilibrium value, as we found for bromine.

Because of the low value for [Cl], the rate of termination, which depends on $[\text{Cl}]^2$, is slow. Step (**3**) ($\text{Hal} + \text{H}_2$), on the other hand, is comparatively fast. Propagation is thus much faster than termination, and the links in the chain are

Table 9.2 Dissociation energies of the halogens

Halogen	Dissociation Energy/kJ mol^{-1}
Cl_2	242
Br_2	193
I_2	151

Table 9.3 Rate parameters of $\text{Hal} + \text{H}_2$ and $\text{H} + \text{Hal}_2$

Reaction	A/dm^3 mol^{-1} s^{-1}	E/kJ mol^{-1}	k/dm^3 mol^{-1} s^{-1} (at 500 K)
$\text{Cl} + \text{H}_2$	4×10^{10}	23	1.6×10^8
$\text{Br} + \text{H}_2$	1.4×10^{11}	82	3.8×10^5
$\text{I} + \text{H}_2$	2.4×10^{11}	142	3.5×10^{-4}
$\text{H} + \text{Cl}_2$	3×10^{11}	13	1.3×10^{10}
$\text{H} + \text{Br}_2$	1.5×10^{11}	4	5.7×10^{10}
$\text{H} + \text{I}_2$	1.5×10^{10}	0	1.5×10^{10}

repeated many times before a break occurs, leading to a longer chain length than is found for $H_2 + Br_2$.

In contrast, the $I + H_2$ reaction is very slow, and the $H_2 + I_2$ reaction does not show chain characteristics. Instead the kinetics are second order:

$$\rho = k[H_2][I_2] \tag{E 13}$$

and for a long time the reaction was thought to be an exemplary bimolecular reaction. However, an alternative mechanism is:

$$I_2 \rightleftarrows 2I \tag{R 7}$$

$$2I + H_2 \rightarrow 2HI. \tag{R 8}$$

Assuming that the equilibrium is maintained in reaction (7), then $[I]^2 = K_7[I_2]$ and the overall rate of reaction is:

$$\rho = k_8[H_2][I]^2 = k_8K_7[H_2][I_2] \tag{E 14}$$

which conforms with the observed behaviour with the overall rate coefficient being the product k_8K_7. This mechanism has been tested photolytically, as is discussed in the study notes.

Study notes
The experimental investigation of the above mechanism is described in a paper by Sullivan, J. H. (1967). Mechanism of the 'bimolecular' hydrogen iodine reaction. *Journal of Chemical Physics,* **46**, 73–8 and almost the whole of the paper can be read, and understood, with profit. The basic idea that lies behind the paper is that the observed activation energy for the overall reaction is 170 kJ mol^{-1}, which is greater than the dissociation energy of I_2 (Table 9.2). If the thermal reaction applies then the equilibrium concentration of I is small at low temperatures, and the reaction is consequently slow. However, the concentration of I can be increased significantly by irradiating the reaction mixture with light in the 500–600 nm (visible) region and this should lead to a significant increase in the rate of formation of HI. No such increase would be observed if the reaction were an elementary bimolecular reaction involving the lowest electronic states of H_2 and I_2. Sullivan conducted such an experiment and demonstrated the increase in rate expected for an atomic mechanism. He went further however, and showed that the high temperature reaction conducted in the absence of light is quantitatively compatible with the low temperature photolytic mechanism. The paper provides an interesting insight into the care which has to be exercised in such investigations.

The experimental section is largely composed of an account of how the light absorbed by the reaction mixture was determined. These are standard methods in photochemical experiments and a general discussion may be found in *Photochemistry* by Okabe, H. (1978) (Wiley, Chichester). The most difficult section of the Sullivan paper discusses how reflection of the light from the various optical surfaces was allowed for. This section can be omitted but its presence indicates the care which needs to be taken.

The first paragraph of the results section requires some knowledge of the behaviour of electronically excited iodine. The gist of the argument is that, under the experimental conditions, over 98 per cent of the light absorbed led to I_2 photodissociation, and hence that reactions involving electronically excited iodine molecules can be ignored.

The next paragraph describes how iodine atom recombination was allowed for, since the production of HI via reaction (8) competes with I_2 regeneration, with I_2 and H_2 as third bodies. Finally k_8 is evaluated firstly from the thermal data, since $k_8 = \rho/[H_2][I]^2$ and [I] can be calculated from the equilibrium constant, and then from the lower temperature photolysis experiments from a knowledge of the HI production rate, the number of photons absorbed per second and the I + I recombination rate. Figure 1 in the paper shows an Arrhenius plot for k_8 and shows that the two sets of data lie close to the same straight line.

Finally the discussion section examines the nature of the reaction. The favoured mechanism involves formation of a van der Waals complex H_2I which then reacts further with I. Such complexes have been widely invoked in descriptions of iodine atom recombination. A further possible mechanism, involving excited iodine molecules is rejected, although this mechanism has been invoked by others since Sullivan's original publication (see Hammes, G. C., and Widom, B. (1974). On the mechanism of the hydrogen iodine reaction. *Journal of the American Chemical Society*, **96**, 7623–4). Anderson, J. B. (1996). The hydrogen-iodine reaction: 100 years later. In *Gas phase chemical reaction systems* (ed. J. Wolfrum, H.-R. Volpp, R. Rannacher, and J. Warnatz), pp. 167–76. Springer, Berlin.

Question 9.3 develops some of the themes discussed above.

9.3 A general strategy for analysing the kinetics of straight chain reactions

We can generalize the method of analysis we used to derive the formulae controlling the Br_2/H_2 chain reaction in the following way. The aim in the analysis is the derivation of an expression for product formation (or reactant consumption) involving the rate coefficients for the component elementary reactions and the concentrations of *stable species*. Since radicals are involved in the propagation steps consuming reactants or generating products, we need also to develop expressions for these chain carrier concentrations and, in general, expressions for radical concentrations can be found via the steady-state approximation. The rates of the propagating steps will have much higher values than the initiation or termination reactions and, at least to a first approximation we can ignore these latter terms when calculating the rate of product formation , see p. 229 on ethane pyrolysis for an example of such simplifications.

After postulating a reaction mechanism, based on experimental observation, rate data, or chemical intuition (e.g. Table 9.1), the steady-state equations are set up, to give a set of simultaneous differential equations, each containing the concentrations of several radicals (e.g. eqn (1) and (2)). The sequential nature of the propagation steps means that the rate of a given reaction occurs more than once in these equations, as both a formation and a removal step. For example the term $k_2[Br][H_2]$ appears in both equation (1) and (2) and determines the consumption of Br and the production of H in the propagation steps. Thus, by adding or subtracting the simultaneous equations, simple expressions can be derived containing the concentration of a single radical. Finally these expressions can be substituted into the equation controlling product formation (e.g. eqn (4) and (5) determining [Br] and [H] respectively, are substituted into eqn (6), the expression for the rate of production of HBr). This strategy is illustrated and developed in the next two sections and in the problems at the end of this chapter.

9.4 Alkane pyrolysis

A central process in the chemical industry is the so-called cracking of petroleum based feedstocks (e.g. naphtha, a C_6–C_{10} fraction) to produce smaller hydrocarbons and notably ethene. Ethene and its derivatives are primarily used as the raw materials for polymer production, e.g. ethene → polythene. The basis of cracking is very simple, since it depends on heating the hydrocarbon to a high temperature where it undergoes unimolecular decomposition to produce alkyl radicals, which then take part in a chain reaction. The simplest alkane to which this process is commercially applied is ethane and we shall consider this as our initial example.

9.4.1 Ethane pyrolysis

Table 9.4 shows a simple reaction scheme for the pyrolysis of ethane. The major products are ethene and hydrogen. The reaction is initiated by cleavage of the C–C bond in ethane (which is weaker than the C–H bond) to generate methyl radicals. The resulting propagation reactions are more complex than we found for the hydrogen halogen reactions and comprise several cycles.

CH_3 radicals are produced by the unimolecular decomposition of ethane; however, methyl radicals are not the major chain carrying species; they react with ethane in a hydrogen abstraction reaction to generate one of the chain carriers, the ethyl radical. C_2H_5 is comparatively unstable at the temperature employed commercially (≈ 1100 K) and rapidly decomposes to produce ethene and H. This reaction is a central one in the overall scheme and represents the major route to ethene. Ethyl radicals are regenerated via reaction (12) in which H atoms abstract another hydrogen atom from the primary reactant, C_2H_6. The relative rates of reactions (9–12) mean that $[C_2H_5] \gg [CH_3]$ and [H] and hence only termination steps involving ethyl radicals (reactions (13) and (14)) need to be considered. We can set up the following analysis for the initial conversion of ethane to ethene from our postulated mechanism of Table 9.4.

Firstly, the steady-state equations determining radical concentrations:

$$d[CH_3]/dt = 2k_9[C_2H_6] - k_{10}[CH_3][C_2H_6] = 0 \qquad (E\ 15)$$

Table 9.4 Simple mechanism of ethane pyrolysis

Reaction	Classification	Reaction no.
$C_2H_6 \rightarrow 2CH_3$	Initiation	(R 9)
$CH_3 + C_2H_6 \rightarrow CH_4 + C_2H_5$	Propagation (minor)[a]	(R 10)
$C_2H_5 \rightarrow H + C_2H_4$	Propagation (major)	(R 11)
$H + C_2H_6 \rightarrow H_2 + C_2H_5$	Propagation (major)	(R 12)
$C_2H_5 + C_2H_5 \rightarrow C_4H_{10}$	Termination[b]	(R 13)
$C_2H_5 + C_2H_5 \rightarrow C_2H_4 + C_2H_6$	Termination (minor)[b]	(R 14)

[a] The major chain reactions are (11) and (12). In some way, therefore, reaction (10) can be considered as an initiation step as it is generating a major chain carrying species. However, it is also a propagation reaction as one radical (CH_3) is being consumed and another generated (C_2H_5).

[b] Disproportionation and recombination are, in general, small in comparison with chain propagation, so that the kinetic chain length is long. This fact will simplify our steady-state analysis.

$$d[C_2H_5]/dt = k_{10}[CH_3][C_2H_6] + \boldsymbol{k_{12}[H][C_2H_6]} - \boldsymbol{k_{11}[C_2H_5]} -$$
$$2k_{13}[C_2H_5]^2 - 2k_{14}[C_2H_5]^2 = 0 \qquad \text{(E 16)}$$

$$d[H]/dt = \boldsymbol{k_{11}[C_2H_5]} - \boldsymbol{k_{12}[H][[C_2H_6]]} = 0. \qquad \text{(E 17)}$$

Equation (15) is easily solved giving $[CH_3] = 2k_9/k_{10}$ and in turn this value can be substituted into eqn (16). The emboldened terms are common to both eqn (16) and (17) and manipulation of the two equations leads to the following expressions for the steady-state concentrations of $[C_2H_5]$ and $[H]$.

$$[C_2H_5] = \{k_9/(k_{13} + k_{14})\}^{\frac{1}{2}}[C_2H_6]^{\frac{1}{2}} \qquad \text{(E 18)}$$

$$[H] = (k_{11}/k_{12})\,\{k_9/(k_{13} + k_{14})\}^{\frac{1}{2}}[C_2H_6]^{-\frac{1}{2}}. \qquad \text{(E 19)}$$

In order to determine the overall rate of product formation or reactant consumption we finally need to look at the equations governing these species and then substitute in our expressions for the radical concentrations. For product formation we see that ethene is produced in reactions (11) and (14). Because of the significant chain length, formation of ethene via the disproportionation reaction is very small and the rate of ethene formation can be approximated to:

$$d[C_2H_4]/dt = k_{11}[C_2H_5]$$
$$= k_{11}\{k_9/(k_{13} + k_{14})\}^{\frac{1}{2}}[C_2H_6]^{\frac{1}{2}} \qquad \text{(E 20)}$$

and the reaction is half-order in ethane. A similar analysis can be performed for ethane consumption, once again ignoring the minor contributions to ethane consumption or production in the initiation or termination steps:

$$d[C_2H_6]/dt = -k_{12}[H][C_2H_6]$$
$$= -k_{11}\{k_9/(k_{13} + k_{14})\}^{\frac{1}{2}}[C_2H_6]^{-\frac{1}{2}}[C_2H_6]$$
$$= -k_{11}\{k_9/(k_{13} + k_{14})\}^{\frac{1}{2}}[C_2H_6]^{\frac{1}{2}}. \qquad \text{(-E 20)}$$

As material balance in this comparatively simple mechanism requires, the rate of product formation is equal to the rate of reactant consumption.

Whilst this analysis agrees with experimental observations at low conversions, chemical companies are obviously interested in substantial conversions of ethane into ethene and hydrogen, and therefore we must also consider reaction of the central radicals with these product molecules, e.g.

$$H + C_2H_4 \rightarrow C_2H_5 \qquad \text{(R -11)}$$

$$H + C_2H_4 \rightarrow C_2H_3 + H_2 \qquad \text{(R 15)}$$

$$CH_3 + C_2H_4 \rightarrow C_2H_3 + CH_4 \qquad \text{(R 16)}$$

$$CH_3 + H_2 \rightarrow H + CH_4. \qquad \text{(R 17)}$$

We could, if we wished, successfully apply a steady-state treatment to the extended reaction scheme, although the algebra would be a little complicated. Generally, numerical methods of analysis are employed, (see p. 201). An eventual goal of such a process would be the development of an 'on-line' kinetic model that would aid the efficient operation of a commercial hydrocarbon cracker.

It is instructive to examine a cycle diagram of the scheme shown in Fig. 9.2. The reactions are represented as sections of loops between radicals and all the reactions are shown with the exception of the termination steps. Reactants and products are also shown, and the numbers in parentheses refer to the reaction numbers given in this chapter. For clarity the termination reactions are not shown. The diagram also shows, in circles, the reaction *fluxes* i.e. the rate of each elementary reaction under the conditions studied (the example shown refers to 1118 K and 70 per cent conversion). For example, the flux of reaction (9) is $k_9[C_2H_6]$. Initial reaction produces CH_3, which then either reacts with C_2H_6 to generate C_2H_5 (R 10) or with H_2 to form H. At high conversions, the latter will dominate. Once the ethyl radical or hydrogen atom has been generated, we enter a rapid loop in which H and C_2H_5 are interconverted via reactions (11, −11 and 12). The reaction consists of a rather laboured approach to either C_2H_5 or H and then very rapid interconversion between these species, with [H] maintained at a lower value than $[C_2H_5]$. This rapid loop is responsible for most of the consumption of ethane and , at low conversions, the generation of the major products ethene and hydrogen. However, the loop becomes less efficient in ethene generation, because of the increasing importance of reactions (−11) and (15).

As the temperature of the system is increased, the rate of conversion from ethane to ethene increases, but secondary reactions become progressively more important. In addition to adding to ethene, H may abstract from it, via a reaction with a much higher activation energy, (15) to generate vinyl radicals. Vinyl is a source of useful by-products (e.g. butadiene, C_4H_6) and of less useful ones such as ethyne. It is probably the route to larger unsaturated radicals, which eventually produce aromatic products.

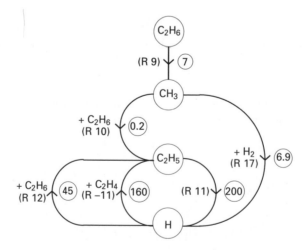

Fig. 9.2 Cycle diagram for ethane pyrolysis at 1118 K and 70 per cent conversion. The initial ethane pressure is 38 torr and the diluent gas is nitrogen. The numbers in parentheses refer to the reaction numbers and those in circles to the reactive flux in units of 10^{16} molecule $cm^{-3} s^{-1}$.

9.4.2 Pyrolysis of higher hydrocarbons

The development of a kinetic model for the decomposition of higher hydro-carbons might be thought to become exponentially more complex with increasing numbers of carbon atoms. However, whilst the complexity does increase, the problem is not quite as difficult as one might expect because larger radicals are very unstable, rapidly breaking down into smaller radicals and molecules such as ethyl, propyl, ethene, propene and hydrogen. Therefore we do not need to know the rates of each radical decomposition, only the pattern of products produced. The rate-determining steps are still the loops and cycles of the sort discussed in ethane pyrolysis.

We need to know two important pieces of information before we can develop our model. Firstly we require the relative efficiencies of the different H atom abstractions from the parent alkane and hence which alkyl radicals are formed and, secondly, the cracking pattern of these larger radicals into the $C_1 \rightarrow C_3$ species. We can illustrate the various possibilities by considering propane pyrolysis. The initial abstraction from propane can occur either at a primary or secondary C-H bond generating 1- or 2-propyl (R 18a or b)

$$H + C_3H_8 \rightarrow 1\text{-}C_3H_7 + H_2 \qquad \text{(R 18a)}$$

$$\rightarrow 2\text{-}C_3H_7 + H_2. \qquad \text{(R 18b)}$$

Isopropyl decomposition can only occur via one channel

$$2\text{-}C_3H_7 \rightarrow H + C_3H_6 \qquad \text{(R 19)}$$

the overall reaction for this route being

$$C_3H_8 \rightarrow H_2 + C_3H_6. \qquad \text{(R 20)}$$

The main decomposition route for the 1-propyl radical involves C-C fragmentation leading eventually to methane and ethene (R 21–23)

$$1\text{-}C_3H_7 \rightarrow CH_3 + C_2H_4 \qquad \text{(R 21a)}$$

$$CH_3 + C_3H_8 \rightarrow CH_4 - C_3H_7 \qquad \text{(R 22)}$$

and, overall:

$$C_3H_8 \rightarrow CH_4 + C_2H_4. \qquad \text{(R 23)}$$

However, a minor decomposition pathway for the n-propyl radical leads to H + propene

$$1\text{-}C_3H_7 \rightarrow H + C_3H_6 \qquad \text{(R 21b)}$$

and the overall reaction represented by (R 20). Thus the results of the model will depend strongly on the input branching ratios for reactions (R 18) and (R 21) and much experimental effort has been expended in trying to determine accurately these values as a function of temperature.

9.4.3 Laboratory studies of alkane pyrolysis

Much of the commercial alkane pyrolysis that occurs in the UK utilizes naphtha (C_5–C_8) as a feedstock and significant experimental and theoretical effort has been put into trying to understand the overall cracking reaction of this complex feedstock. As kineticists, the first thing that we want to do is to 'distil out' and separate the chemistry involved from other (equally important) processes such as heat exchange or fluid dynamics which will effect the conditions in the cracker and hence the final cracking pattern. Experimentally this can be done by performing the reaction with lower alkanes (we have seen above that it is the chemistry of small radicals that is important) under isothermal conditions in a small scale jet stirred reactor (Fig. 9.3). The pure alkane and a diluent gas are preheated and then sprayed into the reactor from a jet orifice. The rapidly recirculating gas ensures good mixing of the incoming gases and we measure the resulting product distribution by gas chromatographic analysis. Under these conditions it is easy to observe how the product spectrum varies with parameters such as temperature, residence time in the reactor and partial pressure of alkane. Figures 9.4 and 9.5 show some typical results for ethane and propane pyrolysis.

The next step is to see whether we can reproduce the experimental observations from our postulated reaction mechanisms. Using numerical integration techniques of the type described in Section 8.3, product distributions can be simulated and compared with experiment. The first two columns of Table 9.5 show the level of agreement for propane pyrolysis using literature values (1989) for the rate coefficients of the elementary reactions in the proposed mechanism.

The agreement is less than perfect and, assuming that we have included all the relevant reactions in our mechanism, the discrepancies must be due to the rate coefficients used. Sensitivity analysis (p. 203) of the numerical integration

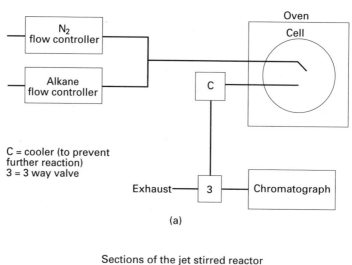

C = cooler (to prevent further reaction)
3 = 3 way valve

(a)

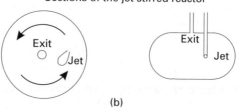

Sections of the jet stirred reactor

(b)

Fig. 9.3 Schematic diagram for laboratory investigations of alkane pyrolysis (a) and sections through the quartz jet stirred reactor (b).

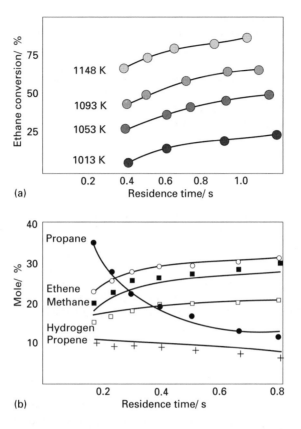

(a)

(b)

Fig. 9.4 (a) Ethane pyrolysis. Variation of ethane conversion with residence time in the reactor. (b) Propane pyrolysis.

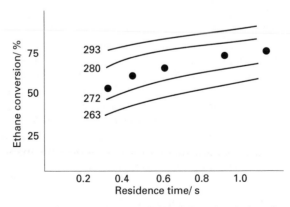

Fig. 9.5 Ethane pyrolysis. Solid lines denote variation of simulated ethane conversion using rate coefficients for vinyl radical reactions based on various heats of formation (kJ mol^{-1}) of the vinyl radical. Filled circles denote experimental values.

Table 9.5 Comparison of experimental results and model predictions of major products for propane pyrolysis at 0.33 s residence time

	Experiment	Simulation with literature values*	Simulation with adjusted values
Overall conversion	64.5 ±1.0 %	73.0%	65.6%
Propane	21.0 ± 1.1%	15.1%	20.3%
Propene	7.8 ± 0.2%	14.4%	8.8%
Hydrogen	17.1 ± 1.0%	23.6%	18.1%
Ethene	26.6 ± 0.4%	23.6%	26.5%
Methane	23.8 ± 0.7%	20.1%	22.9%

*Literature values from Baulch, D. L., Duxbury, J., Grant, S. J., and Montague, D. C. (1981). Evaluated data for high temperature reactions. *Journal of Physical Chemistry Reference Data,* **10**, Supplement 1.

highlights the most important reactions in the mechanism and the values of these rate coefficients can then be investigated in detail. As a first step to improving the agreement with experiment, we can vary the rate coefficients *within their experimentally determined error limits.* If this is ineffective then we need to look at the experimental measurements themselves. Should they be remeasured, perhaps using alternative experimental techniques? However, if the values of the rate coefficients appear to be well characterized then we must re-examine the mechanism.

For many reactions there are no experimental determinations of the elementary rates, or lengthy extrapolations have to be made from low temperature data. In these cases we are reliant on estimates based on comparisons with similar species. For example there is little experimental evidence on the rates of vinyl radical reactions. In one numerical simulation of ethane pyrolysis (P. D. Lightfoot D.Phil thesis, Oxford 1986) the rates of vinyl reactions were estimated by comparison with those of methyl radicals, with the activation energies being scaled to reflect the different heats of formation of the two radicals. Unfortunately at the time of the investigation the heat of formation of the vinyl radical was poorly characterized, leading to large variations in the simulated results, dependent on which heat of formation was used, as shown in Fig. 9.5.

This kind of analysis shows up the strong links between phenomenological studies, such as alkane pyrolysis, and the experimental studies of elementary reaction rates detailed in Chapter 2. Sensitivity analysis of a complex mechanism identifies the reactions requiring further investigation. The new values are then input into the model and the process repeated. In many cases the agreement can be very good showing that we have a broad understanding of the reactions involved. The solid lines in Fig. 9.4 and a comparison of the first and third columns of Table 9.5 show that for the case of propane pyrolysis the numerical model can simulate the observed product spectrum over a range of conditions.

9.5 Free radical polymerization

The final category of chain reaction we shall discuss is free radical polymerization, in which a radical adds to an olefin to form a larger radical which then adds another olefin and so on:

$$(\text{R } 24)$$

The olefin is termed the *monomer* (M) whilst the growing radicals are *oligomers*. The addition reactions are propagation steps, which are eventually terminated e.g. by disproportionation, to form a stable polymer molecule. Typical monomers and polymers are shown in Table 9.6. World-wide commercial production of all plastics and synthetic fibres exceeded 50 billion kilograms in 1989 illustrating the industrial importance of polymerization.

The reaction scheme is shown in Table 9.7. The initiation reaction which generates R is frequently a low activation energy dissociation reaction of a compound R_2 such as benzoyl peroxide to generate the initial link in the chain

$$(C_6H_5CO)_2 \rightarrow 2C_6H_5CO. \qquad (\text{R } 25)$$

Table 9.6 Typical monomers and polymers

Monomer	Polymer
Ethene	Polyethene (polythene)
Chloroethene (vinyl chloride)	Polychloroethene (polyvinylchloride, PVC)
Phenylethene (styrene)	Polyphenylethene (polystyrene)

Table 9.7 Reaction scheme for polymerization

Reaction	Classification	Rate
$R_2 \rightarrow 2R$	Initiation	$2k_d[R_2] = I$
$R + M \rightarrow X_1$	Initiation	$k_1[R][M]$
$X_1 + M \rightarrow X_2$	Propagation	$k_p[X_1][M]$
$X_2 + M \rightarrow X_3$	Propagation	$k_p[X_2][M]$
$\downarrow$	$\downarrow$	$\downarrow$
$X_i + M \rightarrow X_{i+1}$	Propagation	$k_p[X_i][M]$
$X_i + X_j \rightarrow A_i + O_j^a$	Termination	$k_t[X_i][X_j]$

^{a} A_i is a polymer with a saturated end group and O_j one with an unsaturated end group.

The rate of this reaction $(2k_{25}[R_2])$ is conventionally set equal to I. R then adds to the monomer to generate X_1, a radical with one monomer unit incorporated and the propagation reactions progressively incorporate more and more monomer e.g. X_i contains i monomer units. Since the different X_i are similar, it is generally assumed that the propagation steps all have the same rate coefficient k_p. Whilst this may be a significant assumption when i is small, it is more likely to be valid for the remainder of the oligomer growth which may involve several 100 or 1000 propagation steps. Finally, we shall assume that quadratic termination occurs via a disproportionation mechanism. We now set up the steady-state equations for all the radicals

$$d[R]/dt = I - k_1[R][M] = 0 \tag{E 21}$$

$$d[X_1]/dt = k_1[R][M] - k_p[X_1][M] - k_t[X_1]\Sigma_j[X_j] = 0 \tag{E 22}$$

$$d[X_2]/dt = k_p[X_1][M] - k_p[X_2][M] - k_t[X_2]\Sigma_j[X_j] = 0 \tag{E 23}$$

generally:

$$d[X_i]/dt = k_p[X_{i-1}][M] - k_p[X_i][M] - k_t[X_i]\Sigma_j[X_j] = 0. \tag{E 24}$$

Note that (i) a given radical, e.g. X_1, can terminate with *any* other radical, X_j, so that its rate of termination depends on $\Sigma_i[X_i]$, the total radical concentration; (ii) as with other straight chain reactions, the propagation steps occur as formation and removal terms, so that the addition of all of the steady-state equations eliminates all the propagation steps, and leads to the eqn:

$$I = k_t(\Sigma_i[X_i])^2 \tag{E 25}$$

i.e. the rate of initiation is equal to the total rate of termination. Thus

$$\Sigma_i[X_i] = \sqrt{(I/k_t)}. \tag{E 26}$$

The rate of polymerization, ρ, may be set equal to the total rate of monomer consumption. Assuming that the chains are long, so that the second initiation step of Table 9.7 makes a negligible contribution,

$$\rho = -d[M]/dt = k_p[M]\sum_i[X_i] \tag{E 27}$$

i.e.

$$\rho = k_p[M]\sqrt{(I/k_t)}. \tag{E 28}$$

The overall rate of polymerization will therefore be first order in monomer and half order in initiator. Finally, the chain length, v is equal to the rate of propagation over the rate of termination, i.e. the average number of collisions the growing polymer makes with other monomer species before it collides with the radical centre of another growing chain:

$$v = k_p[M]\sum_i[X_i]/2(k_t\{\sum_i[X_i]\}^2) = k_p[M]/2\sqrt{(Ik_t)}. \tag{E 29}$$

For the disproportionation termination mechanism, v is also equal to the average number of monomer units in the final polymers, which is usually termed DP, the average *degree of polymerization*. Thus

$$DP = k_p[M]/2\sqrt{(Ik_t)}. \tag{E 30}$$

For termination by recombination DP will be twice this value.

Study notes

Free radical polymerization provides an interesting vehicle for the application of kinetic principles, however, space precludes a more complete description of this interesting and commercially important field of study. In this study note we shall give a little more background to polymerization and then discuss some possibilities for further reading.

1   *The thermochemistry of addition polymerization*–The general polymerization process involves the breaking of the π system of the monomer unit with a loss of approximately 260 kJ mol^{-1} and its replacement with a single carbon–carbon bond, with a bond strength of approximately 350 kJ mol^{-1} so that the overall reaction is exothermic. However, the loss of the π system means that high energy (and hence reactive) species such as radicals, ions or co-ordination compounds are required for the reaction to proceed. Ethene molecules will never spontaneously polymerize.

2   *The thermodynamics of polymerization*–As we noted above polymerization is exothermic and this provides the major impetus for reaction. Conversely the entropy of the reaction is negative as all the free moving monomer units are being linked together with a severe reduction in the available degrees of freedom. Eventually the $T\Delta S$ term will dominate and at high enough temperatures the polymerization will have a positive Gibbs energy.

3   The bulk properties of the polymer such as elasticity and strength will depend on the chain length, the shape of the chains (straight, branched or crosslinked) and how the monomers are joined together (stereochemistry). These factors are controlled by the kinetics and dynamics of the elementary reactions.

The chain length depends on the mechanism of termination and on the relative rates of propagation and termination. Greater control of chain length can achieved by using a transfer agent designated TH which can donate a hydrogen atom to the growing polymer terminating the reaction.

$$X_i + TH \rightarrow XH + T. \qquad (R\,26)$$

The T radical formed can then initiate a new chain. The degree of polymerization now has to be modified from its simple relationship (E 30) to include the effect of the transfer agent

$$DP = k_p[M]\sum_i [X_i] / \{(2k_t\{\sum_i [X_i]\}^2) + k_{26}[TH]\sum_i [X_i]\}. \qquad (E\,31)$$

Inverting eqn (31) gives

$$1/DP = 1/DP_0 + (k_{26}/k_p)[TH]/[M] \qquad (E\,32)$$

where DP_0 is the degree of polymerization in the absence of the transfer agent. By controlling the concentration of TH and the nature of TH (to vary k_{26}) we can achieve significant control over the degree of polymerization.

It is possible for the polymer to twist around on itself and, by an internal abstraction reaction, transfer the active radical centre from the end to the middle of the polymer. Subsequent growth will lead to branched polymers which may have physical characteristics very different from the straight chain polymers of equal length. Chains can also be linked together to form more rigid structures.

For asymmetric monomers such as styrene the addition reaction can occur head to tail with the phenyl rings separated by one carbon atom, or head to head with the phenyl rings on adjacent carbon atoms. The bulky nature of phenyl groups ensures the steric factor (p. 64–5) for head to head addition is very small, but this may not be the case for smaller substituent groups.

The carbon atom of the vinyl monomer containing the substituent group will be chiral when incorporated into the polymer. In some cases it may be advantageous to have the same chirality for all the monomer units in the polymer, in which case we need to exercise great control over the dynamics of the addition reaction of the growing polymer. By their nature radical reactions tend to be rather indiscriminate and greater stereoselectivity can be achieved by using inorganic coordination compounds as catalytic agents for polymerization.

4   Finally we note that industrial production of polymers may be a multiphase process, with the molten polymer growing from a gaseous monomer, or perhaps a suspension of insoluble polymer in a monomer solution. The simplest process of all would be the growth of a liquid polymer from liquid monomer, however, this is rarely used because of problems of temperature control. As the polymer chain grows the liquid becomes more viscous and it becomes harder to remove the exothermicity of the reaction. The temperature of the mixture rises and along with it the rate of reaction (see also thermal explosions p. 241). Under these conditions it is difficult to control the reaction and properties such as chain length.

5   *Further reading*. An excellent introduction to various aspects of polymerization, such as how the bulk properties relate to chain length and structure, may be found in an *Introduction to Synthetic Polymers* by Campbell, I. M. (1994), Oxford University Press. Chapters 6 and 7 deal with the kinetics of addition polymerization. *Mechanism and Kinetics of Addition Polymerization* (Kucera, M. (1972) in *Comprehensive Chemical Kinetics* (ed. R. G. Compton) Vol. 31. Elsevier, Amsterdam) provides a thorough and logical account of polymerization. Starting from monomer structure, the author works through initiation, propagation and termination. Chapter 8 on polymerization kinetics may be read with great profit and there are many references to the original literature. A slightly older account of free radical polymerization may be found in Volume 14A of the same series.

9.6 Questions

9.1   Use the data in Tables 9.2 and 9.3 to calculate the *initial* overall activation energy for the production of HBr from an H_2/Br_2 mixture. How will this activation energy vary over the course of the reaction?

9.2   When bromine and methane are irradiated with visible light at 500 K, the rate of formation of CH_3Br is proportional to the square root of the light intensity, and to the first power of the methane pressure, for a fixed pressure of bromine vapour. Devise a kinetic scheme to account for these observations.

9.3   Justify eqn (5) on p. 76 of Sullivan's (1967) paper on $H_2 + I_2$ (*Journal of Chemical Physics*, **46**, 73) and show that it can be rearranged to give the eqn on p. 77. Use the following data from the Sullivan paper to determine k_2 at 417.9 K.

$[I_2]/$ $10^{-5}\,mol\,dm^{-3}$	$[H_2]/$ $10^{-3}\,mol\,dm^{-3}$	$[HI]/$ $10^{-7}\,mol\,dm^{-3}$	E_a = light absorbed/ 10^{-7} Einstein $dm^{-3}\,s^{-1}$	t = exposure time/ $10^3\,s$
6.42	10.96	8.35	4.71	33.60
6.80	10.96	9.96	4.99	41.46
8.96	4.66	12.73	5.18	80.88
14.67	4.66	12.12	6.46	79.20

15.75	2.59	7.63	6.78	85.50
22.00	4.50	6.08	4.13	93.60
17.53	5.72	7.30	3.96	79.20

An Einstein $= 6.02 \times 10^{23}$ photons.

The iodine atom recombination rates with I_2 and H_2 as the third bodies are 2.37×10^{11} and 2.88×10^9 dm^6 mol^{-2} s^{-1} respectively. What are the relative probabilities of a termolecular collision complex $[I–I–H_2]$ leading to a) $I_2 + H_2$, b) $2HI$.

9.4 Devise a scheme for the decomposition of ethanal into methane and CO. The weakest bond is the C–C bond and the aldehydic C–H bond is weaker than the methyl C–H bonds. Show that the rate expression is of order $3/2$ with respect to $[CH_3CHO]$

9.5 The following data refer to the polymerization of methyl methacrylate in benzene in the presence of benzoyl peroxide as an initiator.

Initiation $\quad A_d = 6.8 \times 10^{13}$ s^{-1}
$\qquad\qquad E_d = 124$ kJ mol^{-1}
Propagation $A_p = 10^6$ dm^3 mol^{-1} s^{-1}
$\qquad\qquad E_p = 20$ kJ mol^{-1}
Termination $A_t = 10^8$ dm^3 mol^{-1} s^{-1}
$\qquad\qquad E_t = 5$ kJ mol^{-1}

(i) Determine (a) the rate of polymerization and (b) the average degree of polymerization at 323 K and 343 K for [methyl methacrylate] $= 0.2$ mol dm^{-3}, [benzoyl peroxide] $= 0.05$ mol dm^{-3}.
(ii) Show that the effective activation energy for polymerization, E_{eff}, is given by:

$$E_{eff} = E_d/2 + E_p - E_t/2$$

and obtain an expression for the dependence of *DP* on temperature. For the benzoyl peroxide/methyl methacrylate system, plot $\ln\rho$ and $\ln DP$ vs $1/T$, where ρ is the rate of polymerization.

9.6 Imagine that the steady-state radical concentration, $\Sigma[X_i]$, has been established in a photoinitiated polymerization reaction and that the light is instantaneously turned off, so that initiation stops. Show that the half life for the decay of the total radical concentration is given by:

$$t_{1/2} = [2(k_t I)^{\frac{1}{2}}]^{-1}$$

and hence show that measurements of ρ, *DP*, and $t_{1/2}$ enable I, k_p and k_t to be evaluated. (see *Comprehensive Chemical Kinetics* Vol. 14A pp. 5–6 for a discussion of how $t_{1/2}$ can be determined.)

References

Eastmond, G. C. (1976). The kinetics of free radical polymerization of vinyl monomers in homogeneous solution. In *Comprehensive chemical kinetics*, Vol. 14A (ed. C. H. Bamford and C. F. H. Tipper), pp. 5–6. Elsevier, Amsterdam.
Lightfoot, P. D. (1986). Unpublished D.Phil. thesis. University of Oxford.

10 Explosions and branched chain reactions

10.1 Introduction

Explosion occurs when the reaction rate rapidly accelerates so that products are formed at an ever increasing rate until, eventually, the reactants are consumed. Explosions are generally accompanied by an increase in temperature (arising from the self-heating of the system by the exothermic reaction) and of pressure and can have devastating consequences. Two general types of mechanism apply, namely *thermal explosions*, where the reaction exothermicity cannot be dissipated, leading to increasing reaction temperature and ever greater rate coefficients, and *branched chain explosions*. In the latter case certain steps in the chain reaction mechanism create two radicals from one reactant radical leading to an increase in the number of chain carriers and hence of the overall reaction rate. In many cases both mechanisms will be occurring but in order to emphasize the differing mechanisms we shall consider the two cases separately. We begin with a brief discussion of the former category of reaction.

10.2 Thermal explosions

Our treatment of the hydrogen/bromine reaction (Section 9.2) presumed that the reaction took place at a constant temperature, determined by that of the reaction vessel. The reaction is exothermic, however, and, for such a condition to prevail, the heat generated in the reaction must be removed from the system, otherwise the temperature will increase and the reaction rate with it. We can obtain a semi-quantitative description of the consequences of such an effect as follows.

The rate of heat production per unit volume Φ_+, is given by,

$$\Phi_+ = \rho \Delta H \tag{E 1}$$

where ρ is the overall rate and ΔH the molar enthalpy change (n.b. for a constant volume reactor ΔU would be the appropriate value). Heat loss can occur at the vessel walls. The rate determining step is either the heat transfer from the gas to the wall or, if this is very rapid, convection in the gas itself. In this case the gas adjacent to the walls will be kept at the wall temperature (T_0), whilst the gas temperature will increase towards the centre of the vessel, setting up a thermal gradient in the gas. In a real system, both mechanisms occur, but, for the sake of simplicity, we shall assume that the former mechanism predominates with a uniform reactant temperature (T) throughout the reaction vessel. Under these circumstances, the rate of heat loss Φ_- is given by

$$\Phi_- = \alpha(S/V)(T - T_0) \tag{E 2}$$

where α is a heat transfer coefficient and S and V are the surface area and volume of the reaction vessel. The important point to note is that the rate of heat loss increases *linearly* with reaction temperature. Figure 10.1 shows schematic plots of Φ_+ and Φ_- vs. T for three values of the reactant concentration. Φ_- is linear in temperature, with a slope equal to $\alpha S/V$ and an intercept determined by T_0. Φ_+, on the other hand, depends exponentially on temperature, because it contains rate coefficient terms.

In each case, we assume that the gas starts off at the vessel temperature. In case *a*, Φ_+ is initially greater than Φ_- and the gas self heats until A is reached, at which point $\Phi_+ = \Phi_-$ and thermal equilibrium is established, with the rate of heat loss exactly balanced by that of heat production, and the temperature of the reactants is held constant at a value somewhat in excess of T_0. Under these circumstances, we may be justified in applying an isothermal model, although we ought to measure the temperature of the gas itself and not rely on measurements of the vessel temperature. In case *c*, where we have increased the reactant concentration, Φ_+ is always greater than Φ_- so that the temperature of the reaction mixture rises and the rate increases (because of the exponential dependence of the rate coefficient on T) with a limitation set simply by complete consumption of the reactants − the system undergoes a thermal explosion. Case *b* represents the critical case and establishes the *explosion limit*, which is the combination of vessel temperature and critical concentration for reacting gas in that *particular* vessel. At higher concentrations a thermal explosion takes place, whilst at lower concentrations an isothermal mechanism pertains. The critical conditions may be determined by recognising that at point C not only will the rate of heat loss and gain be equal, but also that the gradients of the two curves will be equal:

$$\Phi_+ = \Phi_- \text{ and } d\Phi_+/dT = d\Phi_-/dT \tag{E 3}$$

as is illustrated in Questions 10.1–10.3. Although we have illustrated thermal explosions with examples and problems related to the hydrogen/halogen systems, thermal explosions do not require chain reactions.

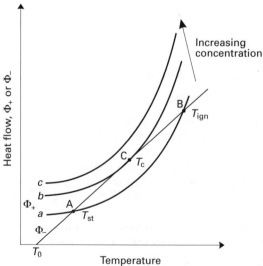

Fig. 10.1 Heat flow contributions for an exothermic reaction in a closed vessel. Curves *a*, *b*, and *c* represent the heat production terms for increasing reacta t concentration.

10.3 Branched chain reactions

In the straight chain reactions considered in Chapter 9, the propagation steps consist of the replacement of one radical species by another, so that, in the steady-state, the concentration of radicals is constant, as is the overall rate of reaction. A branched chain mechanism, on the other hand, involves elementary reactions in which one reactive centre (atom or radical) reacts to produce two chain carriers, so that the concentration of radical species and, in consequence, the overall rate of reaction increases with time. We shall illustrate this type of reaction by reference first to the hydrogen/oxygen reaction and then to hydrocarbon oxidation. For the system to be spontaneously explosive the overall chain reaction must be exothermic, and therefore most real chain branching explosions will have a thermal component as well.

10.3.1 The hydrogen–oxygen reaction

It is interesting to compare the hydrogen–oxygen reaction with the hydrogen halogen reactions discussed in Chapter 9. The bond energy in oxygen (≈ 493 kJ mol^{-1}) is much higher than in the halogens and the initiation reaction is probably a bimolecular reaction rather than a simple bond fission. Initiation generates OH radicals which react with H_2 to produce H (see Table 10.1). The subsequent reactions, (3) and (4), differ from the equivalent halogen reactions because the diatomic product OH, is also a radical. Thus unlike the propagation steps of the hydrogen-halogen reactions, which maintain the total radical concentration, steps (3) and (4) serve to increase it and they are known as *branching steps* (two radicals are produced from one). The accelerating effect of branched chain reactions can be seen by considering the fate of the products of reaction (3):

Table 10.1 Basic reaction mechanism for the H_2/O_2 reaction

Reaction		Reaction Type	A/cm^3 molecule^{-1} [a]	E/kJ mol^{-1} [a]
(1)	$H_2 + O_2 \rightarrow + HO2H$	Initiation	6.8×10^{-8}	200
(2)	$OH + H_2 \rightarrow H + H_2O$	Propagation	4.2×10^{-11}	22
(3)	$H + O_2 \rightarrow OH + O$	Branching	3.8×10^{-10}	70
(4)	$O + H_2 \rightarrow OH + H$		4.8×10^{-11}	42
(5)	$H + O_2 + M \rightarrow HO_2 + M$	Termination	$5.8 \times 10^{-30}\, T^{-1}$	[b]
(6–8)	H, O, OH $\rightarrow$ wall		[c]	[c]
(9)	$HO_2 + H_2 \rightarrow H + H_2O_2$	Restore the chain at high temperature	1.0×10^{-12}	77
(10)	$2HO_2 \rightarrow H_2O_2 + O_2$		7.0×10^{-10} [d]	50
(11)	$H_2O_2 \rightarrow 2OH$		1.0×10^{-7} [d]	1110

[a] Several of the reactions, and especially those with low activation energies, have rate coefficients with a complex temperature dependence. The data given above are approximate and should only be used at temperatures reasonably close to 800 K.
[b] Third order reaction, with units cm^6 molecule^{-2} s^{-1}. Note the negative T^{-1} temperature dependence.
[c] The rates of these reactions depend on diffusion coefficients (and hence on pressure) and on the vessel dimensions and surface characteristics.
[d] Unimolecular reaction in the second-order regime, hence the large A factor.

$$H + O_2 \rightarrow OH + O \qquad\qquad (R\ 3)$$

$$OH + H_2 \rightarrow H + H_2O \qquad\qquad (R\ 2)$$

$$O + H_2 \rightarrow H + OH \qquad\qquad (R\ 4)$$

$$OH + H_2 \rightarrow H + H_2O \qquad\qquad (R\ 2)$$

$$\text{net: } H + O_2 + 3H_2 \rightarrow 3H + 2H_2O$$

After several reactions with stable reactants we have trebled the H atom concentration. The reaction mechanism is completed by including the termination steps (5)–(8). Reaction (5) is a gas phase termination reaction which involves formation of the hydroperoxide radical HO_2. Although a radical, HO_2 is very unreactive at low temperatures. We shall return to its important high temperature behaviour later. Reactions (6)–(8) are linear terminations at the wall. In this process the radicals diffuse to the wall of the reaction vessel and are removed.

As we shall see, the branching steps mean that the steady-state approximation is not always applicable and we employ instead a rather crude analysis which examines the time dependence of the total radical concentration, n.

There is a net generation of chain centres in the initiation reaction and in the branching reactions. The rate of the latter depends on the concentration of the chain centres, and for simplicity we shall consider them all to be equivalent. There is no net production or removal of chain centres, n, in the propagation reactions, and these may be omitted from our analysis. Considering only linear termination,

$$dn/dt = i + gn - fn \qquad\qquad (E\ 4)$$

where i is the rate of initiation, and gn and fn are the rates of branching and linear termination respectively. Solving equation (4)

$$n = i\{\exp(\phi t) - 1\}/\phi \qquad\qquad (E\ 5)$$

where $\phi = g - f$, the net branching factor.

We now see under what conditions the steady-state approximation is inapplicable. If ϕ is positive, i.e. if branching occurs more rapidly than termination, the chain centres accumulate, generating more chain centres, and this leads to an exponential growth in n (see Fig. 10.2). This growth can be very rapid; it has been calculated that, in the hydrogen/oxygen reaction, each initial hydrogen atom generates 10^{15} hydrogen atoms in 0.3 s at 700 K and an oxygen pressure of 11 kPa. In a word the reaction mixture *explodes*!

For $\phi < 0$, termination dominates, and the exponential term tends rapidly to zero and a steady-state is set up, with $n = i/|\phi|$. The relative importance of termination and branching will depend on a number of factors and the region in which the H_2/O_2 reaction is explosive, the so-called explosion peninsula, is shown in Fig. 10.3. At very low reactant pressures, (near A, where the total pressure is about 10^{-3} atm or 10^2 Pa) the branching reactions are slow, and the chain centres diffuse unhindered to the vessel walls, where they are removed. Linear termination is faster than branching, ϕ is negative, and the reaction is slow. As the pressure is raised, the branching reactions accelerate. Eventually g becomes greater than f, and explosion occurs (B in Fig 10.3). A further substantial increase in pressure

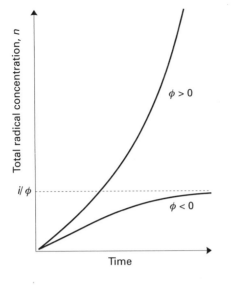

Fig. 10.2 Growth of chain-carrier concentration, n, for different values of the net-branching factor, φ.

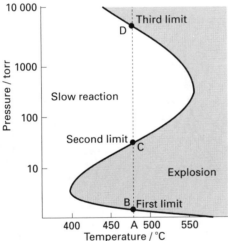

Fig. 10.3 The explosion boundary for the hydrogen–oxygen reaction as a function of temperature and pressure.

enhances the rate of the termolecular recombination reaction (5) sufficiently for quadratic termination to exceed branching, and we re-enter the slow region (**C**).

As Table 10.1 shows, the branching reactions have non-zero activation energies, and their rates increase with temperature. Diffusion is only weakly temperature dependent, and the rate of termolecular recombination decreases as the temperature is raised. As a result, the branching reactions are favoured by increasing the temperature, and the explosion limits become wider, giving the peninsula its characteristic shape.

Figure 10.4 shows the dependence of the reactants, intermediates and products in the non-explosive and explosive regions. In Fig. 10.4(a), the radical concentrations, with the exception of HO_2, are all small, although they do vary somewhat with time. In Fig. 10.4(b), the radical concentrations rapidly build up to very high values and, at its peak, the hydrogen atom concentration is about one tenth of $[H_2]$. The reactant consumption becomes very fast as the explosion

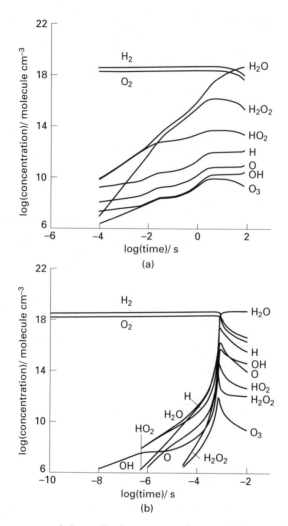

Fig. 10.4 Calculated concentration profiles for the major species in the hydrogen–oxygen reaction. The figures show the sensitivity of the system to temperature (i.e moving to the right in Fig. 10.3). $[H_2] = 2.8 \times 10^{18}$ and $[O_2] = 1.4 \times 10^{18}$ molecule cm^{-3}. (a) $T = 825$ K, non-explosive behaviour, (b) $T = 875$ K explosive behaviour with rapid consumption of reactants. Note the different timescales for the two processes.

occurs and the radical concentrations and the rate then die away as H_2 and O_2 are replaced by H_2O.

The hydrogen–oxygen reaction also shows a third explosion limit, which is more complex. It arises because, at the higher pressures, reactions (9) and (10) compete with diffusion of HO_2 to the wall; thus H_2O_2 is formed and generates OH via reaction (11). However, the explanation to the third limit is not quite so simple, since here the H_2/O_2 reaction shows some features of a thermal explosion. The heat generated by the reactions cannot escape and thus the temperature of the system rises and the rates accelerate.

Study notes

An extensive numerical integration and sensitivity analysis of the H_2/O_2 reaction has been performed by Dougherty, E. P. and Rabitz, H. (1980). *Journal of Chemical Physics*, **72**, 6571–86. Although the paper contains some mathematical details, these can be avoided, leaving a detailed but comprehensible discussion of the reaction. Sections I, II, and IV should be studied. Analysis of this paper will provide the background for questions 10.4–6.

10.4 Hydrocarbon oxidation

The oxidation of hydrocarbons provides a source of energy which is central to many processes on which we depend (automobile engines, heating, power generation etc). On a molecular scale the processes are complex, and on a macroscopic scale they provide an almost bewildering variety of phenomena. We cannot, unfortunately, do justice to them all, but a brief examination provides us with some interesting kinetic insights.

Figure 10.5 shows an ignition diagram for a typical hydrocarbon. The term ignition is used rather than explosion because strictly speaking an explosion refers to a rapid increase in pressure brought about by increasing rates of reaction. In form hydrocarbon oxidation has similarities to the region of the H_2/O_2 system around the first explosion limit, i.e. a competition between chain branching via the $H + O_2$ reaction, (increasing exponentially with temperature) and quenching by diffusion to the walls (increasing slightly with temperature and inversely proportional to pressure–low pressure allows for rapid diffusion and quenching of the reaction). However, there are differences—a region of negative temperature dependence and an additional lobe occurring at low temperatures corresponding to the so-called *cool flame* region. Cool flames are associated with a sudden temperature rise ($\approx 100\ °C$) and an associated increase in reaction rate which is quenched (i.e. the reaction rate falls) before reaction is complete. Sometimes, multiple cool flames can occur, leading, under certain conditions, to well defined oscillatory behaviour and we shall consider negative temperature dependences and cool flames in detail in Chapter 11. In the following sub section on the mechanism of high temperature hydrocarbon combustion we shall see that there are some similarities between hydrocarbon and hydrogen combustion, at least as far as chain branching mechanisms are concerned.

10.4.1 High temperature hydrocarbon combustion (T >1000 K)

Figure 10.6 shows a flow diagram for the combustion of methane (the major constituent of natural gas) in a stoichiometric mixture of fuel and air. At high temperatures H atom abstraction reactions with the major radical species, H, OH, and O occur rapidly and the hydrogen atoms of the fuel are stripped off one by one until we are left with carbon monoxide. The final step in the oxidation reaction is the reaction of CO with OH to generate CO_2.

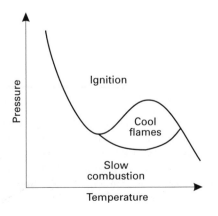

Fig. 10.5 Schematic ignition diagram for a hydrocarbon–oxygen mixture.

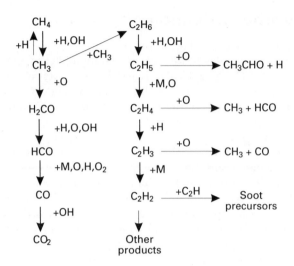

Fig. 10.6 Flow diagram for methane combustion in a stoichiometric methane–air flame at atmospheric pressure.

Depending on the fuel/air ratio and conditions, hydrocarbon radical recombination may compete with abstraction and subsequent degradation, leading to the formation of higher hydrocarbon species. Under conditions of incomplete oxidation compounds such as ethyne (acetylene) and the ethynyl radical (C_2H) can react to give the basic building constituents for soot formation. The picture for methane combustion looks relatively complex and we might imagine that the oxidation of higher hydrocarbons and mixtures of hydrocarbons (petrol is a complex mix of C_5–C_8 hydrocarbons) at high temperatures would be even more convoluted. However, in a similar manner to the development of alkane pyrolysis, considered in the previous chapter, the extremely rapid decomposition of higher hydrocarbon radicals means that the there are only a limited number of reactions that we need consider.

In fact sensitivity analysis (p. 203) shows that only a few reactions control the overall rate of reaction and that none of these reactions are specific to the fuel used. Figure 10.7 shows the influence of varying a single rate coefficient by a factor of five on the calculated flame velocity of a methane/air mixture. Examples of the dominant reactions are –

$$H + O_2 \rightarrow OH + O \tag{R 3}$$

$$OH + CO \rightleftharpoons H + CO_2 \tag{R 12, -12}$$

$$HCO \rightarrow H + CO \tag{R 13}$$

$$CH_3 + H + M \rightarrow CH_4 + M. \tag{R 14}$$

We can rationalize this observation for reactions (3, 13, and 14) by referring back to the flow diagram of methane oxidation. The reaction of oxygen atoms with the various compounds of the degrading fuel is chain branching but relatively unimportant. The major degradation pathways, reaction with H and OH radicals are chain propagating so that the radical concentration remains constant. The major chain branching step is reaction (3) with the H atoms being

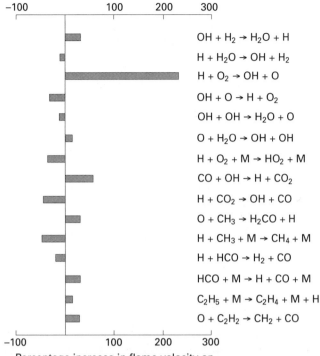

Fig. 10.7 Influence on the flame velocity of a stoichiometric methane–air mixture caused by varying each of the single rate coefficients by a factor of five. Only reactions which produce changes greater than five per cent are considered.

generated from reaction (13). This chain branching reaction along with possible quadratic chain termination reactions such as reaction (14) control the overall radical concentration and hence the rate of reaction. Reaction (12) is of importance because it is a major mechanism for generating H atoms and forms a significant component of the flame exothermicity. We can now appreciate the importance of the hydrogen/oxygen system and why it has received so much study. Many of its elementary reactions are also the crucial steps in all hydro-carbon oxidation.

As kineticists we are interested in predicting observable quantities such as radical concentrations, flame velocities, and ignition delays with a model, based on a series of elementary reactions. The commercial and environmental (see Box 10.1) importance of combustion processes has provided the impetus for significant research in this field. As in previous systems we set up a series of differential equations which control the temporal profile of each of the radical species and solve the set of simultaneous equations using numerical integration techniques. In our simulations of alkane pyrolysis (pp. 231–4) we were able to separate the chemistry from physical processes such as mixing and heat transfer. This is not so easily achieved in flame simulations and any model must take into account a number of complex physical processes. However, this is not the place to discuss such methods. Further information can be found from the references in the study notes. Chemically, the success of the model depends on two factors, the accuracy of the mechanism and the accuracy of the elementary rate coefficients.

The requirement of accurate and precise rate coefficients for combustion modelling has been the driving force behind many of the experimental and theoretical studies of elementary reactions described in Chapters 2 – 4. Many elementary reactions can give several different sets of products an example being the reaction of NH radicals with NO, important in controlling NO emissions

$$NH + NO \rightarrow N_2O + H \tag{R 15a}$$

$$NH + NO \rightarrow N_2 + OH. \tag{R 15b}$$

For our model to be valid we need to know not only the rate at which NH and NO are removed, i.e. the overall rate coefficient, but also the branching ratio between the two channels. These latter experiments are much harder to perform (mainly because many sensitive detection techniques such as laser induced fluorescence (p. 34) do not measure absolute concentrations so that some form of calibration is required) but the information that they provide is essential for the model and can reveal a lot about the molecular mechanism of the reaction.

Box 10.1 Combustion emissions

There are basically three types of combustion emissions. Hydrocarbons and CO result from incomplete combustion within the reaction chamber and can generally be reduced by more efficient engine design so that all the fuel and air mix, and by adjusting the fuel:air ratio. Nitrogen oxides are the third type of emission and these can arise from three different mechanisms.

1   Fuel nitrogen oxidation—fossil fuels contain small quantities of nitrogen based compounds. This is not surprising as the fuels were derived from living matter. These compounds are released during the combustion process and are rapidly oxidized to NO.

2   $N_2 + O_2 \rightleftarrows 2NO$ equilibria—Combustion generally occurs in nitrogen containing air rather than pure oxygen. At typical combustion temperatures NO may be 2–3 per cent of the equilibrium mixture. The $N_2 + O_2 \rightleftarrows 2NO$ equilibrium was thought to be established by reactions such as

$$N + O_2 \rightleftarrows NO + O \tag{R 16, -16}$$

$$O + N_2 \rightleftarrows NO + N. \tag{R 17, -17}$$

3   However, it was found that this mechanism only applies at high temperatures and these reactions are too slow to establish the observed equilibium concentrations of NO. Alternative mechanisms of NO formation involve reactive hydrocarbons such as CH reacting with atmospheric nitrogen

$$CH + N_2 \rightarrow HCN + N \tag{R 18}$$

with the products of this reaction being rapidly oxidized to NO.

As the exhaust gases rapidly cool the radicals are quickly consumed and the high temperature equilibrium concentration of NO is frozen into the mixture. It is the simultaneous prescence of nitrogen oxides and hydrocarbons that lead to the dramatic photochemical smogs now seen in many urban areas.

Nitrogen oxide control strategies focus either on preventing NO formation by gradually controlling the addition of oxidant or by removing NO from the exhaust gases. The mechanism of NO removal will depend on the combustion application. Mobile sources such as automobiles require compact abatement packages such as catalytic converters which reduce the NO in the exhaust, whereas larger fixed plants, such as power stations, can use a wider variety of control measures.

Study notes

1   The field of combustion research is enormous and crosses many disciplinary boundaries. As kineticists we tend to concentrate on just one aspect of the overall process—the chemistry—and largely ignore other crucial topics such as flame structure or fluid dynamics. A broader overview of the field (but still with considerable chemical emphasis) can be found in *Flame and Combustion* (2nd edn) (Griffiths, J. F. and Barnard, J. A. (1995). Chapman and Hall, Glasgow.). Flame structures, detonation, liquid and mixed phase reactions, high explosives, rocket propulsion and effects on the environment are some of the wider and more applied topics considered. In addition, the book contains further information on both thermal and branched chain reactions and the chemistry of combustion. An alternative undergraduate text is Warnatz, J., Mass, U., and Dibble, R. W. (1996) *Combustion*, Springer, Berlin, which covers the subject very comprehensively. At graduate level, Glassman, I. (1987) *Combustion*, 2nd edition, Academic, Orlando, provides more detailed coverage but with an emphasis on combustion chemistry.

2   Two relevant reference books on combustion are Gardiner, J. C. (ed.) (1984) *Combustion Chemistry*, Springer-Verlag, New York, and Hucknall, D. J. (1983) *Chemistry of Hydrocarbon Combustion*, Chapman and Hall, London. Both of these books provide extensive background information.

3   A recent review of combustion modelling can be found in Miller, J. A., Kee, R. J., and Westbrook, C. K. (1990). Chemical kinetics and combustion modelling, *Annual Reviews of Physical Chemistry*, **41**, 345–88.

10.5 Questions

The following problems develop the treatment of thermal explosions considered on pp. 241–2 and apply it to the $H_2 + Cl_2$ reaction system.

10.1   By setting $\rho = Ac^n \exp(-E/RT)$, where A is a pre-exponential factor, E an effective activation energy and c^n describes the dependence of the rate on the concentration of the reactants, use the criticality conditions (p. 242) to show that T_c may be found by solving the quadratic equation

$$T_c^2 - ET_c/R + ET_0/R = 0.$$

10.2   The lower limit is the one which is physically realistic; in addition $E \gg RT_0$. By expanding the square root in the solution to the quadratic equation and neglecting all but the first two terms, show that

$$T_c = T_0(1 + \theta_c)$$

where $\theta_c = RT_0/E$. Hence find an expression for the critical concentration.

10.3   These equations may now be applied to the H_2/Cl_2 reaction. Neglect the inhibition term, so that the reaction scheme is:

$$Cl_2 + M \rightleftarrows 2Cl + M \qquad (1, -1)$$

$$Cl + H_2 \rightarrow HCl + H \qquad (2)$$

$$H + Cl_2 \rightarrow HCl + Cl \qquad (3)$$

Show that, for an equimolar system, $\rho = 2k_2(k_1/k_{-1})^{\frac{1}{2}} \cdot c^{\frac{3}{2}}$, where $c = [H_2] = [Cl_2]$ and hence evaluate the critical conditions for the following parameters:

- $k_2 = 1.44 \times 10^{13} \exp(-2200/T) \text{ cm}^3 \text{ mol}^{-1}\text{s}^{-1}$
- $(k_1/k_{-1}) = 1.04 \times 10^{-1} \exp(-24\ 540/T) \text{ mol cm}^{-3}$
- $T_0 = 500 \text{ K}$
- $(\alpha S/V) = 0.27$
- $\Delta H = -184 \text{ kJ mol}^{-1}$

Problems 10.4–6 refer to the H_2/O_2 reaction. You will find it necessary to consult the reference given in the study note on p. 247.

10.4   In the slow region of the H_2/O_2 reaction, sensitivity analysis leads to the same mechanism as that proposed using the steady state approximation. The latter, however, is invalid in the explosive regions and additional reactions, particularly those involving H, become important. The sensitivity analysis is particularly useful in this region. Compare Tables III and IV and Figs 6 and 7c of the Dougherty and Rabitz reference and suggest reasons for the differences in the mechanisms.

10.5   The competition between the two reactions

$$H + O_2 \rightarrow OH + O \tag{1}$$

$$H + O_2 + M \rightarrow HO_2 + M \tag{2}$$

is of central importance, because the former is a branching reaction, whilst the latter is the major terminating step at the second limit. The rates of reactions 1 and 2 become equal at a third body concentration, [M], given by $[M] = k_1/k_2$ and this value gives a crude estimate of the second limit. Using the data in Table 10.1, plot $\ln[M]$ vs T over the temperature range $650 - 800$ K.

10.6   The following simplified mechanism applies in the slow regime, at very low conversions and high pressures:

$$H_2 + O_2 \rightarrow 2OH \tag{1}$$

$$OH + H_2 \rightarrow H + H_2O \tag{2}$$

$$H + O_2 \rightarrow OH + O \tag{3}$$

$$O + H_2 \rightarrow OH + H \tag{4}$$

$$H + O_2 + M \rightarrow HO_2 + M \tag{5}$$

(a) Assuming HO_2 to be a stable species, show that the steady-state concentrations of the radical intermediates are given by:

$$[H] = 2k_1[H_2]/(k_5[M] - 2k_3) \approx 2k_1[H_2]/k_5[M]$$
$$[OH] = k_5[H][O_2][M]/(k_2[H_2])$$
$$[O] = k_3[H][O_2]/(k_4[H_2])$$

(b) Using the rate data given in Table 10.1, plot [OH]/[H] over the temperature range 400–700 K and explain why [OH] is maintained at such a low relative value.
(c) As k_3 increases, [H] increases despite the fact that reaction (3) consumes H. Why?
(d) The present steady-state analysis is only valid provided $k_5[M] > 2k_3$, otherwise [H] becomes negative! At what temperature is $k_5[M] = 2k_3$, for $[M] = 1.5 \times 10^{18}$ molecules cm^{-3}? Which other reactions might reduce [H] and so avoid the break down in the analysis?

Negative feedback and oscillatory behaviour

11.1 Introduction

In all of the reactions we have considered so far, the concentrations of *intermediates* have shown, at most, a single maximum. In some reactions, however, they have been found to exhibit *oscillatory* behaviour. The most celebrated of these involves the cerium catalysed oxidation of malonic acid by bromate in sulphuric acid and was first reported by Belousov in 1959. Later, in 1967, Zhabotinskii showed that if the reaction mixture is left unstirred, spatial periodic variations of the intermediate concentrations can also occur. Figure 11.1 is a plot of the concentrations of Ce^{3+} and Ce^{4+} and of Br^- in a stirred reaction mixture initially containing malonic acid, bromate, bromide, Ce (III) and sulphuric acid. After an initiation period, $[Ce^{3+}]/[Ce^{4+}]$ and $[Br^-]$ show oscillatory behaviour.

Central to oscillatory reactions is a *negative feedback* mechanism (by feedback we mean that the products of the mechanism influence the rates of earlier steps in the mechanism) that checks the runaway acceleration of an intermediate concentration and differentiates the oscillatory reaction from an explosion, where the feedback is always positive. Negative feedback can be caused by a number of different phenomena, some purely physical, others of a more chemical nature. In this chapter we shall briefly consider a number of different oscillatory systems, involving combustion, oxidation on surfaces, cool flames and the Belousov–Zhabotinskii (BZ) reaction already referred to in the opening paragraph.

It is crucial at this point to emphasize that the reaction is *not oscillating between the reactants and final products*, but rather, it is the concentrations of the intermediates that are involved in the various component elementary reactions, that are fluctuating. The overall reaction is always moving towards the minimum free energy.

11.2 Oscillatory behaviour in a well stirred reactor

The first type of oscillatory behaviour that we shall briefly consider is that of the H_2/O_2 reaction in a continuous well stirred tank reactor (CSTR) of the type

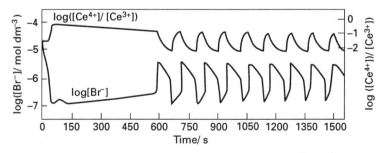

Fig. 11.1 The periodic behaviour of $[Ce^{4+}]/[Ce^{3+}]$ and $[Br^-]$ in a stirred reactor.

shown schematically in Fig. 9.3(b). The mixture can be stirred by the use of one or more gas jets, or with a paddle mechanism. The oscillatory behaviour observed is not only chemical in origin, but also arises from the timescales of reaction induction, ignition and the throughput of the reactants and products.

Hydrogen and oxygen are introduced into the heated reactor and the conditions of temperature and pressure are chosen close to the second explosion limit. If the average residence time of the gas within the reactor (given by reactor volume/volume flow rate) is greater than the time for the radical concentrations to build up to their critical level (see Section 10.3 on chain branching reactions) an ignition will occur on a millisecond timescale and the hydrogen and oxygen reactants will be consumed. The reaction products are swept out of the reactor and replaced by the fresh incoming reactants and the process can begin again.

The physical exchange of products and reactants is in itself is not enough to create oscillatory behaviour. Fresh incoming reactants would be expected to maintain the rapid combustion. A second, chemically more significant feature, ensures that rapid combustion cannot be maintained. The crucial reaction limiting the rate of combustion around the second explosion limit of the hydrogen/oxygen system (see Section 10.3) is the termolecular combination of hydrogen atoms and molecular oxygen

$$H + O_2 + M \rightarrow HO_2 + M. \qquad (R\ 1)$$

The efficiency of this reaction is dependent on the molecular structure of M (see Sections 5.11 and 5.12). The product of the hydrogen/oxygen reaction is water, a polyatomic molecule and, because of its many internal modes of vibration and rotation, a far more efficient third body at stabilizing HO_2 than either hydrogen or oxygen. The increased efficiency of reaction (1) as the overall reaction proceeds ensures that the reaction moves into the slow combustion regime.

Pressure and temperature excursions in the ignition region can be extreme, with temperature increases of several thousand Kelvin possible. Figure 11.2 shows some typical experimental results for the H_2/O_2 reaction in a CSTR. In some cases the response of the thermocouples is too slow to give an accurate indication of the temperature excursion and techniques of higher temporal

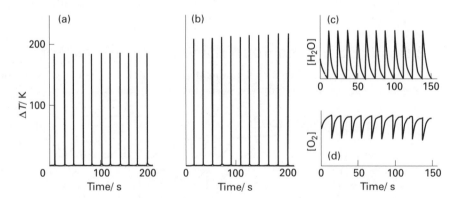

Fig. 11.2 Example experimental records for oscillatory ignition in a hydrogen/oxygen system: (a) temperature rise, (b) light emission intensity, (c) $[H_2O]$ and (d), $[O_2]$. The slow decays or rises of water and oxygen correspond approximately to the rate at which the reactor is recharged with fresh reactant. (From Scott, S. K. (1993). *Chemical Chaos*, p. 286. Oxford University Press.)

resolution such as the spectroscopic observation of relative rotational populations are required (Chapter 2 p. 54).

11.3 Oscillatory behaviour on well characterized surfaces

A further example of oscillatory behaviour where negative feedback is a mixture of chemical and physical processes is the low pressure ($\approx$ 10 Pa) oxidation of carbon monoxide on a Pt(100) surface and provides an interesting link with the topics introduced in Chapter 7. The general mechanism for CO oxidation is shown below:

$$CO_{(g)} + S \rightleftarrows CO_{(ads)} \qquad\qquad (R\ 2, -2)$$

$$O_{2(g)} + 2S \rightarrow 2O_{(ads)} \qquad\qquad (R\ 3)$$

$$CO_{(ads)} + O_{(ads)} \rightarrow CO_{2(g)} + 2S \qquad\qquad (R\ 4)$$

where S represents a vacant surface site. Oscillatory behaviour occurs due to surface coverage induced phase transitions on the surface of Pt(100) and a competition for surface sites.

Bulk Pt(100) exists as a square (1 × 1) lattice, with a high sticking coefficient (p. 181) for dissociative adsorption of molecular oxygen. However, at a clean *surface* a quasi hexagonal phase, where the surface atoms have a higher coordination number, becomes the energetically favourable packing arrangement. This hexagonal phase has a very low propensity for dissociative adsorption of oxygen. Adsorption of CO on to the hexagonal surface changes the energetics of the system and at a critical surface coverage the (1 × 1) arrangement becomes energetically more favourable. This change is reversible; decreased CO coverage or heat drives the surface back to the hexagonal arrangement.

Starting with a clean hexagonal surface and a 10:1 O_2:CO gas mixture, CO adsorption is the only process occurring until a critical coverage (θ = 0.08 monolayer) is reached at which point the Pt(100) surface phase transition occurs. The surface is now activated for O_2 dissociative adsorption. Vacant sites are rapidly filled by O atoms which react with the adsorbed CO molecules releasing fresh sites for further oxygen adsorption. With a rapidly decreasing CO coverage the (1 × 1) arrangement becomes increasingly unfavourable and eventually reverts to the hexagonal structure with a consequent termination of O_2 dissociative adsorption and therefore reaction, until the CO coverage builds and the process can start again. Ertl (1990) has studied this system and using surface techniques such as low energy electron diffraction (LEED) has been able to demonstrate the correlation of surface structure to O_2 uptake.

It should be noted that the surface coverage of CO need not be uniform. In fact there may be local areas of (1 × 1) phase with very high surface coverage. In this situation the autocatalytic nature of the CO and O combination reaction will be emphasized. The reaction of *one* oxygen atom, which has managed to find a vacant site amongst the CO molecules, releases two surface sites allowing the adsorption of *two* further oxygen atoms. The overall reaction rate, which is proportional to both CO and O surface coverage, will rapidly accelerate, reaching a maximum, before declining due to decreased CO coverage.

11.4 Cool flames

Figure 10.5 showed the ignition diagram for a typical hydrocarbon. We noted that there are similarities to the hydrogen/oxygen system but also that there is an additional lobe occurring at low temperatures (500–800 K depending on the fuel) corresponding to the so-called *cool flame region*. Cool flames are associated with a sudden temperature rise of approximately 100 °C and an increase in reaction rate which is quenched before reaction is complete. There is, therefore, some mechanism for negative feedback which we now seek to explain.

The initiation step, as in the H_2/O_2 case, occurs by reaction of the fuel, RH, with O_2:

$$RH + O_2 \rightarrow R + HO_2. \tag{R 5}$$

The HO_2 radical is unreactive at the temperature of interest and we concentrate on R which can undergo two types of *linear* propagation reaction, both of which occur via the peroxy radical RO_2 with the alkyl radical being regenerated by reaction (7)

$$R + O_2 \rightarrow RO_2 \rightarrow \text{stable products} + OH + \text{heat} \tag{R 6a}$$

$$\rightarrow QO_2H \tag{R 6b}$$

$$OH + RH \rightarrow R + H_2O. \tag{R 7}$$

The route to the hydroperoxide radical, QO_2H, is a minor one, but is of crucial importance, because of the subsequent, slow branching steps:

$$QOOH + O_2 \rightarrow \rightarrow \text{branching agent} \tag{R 8}$$

and

$$QO_2H + RH \rightarrow RO_2H + R \tag{R 9}$$

$$RO_2H \rightarrow RO + OH \tag{R 10}$$

and we shall consider hydroperoxy formation in greater detail in section 11.4.1.

During the induction period, the hydroperoxide concentration builds up until a critical level is reached, where sufficient branching, via reactions (8) and (10), occurs to accelerate the overall reaction rate and so raise the temperature of the mixture. This type of reaction sequence, involving branching via a comparatively stable species, is termed *degenerate* branching. Why, then, does the reaction not proceed to completion? Phenomenologically, this effect can be traced back to a region of negative temperature dependence in the overall reaction rate (Fig. 11.3). As the temperature increases, the overall reaction rate decreases; for propane, this occurs in the temperature range 700–750 K. At a molecular level, it can be attributed to the reversal of the initial combination of R and O_2:

$$R + O_2 \rightleftarrows RO_2. \tag{R 6, -6}$$

The forward reaction has, in general, only a slight activation energy. In contrast the reverse reaction, the dissociation of the peroxy compound, involves the breaking of a C–O bond and is therefore strongly temperature dependent.

In cool flame oscillations it is the rise in temperature and the effect that it has on the equilibrium (R 6, −6) that provides the negative feedback. As the temperature rises, the equilibrium shifts to reactants, $R + O_2$, and the source of the

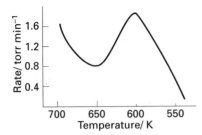

Fig. 11.3 Variation with temperature of the maximum rate of oxidation of propene; total pressure = 160 torr, with a fuel/O_2 ratio of 1:2. Note the negative dependence between 600 and 650 K.

chain branching reagents is terminated. With a marked decrease in chain branching the rate of reaction, and hence heat evolution, decreases. The reaction mixture cools, eventually reaching a point at which the $R + O_2$ equilibrium has a significant RO_2 component and the process can begin again.

Note that all the time the hydrocarbon and oxygen reactants are being consumed, oscillatory behaviour is only observed in the intermediates. Whilst cool flames can be observed in static reactors, oscillatory behaviour is more readily established in a continuously-stirred-tank-reactor (CSTR), where the fuel and oxygen flow separately into a well stirred reaction vessel. The system permits a constant replenishment of the reactants and removal of products, so that the exacting conditions required for oscillation are more easily effected.

Figure 11.4 shows examples of the type of behaviour that can be observed. (Proudler *et al.*, 1991). In these experiments butane and oxygen are flowed into the reactor, different types of oscillatory behaviour are observed depending on the gas and wall temperatures and whether the mixture is being heated or cooled. The rapid temperature fluctuations are measured with a thermocouple and pressure excursions are followed by a pressure transducer. Note that the magnitude of the temperature oscillations is very different than for the H_2/O_2 system, the smaller temperature excursions (typically 20–200 K) indicating that the ignition never proceeds very far before it runs out of chain branching precursor. The period of the oscillations is also much shorter and is considerably less than the residence time in the reactor.

Numerical models (see Section 8.3) have been devised in order to simulate the observed behaviour but to date the agreement is qualitative rather than quantitative. This may be partially due to inaccuracies in the gas phase model, but in addition, Proudler and other workers have observed variations with the composition of the reactor wall surface demonstrating the importance of surface reactions.

11.4.1 Peroxy radical isomerizations

Peroxy radical isomerization is the crucial step leading to chain branching and the mechanisms of these internal abstraction reactions have recently been investigated. The rate of the isomerization process depends on the structure of the radical species and of the ring transition state through which reaction occurs. The activation energy for isomerization tends to be lowered by the internal abstraction of tertiary or secondary hydrogens and ring strain minimized by ring structures of five or more atoms. Conversely the pre-exponential factor for internal isomerization is higher in the case of small ring structures. We can rationalize this

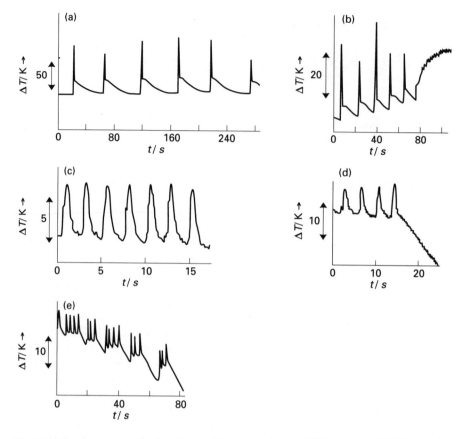

Fig. 11.4 Temperature profiles for a butane/air mixture for an initial pressure of 400 Torr and temperature of 570 K showing the variety of behaviour that can be obtained as the reaction vessel is heated or cooled. The experiments were carried out in a jet stirred reactor where the consumed reactants are slowly being replaced by fresh reactants and the products are gradually swept out by the incoming gas. (a) Oscillatory ignitions, T_{wall} = 577 K, (b) transition from ignitions to cool flames, T_{wall} = 579 K, (c) oscillatory cool flames, T_{wall} = 579 K, (d) cessation of cool flames on cooling, T_{wall} = 542 K, (e) repeated re-ignition of cool flames.

latter observation by considering the part of the radical which is twisting back to internally abstract a hydrogen atom. A 1,6 isomerization, with five atoms in the folding chain, (see Fig. 11.5) will have many more possible conformations in space and hence statistically, will be less likely to be in the correct conformation for the internal abstraction in comparison with the shorter chain of a 1,4 isomerization. A relevant example is the 1,5 isomerization of the neopentyl peroxy radical:

$$
\begin{array}{ccc}
CH_3 & & CH_2 \\
| & & | \\
CH_3-C-CH_2-O-O & \rightarrow & CH_3-C-CH_2-O-O-H \\
| & & | \\
CH_3 & & CH_3
\end{array}
\qquad \text{(R 11)}
$$

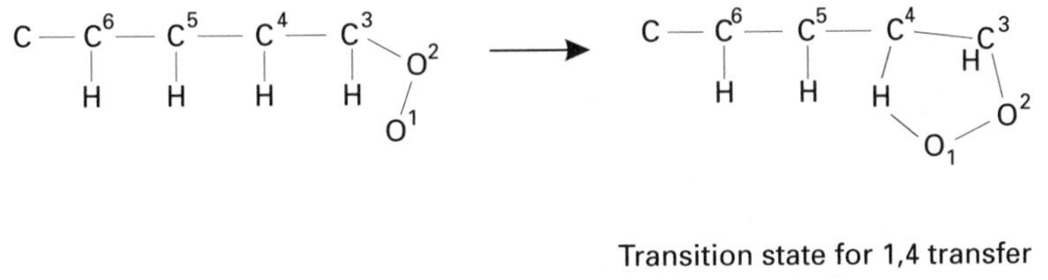

Transition state for 1,4 transfer

Fig. 11.5 Designation of the various sites for an internal abstraction in the peroxy radical isomerization.

A recent paper by Hughes *et al.* (1992) illustrates how the experimental techniques described in Chapter 2 (such as laser flash photolysis/laser induced fluorescence) can be used to study aspects of the cool flame process in isolation and the work of Hughes has lead to a re-evaluation of the barrier heights for alkyl peroxy isomerizations, which have increased by approximately 10 kJ mol^{-1}, a typical value being 122 kJ mol^{-1} for the neopentyl system.

Surprisingly, the simplest reaction of the series

$$C_2H_5 + O_2 \rightleftarrows C_2H_5O_2 \rightarrow C_2H_4OOH \text{ or other products} \quad (R \ 12, -12, 13)$$

is proving to be mechanistically the most controversial and opposing points of view can be found in Wagner *et al.* (1990), R. Baldwin *et al.* (1980), and McAdams and Walker (1987).

Box 11.1 Knock in internal combustion engines

Figure 11.6 shows a cross-section of a cylinder in an internal combustion engine. The piston rises and compresses the fuel/air mixture, the spark plug ignites the mixture and a flame spreads across the gas generating a pressure increase which pushes the piston back. Clearly the timing of the spark initiated combustion with respect to the phase of the piston stroke is of crucial importance if the energy released by the burning fuel is to be efficiently transmitted to the crankshaft. Engine knock arises because of *autoignition*, (i.e. ignition of the fuel before the flame reaches it). The flame front compresses and heats the unburned fuel (called the 'end gas') and a cool flame may occur before the flame front arrives. The compression and heating continues and a second stage ignition takes place leading to a rapid rise in temperature and complete combustion of the end gas, before the arrival of the spark initiated flame front. This is shown clearly in Fig 11.7, where in contrast to normal operation, the ignition in a knocking cycle occurs significantly before the piston has reached the highest point in the cylinder. In consequence, power is lost and engine damage can result. Knock is alleviated by proper design of both engine and fuel, which must respond sensitively to spark ignition, but must be comparatively insensitive to autoignition. Our comparative ignorance of the quantitative and even qualitative nature of the reactions involved means that the kinetics of oxidation is an active area of research.

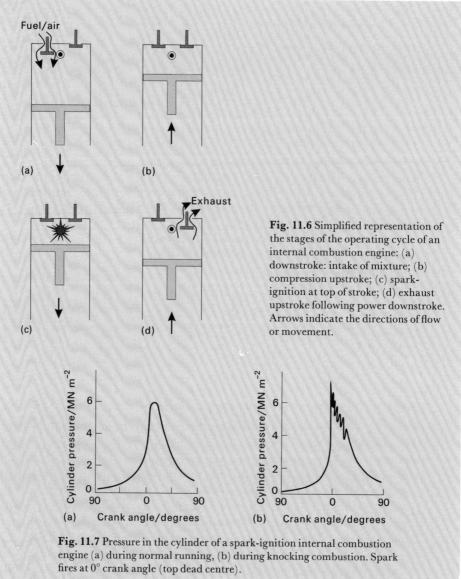

Fig. 11.6 Simplified representation of the stages of the operating cycle of an internal combustion engine: (a) downstroke: intake of mixture; (b) compression upstroke; (c) spark-ignition at top of stroke; (d) exhaust upstroke following power downstroke. Arrows indicate the directions of flow or movement.

Fig. 11.7 Pressure in the cylinder of a spark-ignition internal combustion engine (a) during normal running, (b) during knocking combustion. Spark fires at $0°$ crank angle (top dead centre).

11.4.2 High temperature oxidation of hydrocarbons

What happens at higher temperatures? We know that the deceleration with temperature cannot persist and this is illustrated in Fig. 11.3 by the increase in the overall rate of propene oxidation at temperatures above $750\,K$. A new, higher temperature mechanism is required which cannot, now, involve the peroxy radical, which is unstable under these conditions. We have already discussed the high temperature oxidation of methane in Chapter 10. Higher hydrocarbon oxidation pathways differ from methane in that the initially formed alkyl radical can react with *molecular* oxygen to form the conjugate olefin (e.g. propene from propyl) and HO_2

$$R + O_2 \rightarrow \text{conjugate olefin} + HO_2. \qquad (R\ 14)$$

At these higher temperatures OH radicals are formed from HO_2:

$$2HO_2 \rightarrow H_2O_2 + O_2 \qquad (R\ 15)$$

$$HO_2 + RH \rightarrow H_2O_2 + R \qquad (R\ 16)$$

$$H_2O_2 \rightarrow 2OH \qquad (R\ 17)$$

where the reaction $H_2O_2 \rightarrow 2OH$ corresponds to branching. At higher temperatures still, radical decomposition occurs e.g.

$$HO_2 \rightarrow H + O_2 \qquad (R\ 18)$$

$$CH_3CH_2CH_2 \rightarrow CH_3 + C_2H_4 \qquad (R\ 19)$$

and reaction acceleration proceeds primarily via the branching step

$$H + O_2 \rightarrow OH + O \qquad (R\ 20)$$

as we discussed in Section 10.4

Study notes

1   The complete mechanism of cool flame oscillations is more complex than the brief outline that was sketched out above. Many of the details of the reaction are under current investigation using a variety of approaches and techniques. The basic chemistry which leads, under certain conditions, to oscillatory behaviour is present in the combustion of fuels in an internal combustion engine providing a practical impetus for understanding the subtleties of the reaction.

2   Further details on cool flames may be found in Chapter C2.2 of Cox, R. A. (1987). *Modern Gas Kinetics*, (ed. M. J. Pilling and I. W. M. Smith, Blackwells, Oxford). Cox gives relevant examples of chemical mechanisms and emphasizes that a crucial step in the overall process of cool flames is the isomerization of the peroxy radical, ROO to form a hydroperoxy radical, QOOH as we have discussed above. The chapter also contains brief outlines of higher temperature combustion.

11.5 The Belousov–Zhabotinskii reaction

11.5.1 A schematic representation of the Belousov–Zhabotinskii reaction

The detailed mechanism of this reaction, introduced in Section 11.1, and the origin of the oscillations are complex and are best appreciated by first examining the following schematic mechanism in which A and B are reactants, X, Y, and Z are intermediates and P and Q are products.

$$A + Y \rightarrow X \qquad (R\ 21)$$

$$X + Y \rightarrow P \qquad (R\ 22)$$

$$B + X \rightarrow 2X + 2Z \qquad (R\ 23)$$

$$2X \rightarrow Q \qquad (R\ 24)$$

$$2Z \rightarrow \tfrac{1}{2}Y. \qquad (R\ 25)$$

Note that the concentrations of all the intermediates are inter dependent, the reaction of $A + Y$ generates X, $B + X$ generates Z and finally Z generates Y.

We let the concentrations of A, B, X, Y, Z be a, b, x, y, z and with a and b held constant, apply the steady-state approximation to X, the species involved in the positive feedback step, reaction (R 23):

$$\frac{dx}{dt} = k_{21}ay - k_{22}xy + k_{23}bx - 2k_{24}x^2 = 0$$

$$x = \frac{[(k_{23}b - k_{22}y) \pm \sqrt{\{(k_{23}b - k_{22}y)^2 + 8k_{21}k_{24}ay\}}]}{4k_{24}}$$

$$x = \frac{(k_{22}y - k_{23}b)}{4k_{24}}\left[-1 \pm \left\{1 + \frac{8k_{21}k_{24}ay}{(k_{22}y - k_{23}b)^2}\right\}^{\frac{1}{2}}\right].$$

Thus, because the steady-state equation in x is quadratic, two solutions apply and, under some conditions, these are both real and both accessible to the reaction system. We shall justify this presently through the behaviour of the reaction scheme, but for the present we shall determine approximate values for these solutions by assuming that the concentration of Y may also vary between (comparatively) large and small values. We shall also justify this behaviour in our subsequent analysis. Proceeding in this manner helps us to divide the reaction mechanism into its component parts.

1   If y is large, then $(k_{22}y - k_{23}b)^2 \gg 8k_{21}k_{24}ay$ and we may expand the square root using the binomial expansion

$$x = \frac{(k_{22}y - k_{23}b)}{4k_{24}}\left[-1 \pm \left\{1 + \frac{4k_{21}k_{24}ay}{(k_{22}y - k_{23}b)^2}\cdots\right\}\right]$$

Taking the positive root,

$$x = \frac{k_{21}ay}{k_{22}y - k_{23}b} \approx \frac{k_{21}a}{k_{22}}$$

provided $k_{22}y \gg k_{23}b$. Thus the steady-state concentration of x is determined simply by reactions (21) and (22), which dominate the mechanism.

2   If y is small, then the square root is approximately unity and , taking the negative square root

$$x = \frac{2(k_{23}b - k_{22}y)}{4k_{24}} \approx \frac{k_{23}b}{2k_{24}}$$

provided $k_{22}y \ll k_{23}b$, and now reactions (23) and (24) (and 25 which does not affect x) dominate.

There are, therefore, two limiting mechanisms. Mechanism I applies when y is large,

$$A + Y \rightarrow X \hspace{3cm} (R\ 21)$$
$$\underline{X + Y \rightarrow P} \hspace{3cm} (R\ 22)$$
$$A + 2Y \rightarrow P.$$

This mechanism maintains X in a steady-state but *consumes* Y, so that y falls. Eventually y decreases through a critical value (at which $k_{22}y = k_{23}b$) so that X, instead of reacting mainly with Y via reaction (22), now reacts predominantly with B via reaction (23) and mechanism II takes over:

$$2B + 2X \rightarrow 4X + 4Z \qquad 2 \times \text{(R 23)}$$

$$2X \rightarrow Q \qquad \text{(R 24)}$$

$$4Z \rightarrow 2Y \qquad 4 \times \text{(R 25)}$$

$$2B \rightarrow Q + 2Y.$$

In mechanism II, X reacts with B in an autocatalytic step that *generates more* X so that x begins to rise; as it does so reaction (24), which is *second order* in X becomes more important and eventually the rates balance and X achieves a new steady-state. The final link between the two mechanisms, that allows for oscillations between the two mechanisms, comes from reaction (23) which produces Z, another intermediate, which regenerates Y, so that y increases and eventually exceeds $y_{critical}$ so that mechanism I reasserts itself and the sequence starts again. Thus the concentrations of X, Y, and Z all oscillate between maximum and minimum values.

The salient features of the mechanism are:
- the *competition* between reactions (22) and (23) for X and the dependence of this competition on y.
- the *consumption of* Y in mechanism I.
- the *auto-catalytic generation of* X in reaction (23).
- the *second-order consumption of* X in reaction (24).
- the *regeneration of* Y in mechanism II.

11.5.2 The chemistry of the Belousov-Zhabotinskii reaction

In the Belousov–Zhabotinskii (BZ) reaction, the roles of both A and B are played by BrO_3^- whilst X and Y are $HBrO_2$ and Br^- respectively and Z is Ce^{4+}. Obviously other species take part in the mechanism and some of the apparently simple reactions we examined in the scheme above involve several steps. However, kinetically we can group these steps together in one overall reaction that is equivalent to the reaction in our model scheme. In the following outline of the mechanism, the kinetically important species A, B, X, Y and Z are shown in bold type:

$$\mathbf{BrO_3^-} + \mathbf{Br^-} + 2H^+ \rightarrow \mathbf{HBrO_2} + HOBr \qquad \text{(R 26)}$$

$$\mathbf{HBrO_2} + \mathbf{Br^-} + H^+ \rightarrow 2HOBr. \qquad \text{(R 27)}$$

These reactions are then followed by the conversion of HOBr into Br_2 and the reaction of Br_2 with malonic acid in reactions which do not affect the oscillatory kinetics, but which do generate the product, P, which is bromomalonic acid.

$$3HOBr + 3Br^- + 3H^+ \rightarrow 3Br_2 + 3H_2O \qquad \text{(R 28)}$$

$$3Br_2 + 3CH_2(CO_2H)_2 \rightarrow 3CHBr(CO_2H)_2 + 3Br^- + 3H^+. \qquad \text{(R 29)}$$

Thus the overall reaction in mechanism I is;

$$\mathbf{BrO_3^-} + 2\mathbf{Br^-} + 3CH_2(CO_2H)_2 + 3H^+ \rightarrow 3CHBr(CO_2H)_2 + 3H_2O. \qquad \text{(R 30)}$$

The reaction thus consumes bromide (Y) and maintains $HBrO_2$ (X) at its lower steady-state concentration. As the bromide concentration falls below its critical value, the reaction between $HBrO_2$ and BrO_3^- (now acting as species B) becomes faster than that between $HBrO_2$ and Br^- and mechanism II becomes dominant;

$$2BrO_3^- + 2HBrO_2 + 2H^+ \rightarrow 4BrO_2 + 2H_2O \qquad (R\ 31)$$

$$4BrO_2 + 4Ce^{3+} + 4H^+ \rightarrow 4HBrO_2 + 4Ce^{4+} \qquad (R\ 32)$$

$$2HBrO_2 \rightarrow BrO_3^- + HOBr + H^+ \qquad (R\ 33)$$

$$HOBr + CH_2(CO_2H)_2 \rightarrow CHBr(CO_2H)_2 + H_2O. \qquad (R\ 34)$$

Note that the auto-catalytic reaction (31, 32) takes place in two steps and involves the intermediate formation of BrO_2. The kinetically important second order consumption of $HBrO_2$ gives HOBr as an intermediate product (Q) and this then reacts to form the final product, $CHBr(CO_2H)_2$ in a kinetically un-important step (reaction 34).

Finally Br^- is regenerated in reaction (35) by a step involving the product $CHBr(CO_2H)_2$:

$$4Ce^{4+} + CHBr(CO_2H)_2 + 2H_2O + HOBr \rightarrow$$
$$2f\,Br^- + 4Ce^{3+} + 3CO_2 + 6H^+ \qquad (R\ 35)$$

where $(1 + \sqrt{2} > f > \frac{1}{2})$. This reaction explains the long induction period before oscillation sets in (Fig. 11.1), since the concentration of $CHBr(CO_2H)_2$ must build up to a sufficient level to effect reaction (35). During the induction period $[Br^-]$ is small and Mechanism II dominates, except that the reaction

$$6\,Ce^{4+} + CH_2(CO_2H)_2 + 2H_2O + HOBr \rightarrow 6\,Ce^{3+} + 3CO_2 + 7H^+ + Br^- \quad (R\ 36)$$

slowly converts Ce^{4+} into Ce^{3+} and $CHBr(CO_2H)_2$ is allowed to accumulate via reaction (34).

Clearly the mechanism is a complex one and the original scheme shown on pp. 261–2 is simplified and contains only the salient kinetic features. Thus, for example, all organic reactions were represented by the single step (R 25). Reaction (33) is not recognized in the original scheme, because it does not affect the kinetic behaviour of the intermediates, whose description is the primary aim of the mechanism. It is remarkable that the researchers who devised the model, Noyes, Field, and Körös, working at the University of Oregon (and hence the name of the scheme—Oregonator!), approached the problem from a chemical view-point and devised the detailed reaction mechanism from a consideration of available thermodynamic and kinetic data (Noyes *et al.* 1972). In addition a liberal application of common sense, combined with an acute appreciation of the interplay between the chemical reactions was also of great importance. Once they had devised the overall mechanism, they reduced it to the simplified model described on p. 261—the model known as the Oregonator. The arguments they used make compelling reading and references to their papers are given in the study notes on p. 269.

Box 11.2 Recipe for the BZ reaction

Prepare four solutions: Solution 1: 0.6 M $NaBrO_3$ and 0.6 M H_2SO_4. Solution 2: 0.48 M malonic acid. Solution 3: 1 g NaBr dissolved in 10 cm^3 water. Solution 4: 0.025 M freshly made ferroin solution (Cl^- free). Mix 14.0 cm^3 of solution 1, 7.0 cm^3 of solution 2, 2.0 cm^3 of solution 3 and wait for 5 minutes while the colour of Br_2 disappears. Add 2.0 cm^3 of solution 4. Pouring the mixture to a Petri dish, making a 1–2 mm layer, spatial patterns evolve. In a stirred beaker the mixture demonstrates homogeneous temporal oscillations.

11.6 Representations of oscillatory behaviour

For simple systems, representations of intermediate concentrations such as Fig. 11.1 may be adequate, however, they do not emphasize the links between the two oscillating intermediates. The development of a simple two dimensional representation of such behaviour is outlined in this section.

Figure 11.8(a) shows a graphical representation of a reaction involving two intermediates A and B, as it evolves with time to an equilibrium state. There are no oscillations, the reaction proceeding smoothly to an equilibrium distribution of A and B. This might correspond to formation of a steady-state in A and B. The solid line is known as the trajectory of the reaction. The same steady-state concentration may be attained from a number of different initial concentrations as shown in Fig. 11.8(b). Here trajectories from different starting conditions lead to the same final equilibrium distribution of the intermediates A and B. These three dimensional pictures give a good visualization of the progress of the reaction but are rather cumbersome. If we are not interested at the rate at which these trajectories move then we can project such trajectories onto the *ab* plane to give a two dimensional representation of the reaction and this is depicted in Fig. 11.8(c) where a system spirals down to an equilibrium state from two different starting points.

The essential component of oscillatory behaviour is that the concentration of the intermediates oscillate between two different limits (see Fig. 11.1). The three dimensional trajectory of this process is shown in Fig. 11.8(d) with the trajectory starting from a point on the *ab* plane and moving onto the surface of a cylinder (the induction period). The regular oscillatory motion of the reaction corresponds to the trajectory moving in a helical spiral along the cylinder surface. If we compress this trajectory then we get the two dimensional plot shown in Figure 11.8(e) with the thinner line corresponding to the motion onto the cylinder and the thick line representing the regular oscillation of the reaction between the limiting values of *a* and *b*, known as the *limit cycle*.

We have restricted our discussion to two variables, which give simple limit cycles, however, such depictions are valid for three or more variables with the trajectories occurring in four- (or more) dimensional space. We can still compress the trajectory down onto a two-dimensional surface, to see how two variables behave during a limit cycle, however, now the behaviour is more complex and there may be several apparent crossings on the limit cycle. Further information may be found from the references given in study note on p. 269.

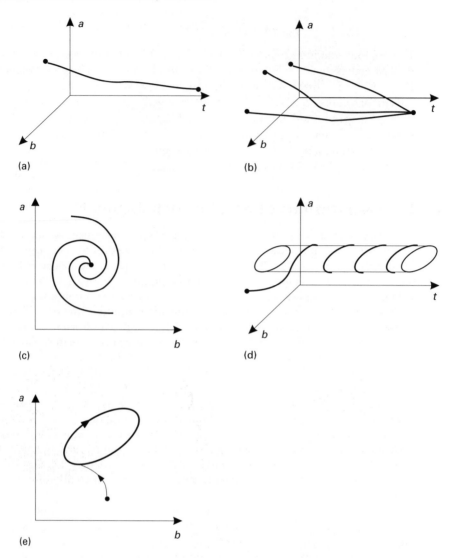

Fig. 11.8 The evolution of concentrations with time. (a) A trajectory corresponding to the development from an initial composition a_i, b_i to a final equilibrium state. (b) trajectories from several different initial conditions converging to a common equilibrium state. (c) a phase-plane portrait for two different initial compositions spiralling down to a common equilibrium state. (d) the trajectory for the establishment of oscillatory behaviour. (e) a phase-plane representation of the establishment of oscillatory behaviour.

11.7 Further developments

The Belousov–Zhabotinskii reaction is not the only system to show temporal oscillations. The Bray–Liebhafsky reaction, involves the catalytic oxidation of hydrogen peroxide by iodate:

$$2IO_3^- + 5H_2O_2 + 2H^+ \rightarrow I_2 + 5O_2 + 6H_2O \qquad (R\ 37)$$

$$I_2 + 5H_2O_2 \rightarrow 2IO_3^- + 2H^+ + 4H_2O. \qquad (R\ 38)$$

The iodate catalyst is consumed in the first reaction and oxygen production ceases until the second reaction regenerates enough of the iodate and consumes the iodine.

Spatial oscillations and the building up of structures via reactions can be even more striking. If the Belousov–Zhabotinskii solution (containing a little less acid than usual and a little more bromide) is placed in a shallow dish and the stable red solution is touched with a hot needle, then a blue ring propagates outwards. If the dish is rocked gently, spirals with a uniform spacing spread out from the point of initial disturbance and quite complex patterns evolve (Fig. 11.9). If a solution is taken in which, say, the $HBrO_2$ concentration is small and the bromide concentration is reduced significantly by adding a drop of solution with a low bromide concentration then the steady-state with high $HBrO_2$ concentration becomes applicable and $[Br^-]$ is depressed even more. A trigger wave (or chemical wavefront) of oxidation then spreads through the solution. This ability to undergo such a change is termed 'excitability' and requires a steady-state which is stable to small fluctuations but becomes unstable to slightly larger variations in conditions. Oscillatory behaviour may not result, but the ability of a system to react to a stimulus by generating a pulse of large chemical change is of great importance in the generation of order in biological systems.

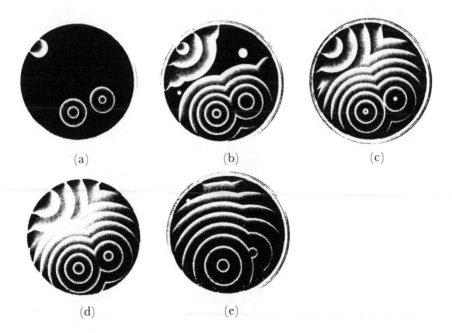

(a) (b) (c)

(d) (e)

Fig. 11.9 Target patterns in a thin layer of BZ reaction in a 9 cm diameter petri dish. (a) 1 min after mixing; (b) after 3 min 30s; (c) after 7 min 15s; (d) after 7 min 35s; (e) after 16 min 20s. Three random pacemaker sites initiate targets, but as the system evolves, the successive annihilations of colliding waves from two adjacent targets occur closer and closer to the lower frequency pacemaker. In time, the higher frequency source entrains the lower one. In a given target, the outermost wave travels at a slightly higher velocity than those inside the target. (Photographs courtesy of M. Pearson.)

Box 11.3 Wavefronts and diffusion

If we assume that the chemical wavefronts propagate by simple diffusion we ought to be able to relate the rate of propagation to calculated or measured diffusional coefficients for typical species in solution (see Section 6.3). The observed velocity of these chemical wavefronts is approximately 1 mm min^{-1} and remains constant but simple diffusional calculations predict rates an order of magnitude smaller and with a decreasing rate. How can we explain these observations as diffusion is the only available method of transportation? A rationale for this contradiction can be found in the autocatalytic behaviour of the reactions and the random nature of diffusion. Diffusion is a statistical process and hence a small fraction of the chain branching species will diffuse at a much greater rate. This vanguard of autocatalytic reagent can then react at an ever increasing rate with the fresh reagents generating significant product concentrations at a much greater rate than diffusional transfer alone. If more than one autocatalytic intermediate is involved and if these intermediates have significantly different mobilities, or diffusion coefficients, then even more complex behaviour can occur!

In the vast majority of this book we have solely considered reactions occurring in uniform distributions of reactants, indeed for experimental determinations of rate coefficients (Chapter 2) this is essential. However, in a number of processes including those discussed above, reaction takes place in non uniform concentrations. An every day example of this is the flame of a bunsen burner, where concentrations of species vary enormously across the flame front. In these circumstances transport phenomena may be as important as the rate coefficients of the elementary reactions in determining the overall rate of product formation. We have largely ignored these phenomena, as they rarely reveal much information about the molecular mechanisms by which two molecules react, but, in many cases (on both a macro and micro scale), transport phenomena and mixing must be allowed for.

The problem of excitability has been widely discussed, particularly by Prigogine, a Russian Nobel Prize Winner. A detailed discussion would be inappropriate here, but an indication of the subject matter is relevant, because it is likely to form an increasingly important aspect of the kinetics of complex systems. We are accustomed to dealing with equilibrium thermodynamics, in which chemical systems move inexorably to their equilibrium position, determined by the achievement of maximum entropy in the system and its surroundings. In some types of system, typified by the Oregonator model, with non linear kinetic equations and a feedback mechanism, then non-equilibrium states called dissipative structures can be excited under the correct stimulus. They are frequently associated with the establishment of some sort of spatial ordering or temporal oscillation. Biological examples include glycolysis, in which oscillatory behaviour is observed in the chemical components of the cycle, and cellular division. Finally, the prebiotic synthesis of macromolecules via an autocatalytic process, and the subsequent pressure on them to evolve through errors in reproduction, may also be discussed, if not fully understood, using the same ideas of dissipative structures.

Study notes
1   The earlier work of Noyes and his co-workers is clearly reviewed in two short articles to be found in *Accounts of Chemical Research*. Noyes, R. M. and Field, R. J. (1977), Oscillations in chemical systems, *Accounts of Chemical Research*, **10**, 214–21, which discusses the Oregonator model in some detail and Noyes, R. M. and Field, R. J. (1977). Mechanisms of chemical oscillators, *Accounts of Chemical Research*, **10**, 273–80, which describes various examples of oscillatory behaviour.
2   The original articles by Noyes. Field, and Körös (1972) on the mechanism of the Belousov–Zhabotinskii reaction can be found in Oscillations in chemical systems, *Journal of the American Chemical Society*. **94**. 1394–5 and the production of spiral structures in shallow dishes is described by Winfree, A. T. (1974) in Rotating chemical reactions, *Scientific American*, **230**, (*June*), 82.
3   The mathematical treatment of oscillations and *bifurcations* (the sudden shooting off to a new steady-state) is complex, but the introductory chapter in *Self-Organisation in Non-Equilibrium Systems* by Nicolis, G. and Prigogine, I. (Wiley, New York 1977) could be read with profit. It is a most stimulating book and some of the later chapters on biological systems and on so-called ecosystems may also be dipped into to obtain a flavour of the subject area—more than this requires some effort!
4   Recently several books have been produced by the Leeds group concentrating on the kinetics and analysis of oscillatory reactions. *Oscillations, Waves, and Chaos in Chemical Kinetics* (Oxford University Press, 1994) by S. K. Scott is a short introductory text to the field and sets the scene for two more detailed texts *Chemical Oscillations and Instabilities; Non-Linear Chemical Kinetics* P. Gray and S. K. Scott (Oxford University Press, 1990) and *Chemical Chaos* by S. K. Scott (Oxford University Press, 1991). Chaos is one of the buzz words of recent times and its application to some complex chemical systems is just one of many areas in which chaos theory is being applied. We should note that the term chaos does not refer to the familiar images of randomness and disorder. Chemically chaotic behaviour is indeed irreproducible and unpredictable, but this arises not from randomness, but rather from the extreme sensitivity of the system to minute changes in conditions. An overall view of the subject of chaos may be found in *The New Scientist Guide to Chaos* (ed. N. Hall) (Penguin, London, 1992).

11.8 Questions

11.1   Plot limit cycles for the oscillations of the two intermediates, α and β, shown in Figs. 11.10(a) and (b).

11.2   The Lotka–Volterra mechanism shows oscillatory behaviour. If the concentration of A is kept constant (by adding more A as is necessary) explain why, starting with small initial concentrations of X and Y, oscillations will occur.

$$A + X \rightarrow 2X \qquad\qquad (R\ 39)$$

$$X + Y \rightarrow 2Y \qquad\qquad (R\ 40)$$

$$Y \rightarrow B. \qquad\qquad (R\ 41)$$

Set up the differential equations controlling the intermediates X and Y and solve them numerically (e.g. using the Runge–Kutta algorithm). Plot the concentration of X and Y and limit cycle. How do the limit cycles vary with the initial starting conditions of X and Y? Use the following data: $k_{39} = 1$, $k_{40} = 1$, $k_{41} = 1$, and A = 1.

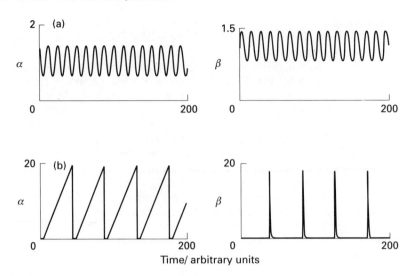

Fig. 11.10 Fluctuations of two intermediates α and β for two different reactions.

11.3 The Brusselator is another set of equations that can lead to oscillatory behaviour.

$$A \rightarrow X \qquad (R\ 42)$$

$$2X + Y \rightarrow 3X \qquad (R\ 43)$$

$$B + X \rightarrow Y + C \qquad (R\ 44)$$

$$X \rightarrow D. \qquad (R\ 45)$$

Set up the differential equations, once again assuming that A and B, the reactants are maintained at a constant concentration, and solve numerically. What do you notice about the limit cycles as you vary the initial starting conditions of X and Y? Use the following data: $k_{42} = 1$, $k_{43} = 1$, $k_{44} = 1$, $k_{45} = 1$, A = 1 and B = 3. This type of behaviour is known mathematically as an *attractor*.

References

Baldwin, R., Pickering, I. A., and Walker, R. W. (1980). Reactions of ethyl radicals with O_2. *Journal of the Chemical Society, Faraday Transactions 1*, **76**, 2374–82.

Ertl, G. (1990). *Spatial inhomogeneities and transient behaviour in chemical kinetics*. (ed. G. Nicoles, F. Baras, P. Birkmans and S. K. Scott). pp. 565–78. Manchester University Press.

Hughes, K. J., Lightfoot, P. D., and Pilling, M. J. (1992). Direct measurements of the peroxy–hydroperoxy isomerization. *Chemical Physics Letters*, **191**, 581–6.

McAdam, K. G. and Walker, R. W. (1987). Arrhenius parameters for the reaction C_2H_5 + $O_2 \rightarrow HO_2$ + C_2H_4. *Journal of the Chemical Society, Faraday Transactions 2*, **83**, 1509–17.

Noyes, R. M., Field, R. J., and Körös, E. (1972). Oscillations in chemical systems. *Journal of the American Chemical Society*, **94**, 1394–5.

Proudler, V. K., Cedarbalk, P., Horowitz, A., and Pilling, M. J. (1991). Oscillatory ignitions and cool flames in the oxidation of butane. *Philosophical Transactions of the Royal Society, London A*, **337**, 211–23.

Wagner, A. F., Slagle, I. R., Sarzynski, D., and Gutman, D. (1990). Experimental and theoretical studies of C_2H_5 + O_2 reaction kinetics. *Journal of Physical Chemistry*, **94**, 1853–68.

12 Photochemistry

12.1 Introduction

In its broadest sense, photochemistry involves the study of the absorption of radiation and the subsequent reactions of the excited species or dissociation products and incorporates aspects of spectroscopy, energy transfer, reaction kinetics and dynamics. Its importance stems from the fact that the sun's radiation is the ultimate driving force behind virtually all the processes occurring on the Earth's surface and in its atmosphere. In this chapter we shall concentrate mainly on the kinetic and dynamical aspects of photochemistry and, as such, accept that we are only covering a fraction of the overall topic. Important areas such as photosynthesis and vision are covered in a number of more specialized texts.

The fate of electronically excited molecules provides an excellent application of the kinetics of parallel reactions which we developed in Section 8.2. In Section 12.2 we consider the various processes by which electronically excited species can be removed and in Section 12.3 how the kinetics of these processes couple together.

Section 12.4 investigates in greater detail one of the possible fates of electronic excitation, that of dissociation. We have already come across dissociation processes in a number of instances (for example generating radical species in the experimental technique of flash photolysis pp. 39–43). The section briefly surveys some dissociation mechanisms and we see what we can learn from the study of the states of the photofragments that are produced. Photodissociation studies have a close relationship with reaction dynamics. In both cases we are interested in the distribution of energy into products. In reaction dynamics energy redistribution occurs from a transition state generated by the combination of two reagents whereas in photodissociation we generate our transition state by the absorption of a photon. Photodissociation can be thought of as a 'half reaction'.

As well as being of mechanistic interest photofragmentation is an important tool in flash photolysis studies for generating reactive species (Section 12.6) and as a driving force behind much of the atmospheric chemistry of the Earth's atmosphere (see, for example, Section 8.4) and that of the other planets (Section 12.7).

Study notes

Photochemistry is too large a subject to cover in any detail in one chapter and we are aware that we have missed out large amounts of material. The interested reader is recommended to consult the following texts for further information

1   Wayne, R. P. (1989). *Principles and Applications of Photochemistry*. Oxford University Press. An excellent undergraduate text covering the full range of the subject.

2   Okabe, H. (1978). *Photochemistry of Small Molecules*. Wiley, Chichester and New York. The first part of this book is an in depth coverage of the principles and practice of photochemical studies related to small gas phase molecules. The

second part is a very useful review of the photochemical studies of two-five atom species.

3   Calvert, J. G., and Pitts, J. N. (1966). *Photochemistry*. Wiley, Chichester and New York, 1966). Now slightly out of date but an excellent introduction to organic photochemistry.

4   Unravelling the mechanism of photosynthesis and other photochemical biological phenomena such as vision is a major field of research and interdisciplinary study. Hader, D.-P., and Trevi, M. (1987). *General Photobiology*. Pergamon Press, Oxford, provides good coverage of these and other fields and the references give access to relevant original literature.

12.2 The production and fate of electronically excited molecules

One aspect of photochemistry is the study of the behaviour of electronically excited atoms and molecules, produced by the absorption of light. We shall concentrate on comparatively large molecules, exemplified by aromatic molecules, which show a rich and very well characterized photochemistry. Atoms and small molecules behave differently in regard to some of the aspects we shall study, especially radiationless transitions. The general processes involved are shown schematically in Fig. 12.1 and are discussed briefly below.

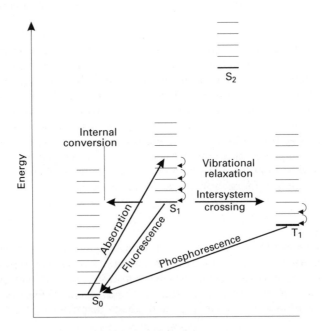

Fig. 12.1 Schematic diagram showing the processes connecting the various electronic states of a polyatomic molecule. This type of diagram is known as a Jablonski diagram.

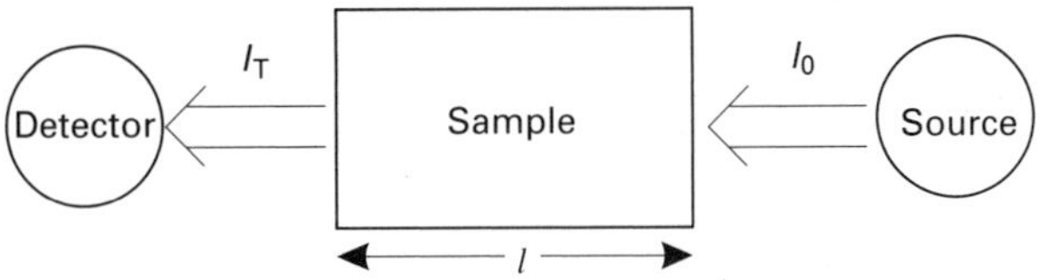

Fig. 12.2 Absorption and the terms involved in Beer's law.

12.2.1 Absorption

For low levels of absorption, the absorption process is controlled by the Beer–Lambert law, eqn (1). When light is passed through a substance (Fig. 12.2) it may be absorbed and the transmitted (I_T) and incident (I_0) light intensities are related via the eqn

$$\ln(I_T/I_0) = -\varepsilon c l \qquad (\text{E } 1)$$

where c is the concentration of the substance (mol dm^{-3}), l is the path length and ε the extinction coefficient. ε depends on the frequency of the light (v), and a plot of ε versus v traces out the absorption spectrum of the substance. For small molecules in the gas phase these spectra may be very structured (Fig 12.3(a)) and spectral analysis will yield important information about molecular structure. With larger molecules and in solution the spectra are generally broad, with fewer structural features (Fig 12.3(b)).

Absorption of light in the visible or ultraviolet region of the spectrum raises the absorbing molecule to an electronically excited state. For example, for benzene, an electron is promoted from a bonding π to an antibonding π orbital (π^*). The spin of the excited electron may be either antiparallel or parallel to the electron remaining in the original orbital, so that a singlet (S) or triplet (T) state may, in principle, be produced. Additional excited states may be produced by excitation to other orbitals, giving rise to a series of excited singlet $(S_1, S_2 \ldots)$ or triplet $(T_1, T_2 \ldots)$ states. For aromatics, the ground state is a singlet and is termed S_0. Note that the triplet state is generally lower in energy than the equivalent singlet state, so that the lowest excited state is T_1.

For a molecule such as benzene, absorption of light occurs without change in the spin of the molecule $(\Delta S = 0$, spin is *conserved*). Thus the S_0–T_1 process is strongly forbidden; the extinction coefficient for such an absorption is very small and, for our purposes, may be taken as zero. Thus absorption of light leads directly to the production of excited singlet states only, the particular state produced depending on the wavelength of the light used.

12.2.2 Emission

Once produced, the excited state (which in general we shall refer to as AB*) can undergo a number of processes, shown schematically in Fig. 12.4. One possibility is for the molecule to emit a photon of light, a first order process, i.e. if only emission occurs,

$$d[AB^*]/dt = -k_f[AB^*] \qquad (\text{E } 2)$$

where k_f is the first-order rate coefficient for emission. If AB* is an excited singlet, then the process is termed *fluorescence* and the reciprocal of $k_f (\tau_f = k_f^{-1})$ is called the

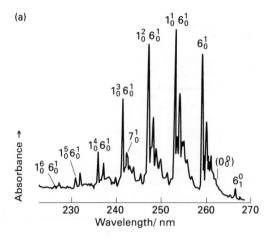

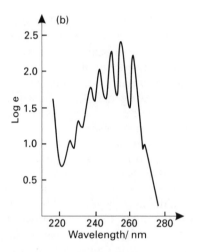

Fig. 12.3 (a) Low resolution absorption spectrum of gas phase benzene showing the well defined vibrational structure. (b) Solution phase spectrum of benzene showing that some of the clear vibrational structure has been degraded.

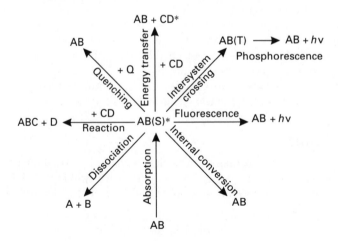

Fig. 12.4 Schematic diagram showing some of the possible fates of the electronically excited molecule AB*.

fluorescence lifetime. Excited states reached via a strong absorption have large values for k_f, i.e. k_f is related to ε. Typically $k_f = 10^8$ s^{-1}. As we shall see below, excited triplet states can also be produced and they emit light by a process known as *phosphorescence.* Because $\varepsilon(S_0–T_1)$ is so small, the phosphorescence lifetime, τ_{ph} is very long – hence the ability of phosphorescent substances to continue glowing after the exciting light source has been turned off. Typically $\tau_{ph} = 10^{-2} – 10$ s.

The emission process described above is termed spontaneous fluorescence and can be represented by reaction (1)

$$A^* \rightarrow A + h\nu. \tag{R 1}$$

An alternative emission process can also occur which is *stimulated* by another photon.

$$A^* + h\nu \rightarrow A + 2h\nu. \tag{R 2}$$

The two photons produced are in phase and propagate in the same direction. Laser action depends on this process and the unique nature of laser light arises from the properties of this process. We can see now why the Beer–Lambert law only applies for low levels of absorption. Once there is a significant excited state population then stimulated emission will compete with absorption. A more detailed discussion of the relationship between absorption, spontaneous and stimulated emission can be found in the initial chapter of *Modern Spectroscopy* by Hollas(1992), in *Lasers in Chemistry* by Andrews (1992), or most texts on photochemistry or spectroscopy. A brief description of dye lasers may be found in Box 12.3.

12.2.3 Vibrational relaxation

The absorption spectrum for the $S_0–S_1$ transition in benzene extends over a wide range of frequencies and at the highest frequency end of this transition, the molecule has been given 64 kJ mol^{-1} of energy in excess of that needed simply to excite the molecule electronically (456 kJ mol^{-1}); this excess of energy leads to vibrational excitation of the electronically excited molecule and the features shown in Fig. 12.3 (a) and (b) correspond to vibrational structure. If the benzene is in solution, then the excited molecules very rapidly lose this vibrational energy to the solvent—this process occurs in $\approx 10^{-11}$ s, well before the molecule has time to emit, so that, immediately following the absorption of light, the molecule is rapidly vibrationally deactivated and we can consider the subsequent processes as involving the vibrationally deactivated S_1 species. The S_1 state can, however, emit to a number of vibrational states in the S_0 level making the fluorescence signal an approximate mirror image of the excitation or absorption spectrum. At low pressures, in the gas phase, vibrational relaxation does not occur and emission from vibrationally excited states can be observed.

12.2.4 Radiationless transitions

The S_1 state may be converted into the T_1 state or the S_0 state:

$$S_1 \rightarrow T_1 + \text{vibrational energy } (k_{isc}). \tag{R 3}$$

$$S_1 \rightarrow S_0 + \text{vibrational energy } (k_{ic}). \tag{R 4}$$

The former, spin-forbidden process, is termed *intersystem crossing* and the latter, spin allowed process *internal conversion*. Both occur with conservation of energy (horizontal transitions in Fig 12.1) and the excess of energy is converted into vibrational energy. Despite its spin-forbidden nature, intersystem crossing often occurs more rapidly than internal conversion. The triplet state, once produced, undergoes rapid vibrational relaxation and then can phosphoresce or inter-system cross to the S_0 state. The rate coefficients for these processes are much smaller than for the S_1 state.

Box 12.1 Energy transfer

We saw in Chapter 5 the importance of vibrational energy transfer in unim-olecular and termolecular reactions. Electronic excitation followed by internal conversion has been used as a means of producing ground state molecules with a very well-defined vibrational energy.

Troe and co-workers are just one of several groups who utilize this method for studying vibrational energy transfer. (Hippler, H., Troe, J., and Wendelken, H. J. (1983). Collisional deactivation of highly vibrational excited polyatomic molecules. *Journal of Chemical Physics*, **78**, 6709–17.) The kind of questions that they are asking are: How much energy is transferred in each collision with a bath gas, $\langle \Delta E \rangle$? How does $\langle \Delta E \rangle$ vary with molecular complexity of both the excited molecule and the bath gas? Does $\langle \Delta E \rangle$ vary with the energy of the excited mole-cule or the temperature of the system?

Flynn and co-workers have looked at the rovibrational levels produced in the *bath gas* following a deactivating collision in order to learn something about the detailed dynamics and mechanism of vibrational energy transfer as it affects the bath gas molecules. (Mullin, A. S., Park, J., Chou, J. Z., Flynn, G. W., and Weston, R. E. (1993). Some rotations like it hot. *Chemical Physics*, **175**, 53–70).

12.2.5 Collisional (quenching) processes

The excited singlet and triplet states can also undergo a number of collisional processes, which are often termed quenching processes because they reduce the fluorescence or phosphorescence intensity. Some of them will be discussed in greater detail below; for the present we content ourselves with a brief classifica-tion:

1. *Energy transfer.* One molecule passes its electronic energy on to another, the excess energy being channelled into vibration e.g.

$$\text{Naphthalene } (S_1) + \text{phenanthrene } (S_0) \rightarrow \text{naphthalene } (S_0) + \text{phenanthrene } (S_1). \quad \text{(R 5)}$$

Alternatively the electronic energy may be transferred into vibrational or rota-tional excitation of the colliding species e.g.

$$\text{I*}(^2P_{\frac{1}{2}}) + \text{HF} \rightarrow \text{I}(^2P_{\frac{3}{2}}) + \text{HF}(v = 2). \quad \text{(R 6)}$$

2. *Chemical reaction.* The excess of electronic energy may enable a reaction to occur which requires a large activation energy for the ground state molecule. The Woodward–Hoffmann addition reactions provide an interesting class of reac-tions, some of which are thermally allowed, others requiring photochemical

excitation. A relevant example for atmospheric chemistry is the generation of OH radicals in the troposphere (lower atmosphere). The photolysis of ozone leads to electronically excited oxygen atoms $(O(^1D))$ which can then react with water vapour generating two OH radicals. The OH radical is perhaps the most important scavenging radical in the troposphere for example, it determines the atmospheric lifetime of CH_4 via the reaction

$$OH + CH_4 \rightarrow H_2O + CH_3. \tag{R 7}$$

Isomerization reactions are another important category of reactions which may be photoinitiated.

12.2.6 Dissociation

In certain cases the high energy state which is accessed by the exciting photon is dissociative. The energy of the photon is greater than that of a particular bond in the molecule and the molecule dissociates into two fragments. The dynamics and mechanism of photodissociation have been the subject of many books and we shall very briefly consider a number of important aspects in Section 12.5.

12.3 Photochemical kinetics

Probably the most important concept in photochemical kinetics is that of the *quantum yield*. The quantum yield for any particular process (e.g. fluorescence, reaction etc) is defined as the fraction of the excited molecules undergoing that process i.e.:

$$\Phi = \frac{\text{number of excited molecules undergoing process}}{\text{number of photons absorbed.}} \tag{E 3}$$

or

$$\Phi = \frac{\text{Rate of process}}{\text{number of photons absorbed per unit volume per unit time.}} \tag{E 4}$$

Its exact evaluation is best appreciated by reference to a few problems.

12.3.1 Quantum yield for fluorescence

Calculations of this type are ideal examples of parallel reactions which were introduced in Section 8.2. The first-order parallel processes which can occur for an excited singlet state (S_1) are:

$$S_0 \rightarrow S_1 \qquad r \tag{R 8}$$

$$S_1 \rightarrow S_0 + h\nu \quad k_f \tag{R 9}$$

$$S_1 \rightarrow T_1 \qquad k_{isc} \tag{R 10}$$

$$S_1 \rightarrow S_0 \qquad k_{ic}. \tag{R 11}$$

Consider the case where the system is irradiated by a steady-state light source. r is the rate of excitation (i.e. $r = I_{abs}$ the number of photons absorbed per unit time per unit volume). The lifetime of S_1 is very short, so we can apply the steady-state approximation (pp. 198–9):

$$d[S_1]/dt = r - (k_f + k_{isc} + k_{ic})[S_1] = 0 \qquad (E\ 5)$$

$$[S_1] = r/\ (k_f + k_{isc} + k_{ic}). \qquad (E\ 6)$$

The rate of fluorescence (number of photons emitted per second) is $k_f[S_1]$, so that the fluorescence quantum yield is (from E 4)

$$\Phi_f = k_f[S_1]/r. \qquad (E\ 7)$$

We can substitute our steady-state value of $[S_1]$ into eqn (7) to give the final expression (E 8)

$$\Phi_f = k_f/\ (k_f + k_{isc} + k_{ic}). \qquad (E\ 8)$$

Thus the quantum yield is, in this case, the fraction of molecules excited which undergo fluorescence or, equivalently, the probability of fluorescence following excitation. Similar expressions may be found for the quantum yield for inter-system crossing and internal conversion.

12.3.2 Unusual quantum yields

For certain reactions the quantum yield may be greater than one. A trivial example would be the quantum yield for the production of methyl radicals from the photolytic dissociation of azomethane. As two methyl radicals are produced for each photon (of wavelength less than 220 nm) absorbed, $\Phi_{CH3} = 2$.

$$CH_3N_2CH_3 + h\nu \rightarrow 2CH_3 + N_2. \qquad (R\ 12)$$

Alternatively, if a dissociation produces a reactive radical species, it is possible that a chain or multi-step reaction could be initiated so that the quantum yield for the production of stable species such as methane or carbon monoxide would be greater than one, e.g.

$$CH_3CHO + h\nu \rightarrow CH_3 + CHO \qquad (R\ 13)$$

$$CHO \rightarrow H + CO \qquad (R\ 14)$$

$$H + CH_3CHO \rightarrow CH_3CO + H_2 \qquad (R\ 15)$$

$$CH_3 + CH_3CHO \rightarrow CH_4 + CH_3CO \qquad (R\ 16)$$

$$CH_3CO \rightarrow CH_3 + CO. \qquad (R\ 17)$$

The exact number of ethanal molecules removed per photon absorbed will depend on the ratio of the rates of ethanal consumption reactions versus radical recombination reactions.

Finally, the quantum yield may be time dependent. We recall that we introduced the idea of a quantum yield on p. 164 in our discussion of iodine photo-dissociation in solution. If we work on a picosecond timescale, then the quantum yield for iodine atom formation is very high (approaching two) but the iodine atom yield is rapidly reduced by geminate recombination, so that measurements on a longer timescale give a smaller quantum yield.

12.3.3 Actinometry

One of the difficulties in measuring a quantum yield is the determination of the absolute value for I_0, the incident light intensity. A detailed discussion of how this can be achieved from first principles may be found in *Photochemistry of Small Molecules* by H. Okabe (1978) (Wiley, Chichester. pp. 125–7). It requires the use of an absolute energy detector and careful consideration of geometrical factors. Comparatively few laboratories are in a position to make such measurements and most rely on an actinometer—a substance whose quantum yield has been previously measured as a function of wavelength.

12.3.4 The lifetime of an excited state

Photochemical measurements are made either with continuous irradiation (e.g. the quantum yield discussion above) where the steady-state approximation may be employed, or with pulsed irradiation (from a laser or flash lamp) where it may not. Lifetime measurements are made under the latter conditions, preferably with a pulsed light source whose duration is much shorter, than the lifetime of the excited state. Under these conditions, immediately following the pulse,

$$-d[S_1]/dt = (k_f + k_{isc} + k_{ic})[S_1] \qquad (E\ 9)$$

subject to the condition that, at time $t = 0$, the S_1 concentration is $[S_1]_0$. Integrating

$$[S_1]_t = [S_1]_0 \exp(-t/\tau) \qquad (E\ 10)$$

where $\tau = (k_f + k_{isc} + k_{ic})^{-1}$ is the natural lifetime of the excited state (the time taken for the concentration to fall to $1/e$ of its initial value). Note that the lifetime is inversely proportional to the sum of the first-order rate coefficients and we can measure it by monitoring any one of the component processes—the easiest one to study is fluorescence and the total light emitted is given by:

$$I_f = k_f[S_1] = k_f[S_1]_0 \exp(-t/\tau) \qquad (E\ 11)$$

k_f can then be determined if Φ_f is known, since $k_f = \Phi_f \tau^{-1}$. We can see from eqn (11) that it is not necessary to measure the total light emitted, but only the time dependence of the relative intensity, since eqn (11) is exponential and we only require τ.

12.3.5 Stern–Volmer plots

Suppose that we add a substrate Q which can deactivate S_1, so that an additional process

$$S_1 + Q \rightarrow \text{products} \qquad (R\ 18)$$

is added to those we considered in our previous discussion of fluorescence quantum yields. Applying the steady-state approximation again, we find

$$[S_1] = r/(\ (k_f + k_{isc} + k_{ic} + k_q[Q]). \qquad (E\ 12)$$

The steady-state rate of light emission is $k_f[S_1]$, so that, provided r is constant, the ratio of the fluorescence intensity in the absence (I_{f0}) and presence (I_f) of quencher is given by

$$I_{f0}/I_f = (k_f + k_{isc} + k_{ic} + k_q[Q])/\ (k_f + k_{isc} + k_{ic}) = 1 + k_q\tau[Q]. \qquad (E\ 13)$$

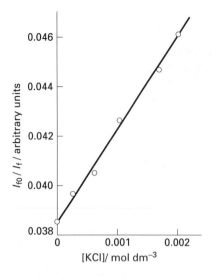

Fig. 12.5 Stern-Volmer plot for the quenching of quinine sulphate fluorescence by the chloride ion.

A plot of I_{f0}/I_f vs. [Q] is a straight line, with slope $k_q\tau$ (Fig. 12.5). Provided τ has been measured, k_q can be evaluated. The Stern–Volmer method provides a facile way of measuring rapid rates: τ provides a very short internal clock typically operating on a 1–10 ns timescale.

Boxes 12.2 and 12.3 detail some applied aspects of photochemical kinetics.

Box 12.2 Delayed fluorescence

Two triplet molecules can pool their energy to produce one excited and one ground state singlet species:

$$T_1 + T_1 \rightarrow S_1 + S_0. \tag{R 19}$$

This process gives rise to fluorescence occurring on long timescales which are more typical of phosphorescence. Let us imagine that we excite the system with a pulsed light source and then examine the fluorescence (from S_1) at long times (long after the directly excited singlets have decayed). We have two differential equations:

$$d[T_1]/dt = -k_T[T_1] - 2k_{19}[T_1]^2 \tag{E 14}$$

$$d[S_1]/dt = k_{19}[T_1]^2 - k_S[S_1] \tag{E 15}$$

where k_T is the sum of the phosphorescence and intersystem crossing rate coefficients for the triplet and $k_S = k_f + k_{isc} + k_{ic}$. Provided the light intensity is low, the triplet concentration is small and $k_T[T_1] \gg 2k_{19}[T_1]^2$, so that:

$$[T_1]_t = [T_1]_0 \exp(-t/\tau_T) \tag{E 16}$$

where τ_T is the lifetime of the triplet. S_1 itself is very short-lived and the steady-state approximation may be applied:

$$[S_1] = k_{19}[T_1]^2/ks = k_{19}[T_1]_0^2 \exp(-2t/\tau_T)/k_s \qquad \text{(E 17)}$$

whilst the rate of fluorescence is:

$$I_f = k_f[S_1] = k_{19}[T_1]_0^2 \Phi_f \exp(-2t/\tau_T) \qquad \text{(E 18)}$$

Thus the delayed fluorescence is observed to decay with a lifetime half that of the phosphorescence.

Box 12.3 Dye lasers

Dye lasers are important because of their ability to be tuned over a wide frequency range. This has led to many applications in spectroscopy and its related fields. The operation of dye lasers and the design of more efficient laser dyes illustrates some of the topics we have discussed in Sections 12.2 and 12.3.

Laser action occurs by the process of stimulated emission (in fact the word laser is an acronym for *L*ight *A*mplification by *S*timulated *E*mission of *R*adiation). If we can produce more excited state species than ground state (a population inversion) and induce these states to decay via stimulated emission, then stimulated emission will be dominant over both absorption and spontaneous fluorescence.

Figure 12.6 shows a schematic diagram of a dye laser cavity. The dye molecules are excited by a short monochromatic laser pulse to an excited singlet state. Vibrational relaxation in the upper electronic state is very rapid and the mole-

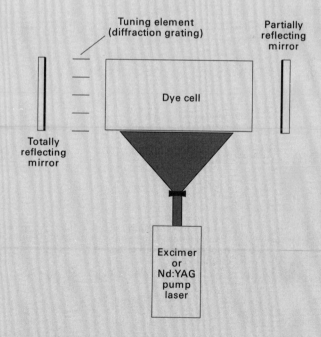

Fig. 12.6 Schematic diagram of a dye laser cavity.

cules will begin to fluoresce down to the ground electronic state. Because photon absorption and emission are 'vertical processes' (the Franck–Condon principle), absorption and fluorescence occur to vibrationally excited levels (Fig. 12.7). Vibrational relaxation is very rapid in the dye solution and hence we can achieve a population inversion between the ground vibrational levels of the upper electronic state and the vibrationally excited levels of the ground electronic state (Fig. 12.8).

Spontaneous fluorescence will occur in all directions and over a wide range of frequencies. A few photons will be produced parallel with the laser cavity and with the correct frequency to pass through the tuning element and be reflected back to the dye cell. Now they can stimulate emission at the same frequency. This stimulated emission will be of identical frequency and direction, so it too can be reflected back along the cavity stimulating more emission. Stimulated emission begins rapidly to dominate over spontaneous fluorescence and we allow out some of the laser light through a partially reflecting mirror to form a monochromatic laser beam. By changing the tuning element we can vary the frequency of the light that we allow to be reflected within the laser cavity.

As well as undergoing stimulated emission (the process we want to promote) the excited singlet state will also undergo all of the other processes that we described in Section 12.2, i.e. fluorescence, intersystem crossing, internal conversion, quenching and reaction. All of these processes will remove molecules from the upper level and reduce the magnitude of the population inversion and hence reduce the power output of the laser. Designers of laser dyes are therefore looking for compounds which exhibit very low rates of intersystem crossing (generally the most important loss mechanism) and investigations of the type described in the section on quantum yields for particular processes, are crucial to laser dye development. In other cases intersystem crossing itself is not the major problem but rather, the triplet states produced may be very efficient at

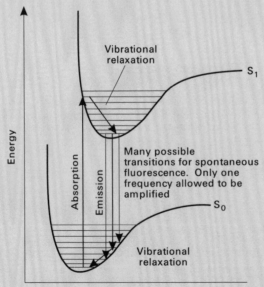

Fig. 12.7 Jablonski diagram for the processes involved in the laser dye.

reabsorbing the laser radiation, decreasing the efficiency of operation.

The other important criterion is the stability of the laser dye. The dye must be stable with respect to photodissociation by the intense excitation (pumping) source and also the singlet or triplet states must have a low reactivity. This reduces any tendency to react chemically rather than to return to the ground state by physical processes, otherwise the laser dye will rapidly be consumed.

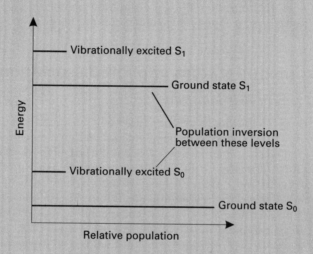

Fig. 12.8 Relative populations of the vibronic states of the laser dye, showing a population inversion between the ground vibrational state of the first singlet state and the upper vibrational levels of the ground electronic state.

12.4 Photodissociation

In the second part of this chapter we shall look in more detail at one of the processes we have already mentioned, namely photodissociation. Photodissociation can occur when the molecule absorbs one or more photons of total energy greater than the bond energy of that particular state and can occur via a number of different mechanisms. In the most simple process the molecule is excited directly into a repulsive state from which dissociation occurs directly. Alternatively, the initial absorption occurs to a bound excited state. Internal conversion from this excited state to a lower state can occur, and if the energy of the lower electronic state is greater than the bond dissociation energy then the molecule may dissociate. Finally, dissociation may occur via the absorption of more than one photon. Infrared multiphoton dissociation is an extreme example of this mechanism. Unlike the other processes, it does not involve excitation to a different electronic state, but rather the rapid sequential absorption of several IR photons until the molecule reaches the top of the Morse well, at which point dissociation may occur.

The remainder of the chapter is set out in the following way: Firstly, we shall look at some experimental examples of photodissociation studies and see what light they cast on the mechanism of photodissociation. Photodissociation can be thought of as a 'half reaction'. Photon absorption 'instantaneously' generates the excited transition state which falls apart to give products. Because we are only working with one reactant and we can selectively excite any one particular transition, such photodissociation experiments can be very specific in the nature of the transition state that we produce especially if we use molecular beam techniques to generate well defined initial states from which the initial absorption takes place. Secondly, we shall review some more applied aspects of photodissociation; the formation of radicals in flash photolysis experiments and the production of reactive species in planetary atmospheres.

12.5 Experimental studies of photodissociation

12.5.1 Direct dissociations

The hydrogen halide molecules provide excellent examples of the detailed information that can be obtained from photofragmentation experiments. Figure 12.9 shows some of the potential energy curves for HI. Photon absorption directly produces the HI molecule on a dissociative excited state and the diatomic falls apart to give H and I atoms. Being a diatomic system there can be no rovibrational energy stored in the photofragments, but Fig. 12.9 shows that for photons of energy greater than 386 kJ mol^{-1} (corresponding to $\lambda < 310$ nm), the iodine atom can be produced in one of two electronic states. The remainder of the energy difference between photon energy and the reaction endothermicity must appear as translational energy of the photofragments. We can calculate the fraction of translational energy carried away by each atom from conservation of momentum and energy. Consider the HI molecule to be initially at rest, when the momentum is zero. This value of the momentum (a vector quantity) must be conserved and therefore the photofragments must move off in opposite directions with equal magnitudes of momentum:

$$0 = m_H v_H - m_I v_I \qquad \text{(E 19)}$$

$$v_H = (m_I / m_H) v_I. \qquad \text{(E 20)}$$

The H atom therefore moves off 127 times faster than the heavier iodine atom. It consequently carries the bulk of the relative translational energy;

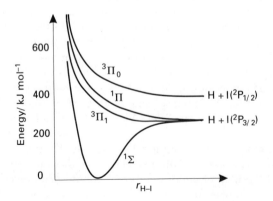

Fig. 12.9 Potential energy surface for HI.

$$E_H = \tfrac{1}{2}m_H v_H{}^2 = \tfrac{1}{2}m_H[(m_I/m_H)v_I]^2 \qquad\qquad \text{(E 21)}$$

$$E_I = \tfrac{1}{2}m_I v_I{}^2 \qquad\qquad \text{(E 22)}$$

$$E_H = 127E_I. \qquad\qquad \text{(E 23)}$$

The kinetic energy of the photofragments can be measured in the following type of experiment known as time of flight mass spectrometry (TOFMS). A short laser pulse is used to dissociate the HI molecules and start the internal clock of the experiment. A fraction of the H atoms pass down a long tube (the flight tube) and into a mass spectrometer. Here they are ionized and detected and the time between photolysis and detection is recorded. As we know the distance between the point of photolysis and detection we can calculate the velocity of the H atoms and hence their kinetic energy. Figure 12.10 shows the type of spectrum obtained in these experiments. Two peaks are observed corresponding to the formation of either of the electronic states of iodine. The area under the peaks is proportional to the fraction of products in each channel and so a precise determination of the branching probability to each electronic state of iodine may be calculated. Photodissociation dynamics is a subject in itself and space precludes taking this discussion much further but before we move on there are a couple of points to note: (1) For larger species, the potential energy surfaces become more complex and we need to consider the possibility that significant amounts of energy can be taken up in the internal states of the polyatomic photofragments. In these cases techniques such as LIF are essential in characterizing the internal energy distributions of the fragments. (2) With the use of polarized laser beams and angular

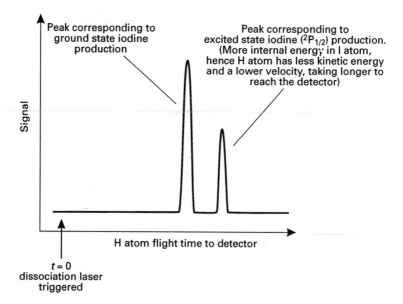

Fig. 12.10 H atom time of flight spectrum following HI photolysis, showing that two different processes are occurring, producing both ground and electronically excited iodine atoms.

detectors (such as those described in Section 4.3) we can begin to map out the angular as well as the energy distributions of the photofragments.

12.5.2 Stepwise dissociations

As an experimental example of the study of a stepwise dissociation mechanism we shall look at the photolysis of acetone in the UV. Gas chromatographic analysis of the stable products from acetone dissociation at 193 nm showed that the major dissociation pathway is:

$$CH_3COCH_3 + h\nu \rightarrow 2CH_3 + CO. \tag{R 20}$$

However, the molecular mechanism of photodissociation was not clear. Does the molecule fall apart in one concerted process, or do each of the methyl radicals come off in sequential processes? Photolysis at longer wavelengths produces the radicals CH_3CO and CH_3 indicating a stepwise process, but experiments performed to try and isolate the intermediate radical CH_3CO at shorter wavelengths have had no success. Leone and co-workers (1988) developed an elegant technique for looking at the products of photofragmentation and their experiments point strongly to a stepwise mechanism.

In these experiments, acetone is photolysed by an excimer laser at 193 nm. The energy of the photon is more than sufficient to dissociate the molecule and some of the excess of energy will be imparted to the vibrational modes of the photofragments which will in turn give out IR fluorescence as they relax. The frequency of the fluorescence will be characteristic of the rovibrational state from which the emission originated. In the experiment the IR emission is analysed by an FTIR spectrometer and much of the paper concerns the relatively complex interface between the acetone dissociation laser and IR collection, however the details of this process need not concern us here. Figure 12.11 shows a nascent spectrum of the CO emission (by nascent we mean that the CO distribution has not been affected by collisional deactivation). One prominent feature is the high degree of rotational excitation of the CO molecule. This was interpreted as strong evidence for a stepwise fragmentation. Were the carbon–carbon bonds to break simultaneously there would be no mechanism for imparting such a high degree of rotational excitation to the CO molecule. However, in a stepwise process, the sequential breaking of carbon–carbon bonds allows significant torques to be imparted onto the CO fragment, especially in the final step, the break up of the CH_3CO radical (Fig. 12.12).

12.5.3 Real time studies of the photodissociation process

The development of sub-picosecond laser pulses has allowed us to probe the time resolved behaviour of processes which had previously to be considered as instantaneous. The results of these cutting edge experiments are giving unprecedented information about the detailed dynamics of dissociation. As an example we shall look briefly at the photodissociation of ICN and NaI.

The experimental apparatus for the study of ICN dissociation is shown in Fig. 12.13. The same femtosecond laser is used both to dissociate the ICN molecule and probe for the *free* CN radical by laser induced fluorescence (pp. 34–5). The aim of the experiment is to measure the time delay between photolysis and probe pulses, before the free CN signal is observed. This will correspond to the time

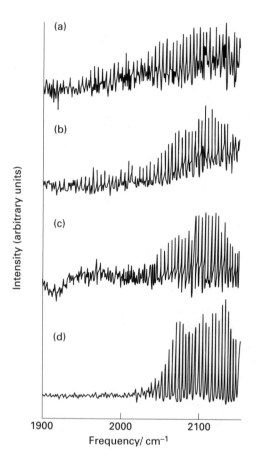

Fig. 12.11 Rovibrational emission spectrum of CO produced from the photolysis of acetone at different delays after the photodissociation. The high rotational levels populated in the nascent photofragments (a) (1.3 Pa, 6 μs delay, ≤0.5 collisions)indicate a mechanism of the type shown in Fig. 12.12. After a number of collisions the rotational distribution relaxes to a 300 K Boltzmann distribution: (b) 2.7 Pa, 6 μs delay, ≤1.0 collisions; (c) 2.7 Pa, 20 μs delay, ≤3.0 collisions; (d) 10.0 Pa, 30 μs delay, 400 Pa Ar, ≤600 collisions (Ar).

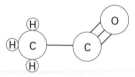

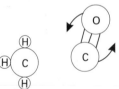

| Acetyl radical formed after the departure of first methyl radical has a bent structure and is highly energetic. | C - C bond cleavage occurs. Fragments start to move apart. The bent structure means that a torque is applied to both fragments. | The torque leads to the observed rotational excitation of the CO fragment. |

Fig. 12.12 Dissociation mechanism of acetone. The two step process involves a bent acetyl intermediate, dissociation of which imparts a considerable torque to the CO fragment and hence high rotational excitation. A simultaneous symmetric rupture of both C–C bonds would leave rotationally cool CO.

taken for the ICN molecule to move down the dissociative potential energy surface to which it is promoted in the initial excitation pulse (Fig. 12.14). The two pulses are delayed by making the probe pulse take a slightly longer path length (which can be varied) and the time for dissociation is recorded as approximately 205 fs.

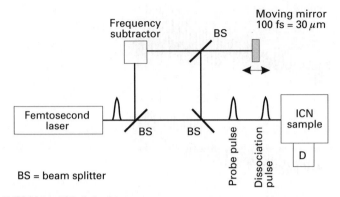

Fig. 12.13 Schematic diagram of the apparatus for studying the real time dissociation of ICN. BS, beam splitter; D, LIF detector.

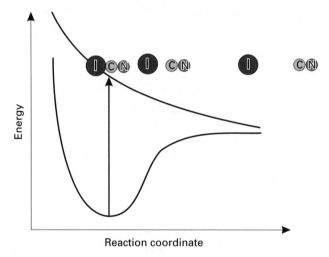

Fig. 12.14 Potential energy diagram for ICN dissociation. The absorption of the dissociating photon raises the molecule into a replusive state. The molecule moves down in energy with the I–CN bond lengthening all the time.

The potential energy for sodium iodide dissociation is shown in Fig. 12.15. The ground state is essentially ionic in character, but the lowest energy products are of course the uncharged atoms. There must therefore be a crossing between the two surfaces. Figure 12.15(b) shows the time-dependent signal for the production of sodium atoms and we can see that instead of an instantaneous production, the growth is more gradual and even shows some structure! To explain these observations we need to consider the motion of the wavepacket which represents the electronically excited NaI molecule initially formed by the absorption of the dissociation photon. The initial vertical transition (Franck–Condon theory) creates a state on the covalent surface. During the course of the dissociation the wavepacket moves across the PES to the crossing of the two surfaces. At the crossing a fraction of the wavepacket leaks through this avoided crossing and out

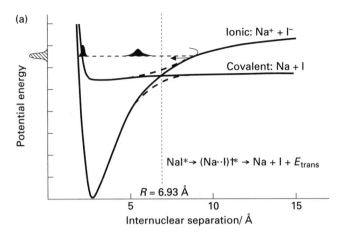

(a)

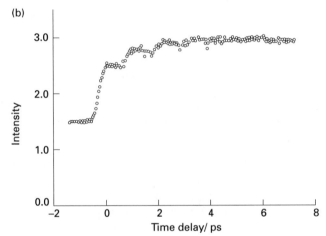

(b)

Fig. 12.15 (a) The potential energy diagram for NaI and Na⁺I⁻ molecules. (b) The observed fluorescence of Na showing traversals within the potential energy well.

onto the covalent surface giving atomic products. However, some fraction is reflected from the crossing. This portion of the wavepacket travels to and fro across the PES before once again reaching the crossing where a further fraction of the wavepacket leaks through. The periodic arrival of the wavepacket at the crossing and subsequent formation of products accounts for the stepwise growth in the product signal, the time between steps being equal to the period of the oscillation on the upper excited state.

Study notes
Further information on the dynamics of photodissociation may be found in the following articles:
1   pp. 133–9 in *Chemistry in Microtime* by Lord Porter (in *The Chemical Bond* (ed. A. Zewail). Academic Press, New York, 1992). The review deals with the historical development of time resolved experiments and the ever decreasing timescales for

which reactions can be studied. It is an excellent introduction to the fields of picosecond and femtosecond studies and their applications in such fields as transition state spectroscopy and photosynthesis.

2 *Real-Time Laser Femtochemistry*—Chapter 9 of *The Chemical Bond* (ed. A. Zewail). This chapter greatly expands on the introduction to the subject given by Lord Porter and provides an entry point for some of the original literature such as Rosker, M. *et al.* (1988). *Science*, **241**, 1200; Imre, D. *et al.* (1984) *Journal of Physical Chemistry*, **88**, 3956; Maguire, T. C. *et al*, (1986) *Journal of Chemical Physics*, **85**, 844; and Zewail, A., (1990). *Scientific American*, **263**, 76.

12.6 Experimental generation of radical species

Throughout this book we have seen the vital role of elementary reactions involving radical species in a number of applications e.g. combustion, atmospheric chemistry. In order to study these important reactions in isolation we need a clean, fast method of radical generation and rapid mixing between the reagent species. In Chapter 2 we described two methods of studying fast reactions, one of which, flash photolysis, involves premixing the radical precursors and reagents and then generating an 'instantaneous', uniform concentration of radical species by photolysis. It is appropriate in this chapter to look in slightly more detail at the different types of photolysis available and at the processes which need to be considered in a flash photolysis experiment.

The first major consideration is the timescale of the reaction to be studied as this will determine the type of photolysis source that can utilized. Ideally we want the generation of the transient species to be very rapid in comparison to reaction, otherwise we will have problems with our signal analysis. If the generation is not rapid we have to deconvolute two different processes; the production of the transient from the photolysis pulse and consumption via reaction. Figure 12.16 illustrates the timescales of a number of kinetic processes that can be studied by flash photolysis. Needless to say the production of shorter pulses requires more complex equipment and greater costs.

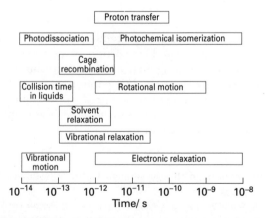

Fig. 12.16 Schematic diagram showing the timescales of various kinetic and dynamic processes.

The photodissociation of acetone at 193 nm (R 20), the molecular mechanism of which we have just discussed above, has been a convenient source of methyl radicals for a number of kinetic studies. Figure 12.17 shows an energy diagram for the process. It is immediately obvious that the photofragments are energetically excited and we saw that much of this excess of energy is contained in the rovibrational excitation of all of the photofragments. We noted in Chapter 4, the significant effect that vibrational excitation could have on reaction rates and therefore, if we are interested in studying the reactions of ground state radicals, it is essential that the radical species are vibrationally deactivated. In general vibrational relaxation is a rapid process, even by the noble gases, and so the addition of large quantities of inert bath gas should ensure that vibrational deactivation occurs before reaction.

Photolysis can also lead in certain circumstances to either translationally or electronically excited species. For dynamical studies this can be a positive advantage (Section 4.3), but for kinetic studies of ground state species we must always ensure complete relaxation on timescales which are rapid in comparison to reaction.

Infrared multiphoton dissociation (IRMPD) provides an alternative method for producing radical species. A schematic figure of the process is shown in Fig 12.18, where the bond to be broken is approximated to a Morse potential. Sequential absorption of IR photons from an intense pulse of laser radiation (generally from a CO_2 laser) raises the molecule above the bond dissociation energy of the weakest bond in the molecule at which point dissociation can compete with the absorption of another photon. The rate of unimolecular reaction (the dissociation process) rises very rapidly with energy once the dissociation barrier has been reached and hence fragmentation occurs close to threshold and the radical species have very little internal energy. Whilst IRMPD has many advantages and has been used in a number of studies there are two major drawbacks to the technique. Firstly the process is rather random. IR absorption occurs on a long timescale in comparison to intramolecular energy redistribution and so fragmentation can occur in bonds other than those targeted for dissociation. Secondly it is difficult (though not impossible) to generate short IR laser pulses of

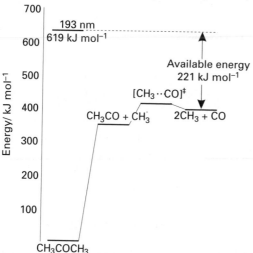

Fig. 12.17 Energy level diagram of the molecular species of interest in acetone photolysis. (Adapted from Fletcher, T. R., Woodbridge, E. L., and Leone, S. R. (1988). *Journal of Physical Chemistry*, **92**, 5390).

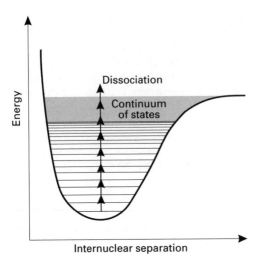

Fig. 12.18 Schematic diagram of infrared multiphoton dissociation. The potential energy diagram has been simplified to a Morse curve but would in reality be more complex for a polyatomic species.

high intensity and so the timescale to which these techniques can be applied is somewhat limited.

The high photon densities available in visible and UV laser pulses can also allow multiphoton processes to occur. Two different processes can exist, firstly the simultaneous absorption of two photons by a single molecule and secondly the sequential absorption of two photons with the formation of an intermediate species. Often such processes are indicated by a squared (or cubic, for three photon process) dependence of the radical concentration with laser intensity but this is not always the case. Multiphoton effects can be advantageous allowing the generation of otherwise unobtainable species but in other cases it may be the target radical that absorbs another photon creating an unwanted or unexpected species which could complicate the reaction kinetics. For example in the photolysis of acetone, a single photon at 193 nm generates the methyl radical which in turn may absorb another photon forming the methylene radical.

$$CH_3COCH_3 + h\nu \rightarrow CO + 2CH_3 \qquad (R\ 20)$$

$$CH_3 + h\nu \rightarrow CH_2 + H. \qquad (R\ 21)$$

Finally, in a flash photolysis experiment the choice of radical precursor is vital. The precursor must be thermally stable at the reaction temperature, and relatively unreactive with the radical species or other reagents. It must yield a significant radical concentration from the light sources available (preferably in the ground state) and produce an unreactive photolysis co-product. Only in a very few cases will such ideal precursors exist for any radical and therefore in any experiment it is important to vary the photolysis conditions or radical precursor to investigate whether any unwanted reactions are affecting the kinetics of the target reaction.

12.7 Photodissociation in atmospheric chemistry

Photodissociation is the driving force behind much of the atmospheric chemistry of this planet and of other planets. We noted the importance of stratospheric ozone in Chapter 8 and saw that two photodissociation steps occurred in the natural cycle between O_3, O_2 and O

$$O_2 + h\nu(\lambda < 242.4 \text{ nm}) \rightarrow 2O \qquad \text{(R 22)}$$

$$O_3 + h\nu(\lambda < 1180 \text{ nm}) \rightarrow O_2 + O. \qquad \text{(R 23a)}$$

We also saw that photolysis of man-made chlorofluorocarbons in the stratosphere leads to catalytic cycles which remove ozone.

Photodissociation is also crucial in the tropospheric chemistry which occurs nearer to the Earth's surface. In comparison to the stratosphere only a limited range of wavelengths is available, as most of the deep UV solar radiation has been filtered out by molecular oxygen or ozone higher in the atmosphere, but still a lot of interesting chemistry is initiated by photolysis. Photolysis of tropospheric ozone leads to the formation of excited oxygen atoms which can react with the abundant quantities of H_2O present in the atmosphere to generate the ubiquitous OH radical, the main oxidizing initiator in the troposphere.

$$O_3 + h\nu(\lambda < 310 \text{ nm}) \rightarrow O_2 + O(^1D) \qquad \text{(R 23b)}$$

$$O(^1D) + H_2O \rightarrow 2OH. \qquad \text{(R 24)}$$

Hydroxyl radical generation occurs in both the natural and polluted troposphere, however OH concentrations are much higher in a polluted environment because of raised ozone levels and a number of pollutant reactions which regenerate OH (see Fig. 8.11).

OH radicals rapidly attack many hydrocarbon species initiating their oxidation to CO_2 via a complex sequence of reactions, an example of which was shown in Fig. 8.11, (in some cases involving photodissociation of aldehydes and ketones formed as intermediates in the oxidation process). Because of its reactivity the lifetime of the OH radical in the atmosphere is very short (≈ 1–10 s). It therefore rapidly disappears after sunset and a completely different type of chemistry takes over during the hours of darkness, initiated by the nitrate radical.

Tropospheric ozone, nitrogen monoxide and nitrogen dioxide interact to give photostationary concentrations:

$$NO_2 + h\nu(\lambda < 400 \text{ nm}) \rightarrow NO + O(^3P) \qquad \text{(R 25)}$$

$$O + O_2 + M \rightarrow O_3 + M \qquad \text{(R 26)}$$

$$NO + O_3 \rightarrow NO_2 + O_2. \qquad \text{(R 27)}$$

Applying the steady-state treatment to both O atoms and ozone

$$d[O]/dt = 0 = k_p[NO_2] - k_{26}[O][O_2][M] \qquad \text{(E 24)}$$

$$d[O_3]/dt = 0 = k_{26}[O][O_2][M] - k_{27}[NO][O_3]. \qquad \text{(E 25)}$$

Combining eqns (24 and 25) gives the steady state $[O_3]$ as

$$[O_3] = k_p[NO_2]/k_{27}[NO]. \qquad \text{(E 26)}$$

In an atmosphere containing NO_2 and NO (common man-made pollutants from fossil fuel combustion, Box 10.1) the concentration of ozone will be determined by the NO_2:NO ratio. Ozone is one of the major components of photochemical smogs causing respiratory inflammation and acting as a strong oxidant.

Interesting photochemistry is not only limited to the Earth's atmosphere but also occurs in the upper reaches of the other planetary atmospheres. The outer planets (Jupiter–Neptune) are characterized by highly reducing atmospheres where it is the photochemistry of reduced species such as hydrocarbons which drives much of the chemistry. One of the main atmospheric components of these atmospheres is methane. Photolysis of methane will lead to hydrocarbon radicals whose subsequent reactions allow the formation of more complex hydrocarbons from methane. The photodissociation of methane has therefore received extensive, but up until 1993, indirect study.

In the deep UV regions of the solar spectrum, characteristic of the radiation reaching the upper stratosphere of the planets, there are four possible dissociation channels for methane photolysis.

$$CH_4 \rightarrow H + CH_3 \tag{R 28a}$$

$$\rightarrow CH + H + H_2 \tag{R 28b}$$

$$\rightarrow 2H + CH_2 \tag{R 28c}$$

$$\rightarrow H_2 + CH_2. \tag{R 28d}$$

Previous indirect studies, involving the gas chromatographic (GC) analysis of the stable products of the photolysis reaction, indicated that the major channels were (28c) and (28d) in approximately equal amounts with a small ($\approx 8\%$) fraction proceeding via channel (28b). Channel (28a) appeared to be insignificant. Atmospheric modellers of the outer planets therefore developed models based on methylene radicals as the basic building blocks for the formation of higher hydrocarbons.

Recently, a direct experiment by Ashfold and co-workers (1993) has indicated a very different mechanism. Using H atom time of flight mass spectrometry they were able to measure the kinetic energies of the H atoms emitted from channels (28a–c). Figure 12.19 shows the measured kinetic energy (KE) of the H atom fragments and also the maximum energies of the H atoms if all of the excess of energy from the dissociation were given to the translational energy of the H atoms. The long tail in the spectrum can energetically only arise from process (28a). Detailed analysis indicates that the dominant pathway producing H atoms is process a, with process b also being of significance. Approximately 25 per cent of the methyl radicals are produced so vibrationally excited that they dissociate further either to $CH + H_2$ or to $CH_2 + H$. Further experiments are required to quantify the branching ratios as the TOFMS experiments are blind to process (28d), however, they do indicate that some significant rethinking is required for higher hydrocarbon formation in the Jovian planets.

12.8 Questions

12.1 Nitrogen dioxide is irradiated with light of a particular wavelength in the presence of xenon buffer gas. The following elementary precesses take place:

$$NO_2 + h\nu \rightarrow NO_2* \tag{R 29}$$

$$NO_2* \rightarrow NO_2 + h\nu' \tag{R 30}$$

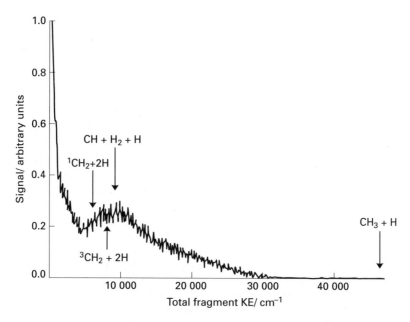

Fig. 12.19 Kinetic energy spectrum of the H atoms resulting from photolysis of CH_4 at 121.6 nm (82 259 cm^{-1}). Threshold energies for the various H atom forming channels are indicated with arrows. The long tail of the distribution shows that a significant portion of the H atoms arise from process (28a).

$$NO_2^* + NO_2 \rightarrow 2NO_2 \qquad \text{(R 31)}$$

$$NO_2^* + Xe \rightarrow NO_2 + Xe \qquad \text{(R 32)}$$

$$NO_2^* + NO_2 \rightarrow 2NO + O_2. \qquad \text{(R 33)}$$

(a) Describe briefly the process which occurs in each step.
(b) Derive an expression for the quantum yield of oxygen.
(c) The fluorescence lifetime τ was measured under the following conditions at 298 K by excitation with a pulsed laser.

$[NO_2]$/molecule dm^{-3}	$[Xe]$/molecule dm^{-3}	$\tau/\mu s$
3.2×10^{18}	1.6×10^{19}	3.38
6.4×10^{18}	1.6×10^{19}	1.89
3.2×10^{18}	0.8×10^{19}	3.64

Use these data to determine the true fluorescence lifetime of NO_2 in its excited state.

12.2 Propanone absorbs 313 nm radiation and exhibits a weak blue emission (380 – 470 nm) from a triplet state with a lifetime of $\tau = 2 \times 10^{-4}$ s. Biacetyl does not absorb 313 nm radiation but strongly quenches the triplet emission. Calculate the rate coefficient of this quenching reaction from the data below.

Φ_0/Φ_B	1.21	1.43	1.65	1.84	2.02
[Biacetyl]/mol dm^{-3}	0.006	0.012	0.018	0.023	0.028

Φ_0 and Φ_B are the phosphorescence quantum yields for propanone emission in the absence and presence of biacetyl respectively.

12.3 For aromatic molecules excited to the S_1 state fluorescence is generally observed from only the first excited singlet state, but for azulene, fluorescence is observed from the second excited singlet state as well. Explain this observation and describe an experiment to test your hypothesis. The energies of the singlet states of azulene are $E(S_1-S_0) = 14\,200$ cm^{-1}, $E(S_2 - S_0) = 28\,300$ cm^{-1}.

12.4 The quantum yield of the $S_1 \to S_0$ transition for rhodamine 6G (a common dye used in dye lasers) is 0.87, and the corresponding lifetime is ≈ 5 ns. Calculate the radiative and non-radiative lifetimes (τ_{sp} and τ_{nr} respectively) of the S_1 level.

12.5 Calculate the initial velocity and translational energy of H atoms produced from the photodissociation of HBr at 193 nm. Assume that all the bromine atoms are produced in their ground electronic state $(^2P_{3/2})$. $(D(\text{H-Br}) = 3.75\text{ eV})$.

H atoms produced from diatomic or triatomic hydride photolysis are used to investigate how translational energy may be used to overcome energy barriers to reaction. However, for thermal studies of H atom reactions we need to remove this excess of translational energy by collisions with a bath gas before significant amounts have reacted. By using principles of conservation of momentum and energy show which of the noble gases would be the most effective at relaxing hot H atoms and estimate how many collisions would be required to relax the H atoms to a room temperature thermal energy.

12.6 How could one differentiate between the following mechanisms for methanal photolysis at 308 nm.

$$H_2CO + h\nu \rightarrow H + HCO \qquad\qquad (R\ 34)$$

$$H_2CO + h\nu \rightarrow H_2 + CO \qquad\qquad (R\ 35)$$

with
a) GC analysis of the final products,
b) time resolved spectroscopic techniques.

12.7 Calculate the concentration (in molecule cm^{-3}) of iodine atoms when a excimer laser pulse of 10 mJ at 248 nm is incident on a cubical cell of volume 1 cm^3, containing 3×10^{15} molecules cm^{-3} of ethyl iodide.

$$C_2H_5I + h \rightarrow C_2H_5 + I. \qquad\qquad (R\ 36)$$

What assumptions do you need to make about the quantum yield for this photolysis process? (Hint: use Beers law to determine the number of photons absorbed in the cell.)

12.8 Calculate the rotational period of an H^{79}Br molecule in its first rotational $(J = 1)$ state ($r_{HBr} = 141$ pm). How does this compare with the zero point vibrational period? ($\tau_{HBr} = 2880$ cm^{-1}).

Using polarized lasers (where the electric vector of the electromagnetic radiation is aligned in one direction in space) it is possible to select the laboratory orientation of molecules to be dissociated. Given a molecular beam apparatus and a detector capable of angular resolution devise an experiment to determine the timescale of the dissociation. Why would a molecular beam help to simplify the data analysis?

12.9 Measurements have been made of the mixing ratios of NO, NO$_2$ and O$_3$ in a trajectory across the plume from a power plant (22 km downwind from the

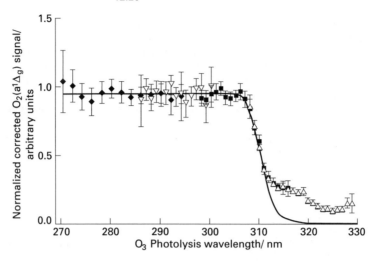

Fig. 12.20 $O_2(a^1\Delta_g)$ signal as a function of the wavelength of the photolysis of ozone. The solid line is the quantum yield of $O(^1D)$ recommended by NASA.

source) as a function of distance (d) from the centre. At a particular time of day the values were as follows:

O_3/ppb	94.0	108.0	129.0
NO_2/ppb	74.5	49.1	30.2
NO/ppb	18.3	10.5	5.4
d/km	0	4.86	12.9

Show that these data are consistent with the establishment of the photostationary state.

Hydrocarbons can increase the ozone concentration in an urban smog (see Fig. 8.11). Propose an explanation why the ozone forming potential of a hydrocarbon is proportional to the rate coefficient for OH attack and the number of carbon atoms in the hydrocarbon.

12.10   The spin allowed processes for ozone photolysis are:

$$O_3 + h\nu \rightarrow O_2(a^1\Delta_g) + O(^1D) \tag{R 37}$$

$$\rightarrow O_2(X^3\Sigma_g^-) + O(^3P). \tag{R 38}$$

What is abnormal about the signal for $O_2(a^1\Delta_g)$ production from ozone photolysis shown in Fig. 12.20 ($\Delta H_f\, O(^3P) = 246.6$, $O(^1D) = 436.7$, $O_2(a^1\Delta_g) = 94.3$, $D(O_2 - O) = 101.3$. All values in kJ mol^{-1})

Fluorescence from the $O_2(a^1\Delta_g)$ state is an important component of the atmospheric air glow, and in conjunction with the $O(^1D)$ quantum yield (assuming only spin allowed processes) and solar fluxes, has been used to estimate atmospheric ozone concentrations. What effect will the observations shown in Fig 12.20 have on the recorded ozone levels derived from such calculations?

12.11   Explain the following observation. When an excitation wavelength of 282 nm is used to monitor OH by laser induced fluorescence (see Fig. 2.8) in moist ozone containing air a higher [OH] is observed in comparison to excitation at 308 nm. (Hint: see question 12.10).

A major problem with LIF observation at 308 nm excitation is that the fluorescence occurs around 308 nm and can be swamped by scattered light from the excitation laser. What effect will lowering the pressure in the LIF cell have on the feasibility of detecting OH at 308 nm?

The fluorescence lifetime of the first excited state of OH is 0.72 μs. Sketch the photomultiplier signal following OH excitation by a 20 ns excitation pulse at 308 nm, at very low pressures, showing the two components of the signal.

References

Andrews, D. L. (1992). *Lasers in chemistry*, (2nd edn). Springer Verlag, Berlin.

Fletcher, T. R., Woodbridge, E. L. and Leone, J. (1988). Photofragmentation of acetone @ 193 nm. *Journal of Physical Chemistry*, **92**, 5387–93.

Hollas, J. M. (1992). *Modern spectroscopy*, (2nd edn), Chapter 1, Wiley, Chichester.

Mourdant, D. H., Lambert, I. R., Morley, G. P., Ashford, M. N. R., Dixon, R. N., *et al.* (1993). Primary products in the photodissociation of methane at 121–6 nm. *Journal of Chemical Physics*, **98**, 2054–65.

Appendices

(I) Fundamental constants

Avogadro constant	L or $\mathcal{N}_A$	$6.022 \times 10^{23}\,\mathrm{mol}^{-1}$
Boltzmann constant	k_B	$1.381 \times 10^{-23}\,\mathrm{J\,K}^{-1}$
Gas constant $(k_B \times L)$	R	$8.315\,\mathrm{J\,K}^{-1}\,\mathrm{mol}^{-1}$
Planck constant	h	$6.626 \times 10^{-34}\,\mathrm{J\,s}$
Speed of light in vacuum	c	$2.998 \times 10^{8}\,\mathrm{m\,s}^{-1}$
Rest mass of electron	m_e	$9.11 \times 10^{-31}\,\mathrm{kg}$
Rest mass of proton	m_p	$1.673 \times 10^{-27}\,\mathrm{kg}$
Charge of an electron	e	$-1.602 \times 10^{-19}\,\mathrm{C}$
Faraday constant $(e \times L)$	F	$9.649 \times 10^{4}\,\mathrm{C\,mol}^{-1}$

(II) Units of energy

As chemists we are probably most familar with the units $\mathrm{kJ\,mol}^{-1}$, however specialist practitioners in various fields often use non-SI units as a matter of convenience or tradition:

$1\,\mathrm{kcal\,mol}^{-1} = 4.18\,\mathrm{kJ\,mol}^{-1}$ [a]
$1\,\mathrm{eV} = 96.5\,\mathrm{kJ\,mol}^{-1}$
$1\,\mathrm{cm}^{-1} = 0.1197\,\mathrm{kJ\,mol}^{-1}$ [b]
Frequency of light: $1\,\mathrm{Hz} = 3.99 \times 10^{-13}\,\mathrm{kJ\,mol}^{-1}$ [c]
Wavelength of light: $1\,\mathrm{nm} = 1.2 \times 10^{5}\,\mathrm{kJ\,mol}^{-1}$ [d]

(III) Conversion factors from pressure to concentration units and for various rate coefficients

Once again tradition and convenience has led to the development of a whole range of pressure and concentration units. Within the book we have tried to limit the number of units used but the reader may need to consult the following tables to compare results from various papers in the literature.

(i) Conversion from pressure to concentration units

$1\,\mathrm{atm}\,(273\,\mathrm{K}) = 2.687 \times 10^{19}\,\mathrm{molecule\,cm}^{-3}$
$1\,\mathrm{atm}\,(273\,\mathrm{K}) = 4.462 \times 10^{-5}\,\mathrm{mol\,cm}^{-3}$
$1\,\mathrm{atm}\,(273\,\mathrm{K}) = 4.462 \times 10^{-2}\,\mathrm{mol\,dm}^{-3}$
$1\,\mathrm{atm}\,(273\,\mathrm{K}) = 1.013 \times 10^{5}\,\mathrm{Pa}$

---

[a] Older kinetic texts and papers will often give activation energies in $\mathrm{kcal\,mol}^{-1}$.
[b] Wavenumbers are often used in vibrational spectroscopy. Typical value for a vibration is $1000\,\mathrm{cm}^{-1} = 11.97\,\mathrm{kJ\,mol}^{-1}$.
[c] Frequency units used in microwave spectroscopy. A typical microwave transition occurs at $\approx 300\,\mathrm{GHz} = 0.4\,\mathrm{kJ\,mol}^{-1}$.
[d] Wavelength is inversely proportional to energy. Wavelength of an ArF excimer laser is $193\,\mathrm{nm} = 1.2 \times 10^{5}/193 = 620\,\mathrm{kJ\,mol}^{-1}$ i.e. in excess of most bond energies.

$$760 \text{ torr } (273 \text{ K}) = 2.687 \times 10^{19} \text{ molecule cm}^{-3}$$
$$1 \text{ torr } (273 \text{ K}) = 3.535 \times 10^{16} \text{ molecule cm}^{-3}$$
$$1 \text{ mtorr } (273 \text{ K}) = 3.535 \times 10^{13} \text{ molecule cm}^{-3}$$
$$1 \text{ torr } (273 \text{ K}) = 5.870 \times 10^{-5} \text{ mol dm}^{-3}$$
$$1 \text{ Pa} = 2.652 \times 10^{14} \text{ molecule cm}^{-3}$$
$$1 \text{ Pa} = 4.403 \times 10^{-7} \text{ mol dm}^{-3}$$

(ii) Conversion factors for bimolecular rate coefficients

To convert from	to	mutiply by
$1 \text{ dm}^3 \text{ mol}^{-1} \text{ s}^{-1}$	$1 \text{ cm}^3 \text{ mol}^{-1} \text{ s}^{-1}$	1000
$1 \text{ dm}^3 \text{ mol}^{-1} \text{ s}^{-1}$	$1 \text{ cm}^3 \text{ molecule}^{-1} \text{ s}^{-1}$	1.66×10^{-21}
$1 \text{ dm}^3 \text{ mol}^{-1} \text{ s}^{-1}$	$1 \text{ Torr}^{-1} \text{ s}^{-1}$	$1.603 \times 10^{-2} \text{ T}^{-1}$
$1 \text{ cm}^{-3} \text{ molecule}^{-1} \text{ s}^{-1}$	$1 \text{ dm}^3 \text{ mol}^{-1} \text{ s}^{-1}$	6.023×10^{20}

(iii) Conversion factors for termolecular rate coefficients

To convert from	to	mutiply by
$1 \text{ cm}^6 \text{ molecule}^{-1} \text{ s}^{-1}$	$1 \text{ dm}^6 \text{ mol}^{-2} \text{ s}^{-1}$	3.628×10^{41}

(IV) Absorption coefficients

A number of different units are used for the absorption coefficient for a molecule. It is important to know this number for example if one needs to calculate the number of radicals produced in a laser pulse. The units will depend on the concentration units used.

To convert from	Base	to	Base	multiply by
$k(\text{atm, 298 K})^{-1} \text{ cm}^{-1}$	e	$\sigma(\text{cm}^2 \text{ molecule}^{-1})$	e	4.06×10^{-20}
$k(\text{atm, 298 K})^{-1} \text{ cm}^{-1}$	e	$\varepsilon(\text{dm}^3 \text{ mol}^{-1} \text{ cm}^{-1})$	10	10.6
$\varepsilon(\text{dm}^3 \text{ mol}^{-1} \text{ cm}^{-1})$	10	$\sigma(\text{cm}^2 \text{ molecule}^{-1})$	e	3.82×10^{-20}
$\varepsilon(\text{dm}^3 \text{ mol}^{-1} \text{ cm}^{-1})$	10	$k(\text{atm, 298 K})^{-1} \text{ cm}^{-1}$	e	0.0942

Index